Statistics for Industry and Technology

Statistical and Probabilistic Models in Reliability

D.C. Ionescu
N. Limnios

Editors

With a foreword by Marius Iosifescu

Birkhäuser
Boston • Basel • Berlin

D.C. Ionescu
Power Engineering Faculty
University Politehnica Bucharest
77206 Bucharest
Romania

N. Limnios
Division Mathématiques Appliquées
Université de Technologie de Compiègne
60205 Compiègne Cedex
France

Library of Congress Cataloging-in-Publication Data

Statistical and probabilistic models in reliability / D.C. Ionescu,
N. Limnios, editors.
p. cm.
Includes bibliographical references and index.
ISBN 0-8176-4068-1 (alk. paper)
1. Reliability (Engineering)—Statistical methods.
2. Reliability (Engineering)—Mathematical models.
3. Probabilities. I. Ionescu, D. C. (Dumitru Cezar) II. Limnios,
N. (Nikolaos)
TA169.S734 1999
620'.00452—dc21 98-29995
CIP

AMS Subject Classifications: 62N05

Printed on acid-free paper

Birkhäuser

ISBN 0-8176-4068-1
ISBN 3-7643-4068-1

Typeset by the editors in LaTeX.
Cover design by Vernon Press, Boston, MA.
Printed and bound by Hamilton Printing Co., Rensselaer, NY.
Printed in the United States of America.

9 8 7 6 5 4 3 2 1

Contents

Foreword

This volume consists of twenty-four papers selected by the editors from the sixty-one papers presented at the 1st International Conference on Mathematical Methods in Reliability held at the Politehnica University of Bucharest from 16 to 19 September 1997. The papers have been divided into three sections: statistical methods, probabilistic methods, and special techniques and applications. Of course, as with any classification, some papers could be as well assigned to other sections.

Problems in reliability are encountered in items in everyday usage. Reliability is an important feature of household appliances, cars, telephones, power supplies, and so on, whether viewed from the vantage of the producer or the consumer. Important decisions are based on the reliability of the product. Obtaining systems that perform adequately for a specified period of time in a given environment is an important goal for both government and industry. Hence study and use of reliability theory, which can be applied in the research, development, and production phases of a system to enable the user to evaluate and improve performance, is a worthwhile venture. If reliability theory is to be useful, it must be quantitative in nature, because reliability must be demonstrable. Subsequently probability and statistics, among others, play an important part in its development.

The increased complexity and sophistication in electronic, mechanical, and structural systems that came into existence in the past four decades required more and more advanced mathematical methods and models in reliability. In connection with such developments, there are very basic and important questions to be considered with respect to implementation of existing theory: To what extent is this applicable to and answers real practical problems? Does not it mainly expand the inapplicable part of what we call applied mathematics?

It is no secret indeed that most of reliability studies start from purely theoretical behavior. As a result, the problems encountered by reliability engineers are not well documented in the structuring of mathematical models. It is clear that, without real communication between model-building theoreticians and practicing reliability scientists, the above questions cannot be avoided.

The aim of the Bucharest conference was to bring together the two kinds of scientists just mentioned, so that they might exchange information, discuss open problems, and fill in the gap between theory and real-life problems. We should consider this conference as a beginning of fruitful cooperation between theoreticians and practitioners. Let us thus hope some others of the same kind will follow it.

The conference was initiated by Politehnica University of Bucharest and the Technological University of Compiegne. The first one is involved in applied research, especially in power engineering, electronics, electrotechnics, and mechanics, while the second is involved in basic and applied research in mathematical reliability. The organization of the conference would not have been possible without the support of the following sponsoring bodies: Romanian Electricity Authority (RENEL), the Bucharest Institute of Power Studies and Design (ISPE), the Pitesti Institute for Nuclear Research (ICN), the Bucharest–Magurele Center of Technology and Engineering for Nuclear Projects (CITON), and the Bucharest Research and Design Institute for Thermal–Power Equipment (ICPET). The editors wish to thank all these institutions for their invaluable help.

Marius Iosifescu April 1998
Honorary President of the
Organizing Committee

Preface

The contemporary society is faced, more than ever, with the unreliability, the insecurity and the risk of the technological plants it devised, used and became dependent upon. Their gradually increasing complexity determined a significant change in the scale of their unreliability's consequences, from the level of personal and local risks, to that of regional or even planetary collective risks.

We are almost used to cosmic or aviation catastrophes, while nuclear accidents are considered possible and we only attempt to diminish their consequences by elaborate emergency plans. Could this mean that technology is out of control, that scientists and researchers have been over taken by the complexity of their own products or that research in this field is falling behind?

In reality, we witness a competition between the more and more advanced results of technological research and the means and methods of endowing them with the desired reliability and safety. This process will never succeed in eliminating risks entirely, only in identifying and decreasing them, but now, the reliability indicators are very strong parameters in the decision process.

In this context, the scientific world, with its preoccupation for research and disseminating information, plays an essential part in creating the culture of every specialty, in opening the way in the new fields and especially in strengthening the newly discovered methods and techniques. This general direction was followed by the recent International Reliability Conference MMR'97 which was focused to "Mathematical Methods in Reliability".

The present work, by means of the problems it deals with, aims at presenting the evolution of the most recent modelling researches based on the use of statistical and probabilistic models, and, at the same time, at pointing out, in new works, the present tendencies of reliability research and its applications. Most of the articles included in the volume have been selected by the editors from the very articles and interventions presented at the conference, or have been adjusted to suit better the problems of modelling.

In spite of the spectacular evolution of the Mathematical models and of the possibilities of calculation offered by the modern computing systems, the reliability analysis of the technical systems is difficult to approach, especially because of their technological complexity and the more and more increased quality requirements.

The main difficulties lie in ensuring the compatibility between "the technological reality with its requirements" and "the reliability models", under the conditions of minimizing modelling and numerical evaluation efforts.

It needs to be specified from the very beginning that modelling and models appear in all stages of the reliability analysis, whether in the structural or functional description of the system, or in the specific description of its random behaviour.

From a theoretical point of view, the simplest approach would consist in devising a global model, including all functional and dysfunctional aspects of the system, and from which the desired results could be extracted, both from a functional perspective and from the point of view of their reliability (random behaviour).

This thorough approach is very difficult to realize and becomes even impossible for complex systems. Therefore, most frequently, the reliability analysis is decomposed and modulated, creating specific submodels, sufficiently precise to ensure their correct integration in the general analysis.

Before passing to a more detailed presentation of the selected articles, a few characteristics of the procedure need to be pointed out, in order to guide the approach to modelling in a more appropriate way and even to explain some of the dissatisfactions and failures that might appear in the domain of reliability analyses, concerning the employed models and their precision.

Essentially, the model is a representation of reality, contrived in Mathematical language, based on a set of hypotheses in relation to which reality is approximated.

If the representation is considered to be accurately achieved, which is generally verifiable, then the model can be perfect in relation to the accepted hypotheses, but quite imperfect in relation to the reality which is subject to modelling.

This implies that it is not enough to achieve very elaborate models, but it is also necessary to find the link between reality and the hypotheses it imposes. Therefore, reality shouldn't be crippled in order to fit the hypotheses of the already known models; what should be done is to discover the particular features of the technical systems and, by their means, to define the minimal hypotheses, developing the correspondent models. As a consequence of the previous statements, the contributions of the different scientists involved in these analyses can be outlined. Mathematicians are generally ascribed the task of building up models and making them correspondent to the accepted hypotheses, they being obviously the ones to look for the most effective ways to achieve their description and numerical evaluation, while the technologists are concerned mainly with defining correctly the hypotheses of modelling in accordance with reality and, of course, with validating the devised models.

It is self-evident that the two aspects cannot be treated separately and that both categories of analysts need to master well enough both the technological and the modelling aspects. In other words, the mathematician will never be able to build up purely theoretical models capable of answering all the problems of the present technical systems, and technologists will never manage to identify the categories of possible hypotheses without knowing the requirements of the modelling techniques.

The selected articles focus better these ideas concerning modelling aspects. Mainly, the following advanced directions were strongly emphasized at the conference: the estimation of accelerating life date, the modelling of the components subject to a random diffuse stress environment and the semi- Markov reliability models. As well important review papers are included : in statistical methods for repairable systems, in accelerating life testing and in asymptotic methods in Reliability analysis of stochastic systems.

This volume is organized as follows.

In Part I, we have a first chapter (B. H. Lindqvist) which gives a survey of statistical methods and models used to fit date of repairable systems. The second chapter (O. Gaudoin) deals with a reliability growth model and in particular a goodness-to-fit test based on the Conditional Probability Integral Transformation (CPIT). Chapter 3 (T. Bedford) concerns competing risk problem, where the use of minimally informative copula with given rank correlation coefficient is developed. Chapters 4, 5 and 6 give estimation methods for accelerating life data. Chapter 4 (V. Bagdonavičius and M. Nikulin) gives a survey on methods in accelerating life testing as well as some new results. Chapter 5 (V. Bagdonavičius and M. S. Nikulin) gives in particular reliability estimators of semiparametric accelerating life data. Chapter 6 (L. Gerville-Reache and V. Nikoulina) gives asymptotic properties of parametric estimators in accelerated failure time model. Chapter 7 (M. S. Nikulin and V. N. Solev) consider the problem of testing the goodness of fit of some parametric family of survival distributions from doubly censored data. Chapter 8 (B. Ouhbi and N. Limnios) proposes nonparametric estimators and derives their asymptotic properties for the estimation of the semi- Markov kernel, of Reliability and Availability of finite semi-Markov systems.

Part II. Chapter 9 (V. S. Korolyuk) gives a survey of asymptotic methods in Reliability analysis of stochastic systems of Ukrainian probabilistic school, based on solution of the problem of singular perturbation for reducible invertible operators. Chapter 10 (V. S. Korolyuk, N. A. Derzko and V. V. Korolyuk) considers a Markov repairman problem and gives diffusion approximation results. Chapters 11 (K. Kolowrocki) and 12 (A. Cichocki, D. Kurowicka and B.

Milczeek) give a review of asymptotic results of reliability function of large series/parallel binary systems. The case of multi-state systems is also considered. Chapter 13 (O. Pourret, J. Collet and J.-L. Bon) gives useful results concerning bounds in stiff Markov chains on state probability as arose in practical reliability problems. Chapter 14 (A. Le Breton and J.-L. Soler) investigates the failure rate behaviour of a component subject to a random diffuse stress environment of an extremely variable random function intensity. Chapter 15 (C. A. Zahalca and M. Chardi) study the Reliability of series and parallel systems which components are subjected to a common point stress process. Chapter 16 (G. Oprisan) presents some investigation about system failure rates and particularly of the finite semi-Markov failure rate. Finally Chapter 17 (B. Ksir) gives asymptotic results concerning the failure time of consecutive-k-out-of-n: systems.

Part III. Chapter 18 (N. Balakrishnan and P. S. Chan) proposes a start up demonstration test performed in two stages. Chapter 19 (B. Harlamov) proposes solution for an optimal control problem for partly observable systems where the optimal times of prophylactic switches are proved. Chapter 20 (C. Lucet and J.-F. Manouvrier) presents a survey of methods to compute network reliability. Chapter 21 (A. Behr and L. Camarinopoulos) presents matroid theory to evaluate reliability of binary coherent systems. Chapter 22 (A. P. Ulmeanu and D. C. Ionescu) proposes a computer-assisted analysis of the semi-Markovian Stochastic Petri nets and an application. Chapter 23 (N. Fota, M. Kaaniche and K. Kanoun) presents a set of guidelines for the efficient and reliable construction of dependability models based on generalized stochastic Petri nets. Finally Chapter 24 (E. Alămoreanu, R. Iatan, R. Chiriţă and R. Ceauşu) gives a detailed method to evaluate reliability of industrial pipes subject to creep and fatigue effects.

Finally, the editors are grateful to Dr. Paul Ulmeanu for his enormous effort to achieve this volume.

D. C. Ionescu April 1998
Bucharest, Romania

N. Limnios
Compiègne, France

Contributors

Alamoreanu, Elena
Mechanical Department, University "Politehnica" Bucharest, 313 Splaiul Independentei, 77 206 Bucharest, Romania
e-mail: *alam@form.resist.pub.ro*

Bagdonavičius, Vilijandas
University of Vilnius, Lithuania
e-mail: *Vilijandas.Bagdonavicius@maf.vu.lt*

Balakrishnan, N.
Department of Mathematics and Statistics, McMaster University, Hamilton, Ontario, Canada L8S 4K1
e-mail: *bala@mcmail.CIS.McMaster.CA*

Bedford, Tim
Department of Mathemathics, Delft University of Technology, Delft, The Netherlands
e-mail: *t.j.bedford@twi.tudelft.nl*

Behr, Andreas
RISA, Sicherheitsanalysen GmbH, 10629 Berlin, Germany
e-mail: *aebehr@berlin.sireco.net*

Bon, Jean-Louis
IUT Sceaux 8, avenue de Cauchy, 92330 Sceaux, France
e-mail: *Jean-Louis.Bon@math.u-psud.fr*

Camarinopoulos, Leonidas
Division of Computational Methods and Computer Programming, Department of Mathematical and Physical Sciences, Faculty of Technology,

Aristotle University of Thessaloniki, Thessaloniki 54006, Greece
e-mail: *risa@eng.auth.gr*

Ceausu, Radu
Mechanical Departement, University "Politehnica" Bucharest, 313 Splaiul Independentei, 77 206 Bucharest, Romania

Chan, P. S.
Chinese University of Hong Kong, Shatin, Hong Kong

Chardi, Mohamed
Research Department, The Royal Naval School, Bd. Sour Idid, Casablanca, Morocco

Cichocki, Adam
Department of Mathematics, Gdynia Maritime Academy, ul. Morska 83, 81-962 Gdynia, Poland
e-mail: *katmatkk@vega.wsm.gdynia.pl*

Collet, Jerome
Electricité de France, Direction des Etudes et Recherches 1, Avenue du Général De Gaulle, 92141 Clamart Cedex, France
e-mail: *jerome.collet@der.edfgdf.fr*

Derzko, Nicolas A.
Institute of Mathematics, Ukrainian Academy of Science, Repine 3 Street Kiev, 252601 Ukraine
e-mail: *ec@publ.freenet.kiev.ua*

Fota, Nicolae
SOFREAVIA 3 Carrefour de Weiden, 92441 Issy les Moulineaux Cedex, France
e-mail: *fotan@smtp.sofreavia.fr*

Gaudoin, Olivier
Laboratoire de Modélisation et Calcul, Centre de Recherche Informatique et Mathématiques Appliquées de Grenoble, BP 53 - 38041 Grenoble Cedex 9, France
e-mail: *Olivier.Gaudoin@imag.fr*

Gerville-Reache, Leo
Département UFR MI2S, Victor Segalen University, Bordeaux 2, B.P. 69, France
e-mail: *gerville@u-bordeuax2.fr*

Harlamov, Boris
Institute of Problems of Mechanical Engineering, Russian Academy of Scince, St. Petersburg, Russia
e-mail: *harlamov@ensure.ipme.ru*

Iatan, Radu
Mechanical Department, University "Politehnica" Bucharest, 313 Splaiul Independentei 77 206 Bucharest, Romania

Ionescu, Dumitru Cezar
Power Engineering Department, Reliability Laboratory, University "Politehnica" Bucharest, 313 Splaiul Independentei, 77 206 Bucharest, Romania
e-mail: *dci@fiab.pub.ro*

Kaaniche, Mohamed
LAAS - CNRS, Toulouse, France
e-mail: *kaaniche@laas.fr*

Kanoun, Karama
LAAS - CNRS, Toulouse, France
e-mail: *kanoun@laas.fr*

Kolowrocki, Krzysztof
Department of Mathematics, Gdynia Maritime Academy, ul. Morska 83, 81-962 Gdynia, Poland
e-mail: *katmatkk@vega.wsm.gdynia.pl*

Korolyuk, Vladimir S.
Institute of Mathematics, Ukrainian Academy of Science, Kiev, Ukraine
e-mail: *ec@publ.freenet.kiev.ua*

Korolyuk, Vladimir V.
Institute of Mathematics, Ukrainian Academy of Science, Repine 3 Street Kiev, 252601 Ukraine
e-mail: *ec@publ.freenet.kiev.ua*

Kurowicka, Dorota
Department of Mathematics, Gdynia Maritime Academy, ul. Morska 83, 81-962 Gdynia, Poland
e-mail: *dorota@vega.wsm.gdynia.pl*

Ksir, Brahim
Institute of Mathematics, University of Constantine, 25000 Constantine, Algeria

Le Breton, Alain

Laboratoire de Modélisation et Calcul, Centre de Recherche Informatique et Mathématiques Appliquées de Grenoble, BP 53 38041 Grenoble Cedex, France
e-mail: *Alain.Lebreton@imag.fr*

Limnios, Nikolaos

Département Génie Informatique, Division Mathématiques Appliquées, Centre de Recherche de Royallieu, Université de Technologie de Compiègne, B.P. 20529, 60205 Compiègne Cedex, France
e-mail: *Nikolaos.Limnios@utc.fr*

Lindqvist, Bo Henry

Department of Mathematical Sciences, Norwegian University of Science and Technology, Trondheim, Norway
e-mail: *bo@imf.unit.no*

Lucet, Corinne

Département Génie Informatique, Heudiasyc UMR CNRS 817, Centre de Recherche de Royallieu de l'Université de Technologie de Compiègne, France e-mail: *Corinne.Luccet@hds.utc.fr*

Manouvrier, Jean-Francois

Département Génie Informatique, Heudiasyc URA CNRS 817, Centre de Recherche de Royallieu de l'Université de Technologie de Compiègne, France e-mail: *JF.Manouvrier@hds.utc.fr*

Milczek, Beata

Department of Mathematics, Gdynia Maritime Academy, ul. Morska 83, 81-962 Gdynia, Poland
e-mail: *katmatkk@vega.wsm.gdynia.pl*

Nikoulina, Valentina

Département Mathematiques, Informatiques et Sciences Sociales, Victor Segalen University, Bordeaux 2, B.P. 69 France and Département of Applied Mathematics, St. Petersburg State University, Russia
e-mail: *M.S.Nikouline@u-bordeaux2.fr*

Nikulin, Mikhail S.

Département UFR MI2S, de l'Université Bordeaux II, B.P. 69, France
e-mail: *M.S.Nikouline@u-bordeaux2.fr*

Oprisan, Gheorghe

Department of Mathematics, University "Politehnica" Bucharest, 313 Splaiul Independentei, 77 206 Bucharest, Romania
e-mail: *oprisan@math.pub.ro*

Ouhbi, Brahim
Département Génie Informatique, Division Mathématiques Appliquées, Centre de Recherche de Royallieu, Université de Technologie de Compiègne, France
e-mail: *brahim.ouhbi@utc.fr*

Pourret, Olivier
Direction des Etudes et Recherches de l' Electricité de France 1, Avenue du Général De Gaulle, 92141 Clamart Cedex, France
e-mail: *olivier@clt50bw.der.edf.fr*

Soler, Jean-Louis
Laboratoire de Modélisation et Calcul, Centre de Recherche Informatique et Mathématiques Appliquées de Grenoble, BP 53X 38 041 Grenoble Cedex, France
e-mail: *Jean-Louis.Soler@imag.fr*

Solev, Valentin
The Laboratory of Statistical Methods, Steklov Mathematical Institute, Fontanka 27, 191000, St. Petersburg, Russia

Ulmeanu, Anatoli Paul
Power Engineering Department, Reliability Laboratory, University "Politehnica" Bucharest, 313 Splaiul Independentei, 77 206 Bucharest, Romania
e-mail: *paul@fiab.pub.ro*

Zahalca, Cristina A.
Power Engineering Department, Reliability Laboratory, University "Politehnica" Bucharest, 313 Splaiul Independentei 77 206 Bucharest, Romania
e-mail: *Cristina-Anca.Zahalca@imag.fr*

List of Tables

List of Figures

Glossary of Terms

Binary System – A system that is only either operating or failed and in which every component is only either operating or failed.

Coherent Binary System – A system in which every component is contained in at least one minimal path of the system.

Cold reserve – A channel in multi-channel system which does not work till a special signal. Without working all technical equipment of the channel are not aging.

Continuous-Time Markov Chain (CTMC) – CTMCs are the simplest stochastic models suitable for representing the state of a system as t varies. Their main application in Reliability Theory is the modelling and assessment of interdependent- component systems.

Counting process – A stochastic process $N(t)$ for which $N(t)-N(s)$ for $s < t$ is the number of events between time s and t.

Diagnostic parameter – Any parameter of a system which depends on a failure time of the system.

Domination of a Binary System – The absolute value of the signed reliability domination.

Enlarged state in space sense – A subset of the state space X of the process. The operation of space enlarging induced by a map $f : X \to X$ means transformation $\xi(t) \to f(\xi(t))$ for all t, where ξ is a random process. An enlarged state in time sense is a finite sequence of states $J = (x_1, \ldots, x_k)$ which is considered like a one unit (for stepped processes). If ξ_1 is such an enlarged process and f is a corresponding map $X \to X$ then $\xi_1(t) =$

$f(\xi(t))$ only for the case when $\xi(t)$ together with some of its neighbors consists the sequence J.

Failure – The termination of a component's ability to perform a required function.

Failure mechanism – The physical or chemical process, design defects, quality defects, or other processes that are the basic reason for failure or that initiate the physical process which deterioration proceeds to failure.

Failure rate (hazard rate) – The rate at which failures occur as a function of time. If T denotes the lifetime of a component, the failure rate $\lambda(t)$ is defined as

$$\lambda(t) = \lim_{\Delta t \to 0} \frac{1}{\Delta t} P(t < T \leq t + \Delta t \mid T > t).$$

The failure rate is sometimes called *force of mortality* (FOM).

Goodness-Of Fit (GOF) test – Let X be a random vector of R^n. A parametric model M for X is a family $\mathcal{P}_M$ of possible probability distributions for X, indexed by a parameter θ:

$$\mathcal{P}_M = \left\{ P_\theta, \theta \in \Theta \subset \mathrm{R}^k \right\}.$$

Let P^* denote the true but unknown probability distribution of X.

Testing the goodness-of-fit of a realization x of X to the model M is testing the null hypothesis H_0 : "$P^* \in \mathcal{P}_M$" versus H_1 : "$P^* \notin \mathcal{P}_M$", from the observation of x.

Homogeneous Poisson process – Let N_t the number of events during $[0, t]$ of a stochastic process in which intervals between successive events are independently distributed according to a common exponential distribution $G_\xi(t) = 1 - e^{-\xi t}$. The stochastic process $\{N_t\}_{t \geq 0}$ is called an homogeneous Poisson process with intensity ξ.

i.i.d. random variables – Stochastically independent and identically distributed random variables.

Interval of constancy of the trajectory ξ covering the time point t – Any interval $[t_1, t_2)$, $t_1 \leq t < t_2$ such that $\xi(s) = \xi(t)$ for all $s \in [t_1, t_2)$. There exists a maximal interval of constancy covering the point t if any.

k-out-of-n configuration – A configuration which performs when at least k of its n components perform, and fails otherwise.

Likelihood function – Function of the parameters of a model which, under these parameter values, measures the chance of obtaining the actually observed data.

Local drift and local variance of a time homogeneous diffusion process ξ at the point x – A vector $a = (a^i)$:

$$a^i = \lim_{t \to 0} \frac{1}{t} \int_B x_1^i \, P_x(\xi(t) \in dx_1),$$

and a matrix

$$b^{ij} = \lim_{t \to 0} \frac{1}{t} \int_B x_1^i x_1^j P_x(\xi(t) \in dx_1),$$

where P_x is a conditional distribution of the process given the initial point x, B is any neighborhood of the point x, x_1^i is the i-th coordinate of the point $x_1 \in \mathrm{R}^d$. For Markov diffusion process these two parameters define all the family of its conditional distributions. There exists analogy of these parameters for a semi-Markov process of diffusion type which are defined in terms of first exit times from a small neighborhood of its initial point.

Marked point process – Stochastic process which records events in time, and in addition a "mark" (a random vector) corresponding to each event.

Matroid Base Family – A set family $\mathbf{B}$ having the property that for every $B_1, B_2 \in \mathbf{B}$ with $x \in B_1$ there exists an element $y \in B_2$ such that $(B_1 - \{x\}) \cup \{y\} \in \mathbf{B}$.

Matroid Base Family System – A binary system whose minimal path set is a matroid base family.

Maximum likelihood estimator – The values of the parameters which maximizes the likelihood function.

Mean Time To Failure (MTTF) – Let T denote the lifetime of a component, with cumulative distribution function $F(t)$ and reliability $R(t)$. The mean time to failure $MTTF$ is the mean (expectation) value of T which is given, if it exists, by

$$MTTF = \int_0^\infty t \, dF(t) = \int_0^\infty R(t) dt.$$

Minimal Path – A minimal set of components such that a given system is operating if the set elements are operating and if it does not contain any other path.

Multivariate exponential distribution – Let $(T_1, T_2, ..., T_n)$ be a random vector of R^{+^n}. The survival function of the multivariate exponential distribution is

$$P(T_1 > t_1, ..., T_n > t_n) = e^{-\sum_{i=1}^{n} \mu_i t_i - \sum_{i<j} \mu_{ij} \max(t_i, t_j) - ... - \mu_{1...n} \max(t_1, .., t_n)}$$

where the $\mu_{i_1, i_2, ..., i_k}$ are positive real numbers.

The survival function of the *symmetrical multivariate exponential distribution* is

$$P(T_1 > t_1, ..., T_n > t_n) = e^{-\mu_1 \sum_{i=1}^{n} t_i - \mu_2 \sum_{i<j} \max(t_i, t_j) - ... - \mu_n \max(t_1, .., t_n)}$$

Non-Homogeneous Poisson Process (NHPP) –

A NHPP with intensity function $\lambda(t)$ is a stochastic counting process on R_+ $\{N_t\}_{t \geq 0}$ defined by:

- $N_0 = 0$.
- For all $t \geq 0$, $P(N_{t+\Delta t} - N_t \geq 2) = o(\Delta t)$.
- $\{N_t\}_{t \geq 0}$ has independent increments.
- For $0 \leq s < t$, $N_t - N_s$ is Poisson distributed with parameter $\int_s^t \lambda(u) du$.

When λ is constant, the process is an Homogeneous Poisson Process (HPP).

Null hypothesis, H_0 – The statement being tested in a test of significance. Usually the null hypothesis is a statement of "no effect".

Parallel configuration – A configuration which fails if all its components fail.

Path – A path is a sequence of consecutive states of the system.

Path of a multicomponent system – A subset of components such that if all the components of this subset perform, and if the other components have failed, then the system performs.

Sometimes the term *link* is used to avoid any confusion with the notion of path in graph theory.

Point stress process – A succession of impulsions (for example shocks) which occur at random times with random amplitudes.

Power of a test – The probability that a test will reject H_0 for given values of the parameters not corresponding to H_0.

P-value – The probability, computed assuming that the null hypothesis is true, that the test statistic would take a value as extreme or more extreme than that actually observed.

Regeneration cycle – A part of trajectory of a random process between two neighboring regenerative times of the process. In wide sense the regenerative time is a random time not depending on future (stopping time or Markov time) such that the future part of the process is independent of the past given the meaning of the process at this time (Markov property). In narrow sense this random time is a hitting time of some fixed point where the sequence of such random times are made not to have limits in finite part of the time axis.

Reliability – The ability of a component to performs an required function, under given environmental and operational conditions and for a stated period of time (ISO 8402).

Let T denote the lifetime of a component. The reliability $R(t)$ of a component is

$$R(t) = P(T > t), \qquad t \geq 0.$$

$R(t)$ is sometimes called the *survival function* at time t of a component and is denoted by $S(t)$.

Reliability (of a) System at time t – The probability that the binary system is operating for all points of time in $[0, t]$.

Semi-Markov transition function of a stepped semi-Markov process – A conditional distribution of the pair: the first jump time and the meaning of the process just after the jump, given the initial state of the process.

Series configuration – A configuration which performs if all its components perform.

Signed (Reliability) Domination of a Binary System – The number of different ways to cover the system's component set by an odd number of the system's minimal paths minus the number of different ways to cover the system's component set by an even number of the system's minimal paths.

Significance level – The value α such that the null hypothesis is rejected if the P-value is less than α.

S-Plus – Computer software for statistical analysis of data (available from MathSoft, Seattle, Washington).

Stepped semi-Markov process – A stepped random process in whom any jump time is a regenerative (wide) time. A random process is called stepped if all its trajectories are piecewise constant and have a finite set of jumps on every bounded time interval.

Stieltjés measure – Let S a positive, nondecreasing, right-continuous, bounded function. The Stieltjés measure m_S is defined by

$$\forall\ (s,t),\ 0 \leq s < t,\ m_S(]s,t]) = S(t) - S(s)$$

where S is called the *cumulative distribution function* (cdf) of m_S.

Stiffness – A CTMC is said to be stiff if some of the states are fast, i.e., their mean sojourn times are low. Stiffness is known to yield computational problems.

Stress environment – External conditions that affect the state of a component.

Strong Markov process – A random process which possesses the Markov property with respect to any Markov time.

Sufficient statistic – Let X be a random vector of R^n. Assume that the distribution of X belongs to the parametric family $\mathcal{P}_M$ defined above. A statistic $T(X)$ is a sufficient statistic for θ if and only if the conditional distribution of X given $T(X)$ does not depend on θ.

The Fisher-Neyman theorem says that $T(X)$ is a sufficient statistic for θ if and only if there exists two measurable functions g and h such that the likelihood function can be written as:

$$\mathcal{L}(\theta, x) = g\left(T(x), \theta\right) h(x).$$

Totally Amenable System – A binary system for which the domination gives an upper bound for the required number of steps to compute system reliability.

Transition rate – Given two states i and j of the system, the probability that the system is in j at time t, knowing that it is in i at time 0, develops into:

$$a_{ij}t + o(t), \qquad \text{as } t \downarrow 0$$

where a_{ij} is called the transition rate from state i to state j.

PART I

Statistical Methods

1

Statistical Modeling and Analysis of Repairable Systems

Bo Henry Lindqvist

Norwegian University of Science and Technology, Trondheim, Norway

Abstract: A repairable system can briefly be characterized as a system which is repaired rather than replaced after a failure. The most commonly used models for the failure process of a repairable system are nonhomogeneous Poisson processes (NHPP), corresponding to minimal repairs, and renewal processes (RP), corresponding to perfect repairs. The paper reviews models for more general repair actions, often called "better-than-minimal repair" models. In particular we study the class of so called trend-renewal processes (TRP), which has both the NHPP and the RP as special cases. Parametric inference in TRP models is considered, including cases with several systems involving unobserved heterogeneity. Trend testing is discussed when the null hypothesis is that the failure process is an RP. It is shown how Monte Carlo trend tests for this case can be made from the commonly used trend tests for the null hypothesis of a homogeneous Poisson process (e.g. the Laplace test and the Military Handbook test). Simulations show that the Monte Carlo tests have favorable properties when the sample sizes are not too small.

Keywords and phrases: Repairable system, nonhomogeneous Poisson process, renewal process, better-than-minimal repair, heterogeneity, trend, Monte Carlo test

1.1 Introduction

A repairable system is a system which, after failing to perform one or more of its functions satisfactorily, can be restored to fully satisfactory performance by any method, other than replacement of the entire system [Ascher and Feingold (1984)].

Much of the literature on repairable systems seems to be motivated from applications to mechanical systems. However, the theory is of course not limited to such situations. An important application area is software reliability. Software systems will usually experience an increasing reliability due to correction of errors, and a number of different models have been suggested for the failure processes observed under testing as well as ordinary use. Finally, the theory of repairable systems will have applications in such areas as medicine and finance, and many other disciplines involving recurrent events.

A particular example of an application area, which has motivated much of the present work, is the oil and gas installations at the Norwegian continental shelf. The oil companies possess large data bases containing survival- and maintenance data for various kinds of equipment. An example of one such database is the OREDA (=Offshore REliability DAtabase), which is a cooperation between the largest oil companies involved in the North Sea. Analyses of the data are used for, e.g., descriptions and comparisons of equipment of different types and manufacturers; predictions for new installations; identification of problems and trends with the aim of improving maintenance procedures; control and improvement of safety and cost-effectiveness; exchange of technology.

The purpose of the present paper is to review some basic models and methods used for the analysis of failure data from repairable systems, with emphasis put on our own research interests. The paper is therefore not particularly intended to give a broad review of the field.

In particular we shall consider "better-than-minimal repair models", i.e. non-Poisson models. Another topic considered is trend testing, when the null hypothesis is that the failure process is a renewal process, and the alternative is a monotone trend in the failure intensity. A brief survey of some important papers in the field is given in the following section.

1.2 "Major Events" in the History of Repairable Systems Reliability

The choice of "events" is of course subjective, and the list is by no means meant to be comprehensive.

A natural point of departure for a brief historic survey is the talk given by Harold Ascher and Harry Feingold at the Office of Naval Research Reliability Workshop in 1983, with the title *Repairable systems: Reliability's stepchild.* The year after appeared their famous and important book, *"Repairable Systems Reliability – Modeling, inference, misconceptions and their causes"*

This is a remarkable book, a statistical book written with passion and anger. Their starting point is the observation that reliability researchers, and practitioners using their methods, have not recognized the crucial difference between treatment of *repairable* systems and *nonrepairable* components. They illustrate this by their "happy" and "sad" systems, with successive times between failures given by, respectively,

15, 27, 32, 43, 51, 65, 177 and 177, 65, 51, 43, 32, 27, 15

Although the two *sets* of numbers are the same, considering them sequentially in time the first system is apparently getting better and better while the second is becoming increasingly worse. Nevertheless, it is believed that even today the method most frequently used by reliability engineers is to treat the numbers as i.i.d. from some distribution and then fitting a suitable parametric model.

The lesson to learn from Ascher and Feingold's book is that repairable systems need to be analysed by nonstationary stochastic point processes rather than the more elementary methods for random samples. However, the book was of course not the first source of such methods. Statistical models and methods appropriate for analysis of repairable systems were known at least from the fifties (see below). But no reliability textbook before 1978 provided *any* test for distinguishing between cases like the "sad" and "happy" systems.

Still, as pointed out by Ascher and Feingold, the world's oldest known formal statistical test, Laplace's test from 1773, would be appropriate to distinguish between the cases. This test is believed to still be the one most used for trend testing (see Section 1.7). Moreover, the book by Ascher and Feingold is still the basic reference on repairable systems reliability.

The next "event" that we shall mention is the talk by David R. Cox, read before the Royal Statistical Society in London in March 1955 and published in Cox (1955). This is one of the first comprehensive treatments of appropriate statistical methods, where reliability applications are explicitly mentioned.

Cox touches a large number of problems, most of them motivated from the clothing industry. Examples of topics of particular importance for reliability applications are: Trend tests, testing whether a failure process is Poisson (which Cox calls *randomness*), autocorrelated intervals, doubly stochastic Poisson processes, heterogeneity (called *components of variance*), correlations between different types of events, mean repair times, availability.

In addition to demonstrating a large number of useful statistical techniques, the paper is special – as expressed by one of the discussants – "because it recognizes that the application of statistics is an *art*; that we cannot merely follow standardized procedures for any situation and always find an answer". Many results from the paper are contained in the subsequent book by Cox and Lewis (1966), which still is one of the most cited sources on the subject.

Another early contribution to the theory, and indeed a very much cited paper, is the 1963 paper by Frank Proschan, "Theoretical explanation of observed decreasing failure rate". Reliability engineers analysing failure data frequently experience an apparent *decreasing* failure rate, often counterintuitive in view of wear and aging effects. Proschan's paper is important because it is presumably the first to point out that such observed decreasing rates could possibly be caused by *heterogeneity* between systems. Proschan presents failure data from 13 airconditioner systems on Boeing 720 airplanes. He argues that failures occur according to a homogeneous Poisson process for each separate plane, but with different rates for each plane. He demonstrates this statistically by using a result from Barlow, Marshall and Proschan (1963) which implies that a mixture of exponential distributions has a decreasing failure rate.

Proschan's test is constructed by pooling all the interfailure times, from all the planes, and then using a test statistic which is sensitive to a decreasing failure rate. In this way he gets a P-value of 0.007 for the null hypothesis of *identical* homogeneous Poisson processes. We also tried the test on "variance components" suggested by Cox (1955), and got approximately the same P-value. We return to the airconditioner data later.

We close the introductory historical remarks by mentioning a definite breakthrough in the analysis of lifetime data. This is the martingale approach to statistical inference in point processes, as introduced by Aalen in his Ph.D. thesis in 1975, and then further refined and applied in the subsequent years. A comprehensive source is the book Andersen et al. (1992). One of the main features is the ability of taking into consideration rather complicated history processes and censoring mechanisms, thus enabling a realistic and intuitive way of modeling.

1.3 Notation and Basic Definitions

We shall in this paper restrict attention to systems which following a failure are immediately restored to a functioning state. For simplicity of notation we shall consider systems observed from time $t = 0$, and let $T_1, T_2, \ldots$ denote the successive *failure times*. The assumption of 0 repair time is reasonable if repair times are negligible compared to times between failures. The assumption can also be justified provided the time scale is taken to be the *operation time* of the system. In the latter case, a marked point process may keep track of the repair times, and possibly additional information such as production conditions, maintenance and external conditions. However, marked processes will not be considered in this paper.

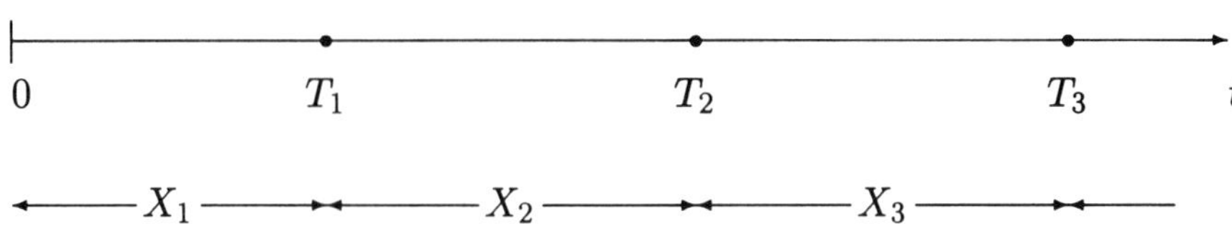

Figure 1.1: Failure times T_i and interfailure times X_i of a repairable system

We shall let the *interfailure times*, i.e. times between failures, be denoted $X_1, X_2, \ldots$. Thus we have

$$X_i = T_i - T_{i-1}, \quad i = 1, 2, \ldots$$

where for convenience we define $T_0 \equiv 0$. (See Figure 1.1). The sequence $T_1, T_2, \ldots$ of failure times and the sequence $X_1, X_2, \ldots$ of interfailure times thus contain exactly the same information about a particular realization of the process. This also holds for the counting process representation

$$N(t) = \text{ number of failures in } (0, t].$$

A plot of $N(t)$ as a function of t (or equivalently, plotting i vs. T_i) is the usual way of depicting data from single repairable systems. An example is given in Figure 1.2 which shows the observed failure process of the airconditioner system of one of the airplanes in the Proschan (1963) data.

The most popular stochastic models for repairable systems are the *renewal process*, RP, and the *nonhomogeneous Poisson process*, NHPP. The defining property of an RP is that the interfailure times X_i are i.i.d. with some distribution F. We shall denote this process by RP(F). The NHPP is defined in terms of an intensity function $\lambda(t)$ such that the probability of a failure in a (small) time interval $[t, t + \Delta t)$ is $\lambda(t)\Delta t + o(\Delta t)$, while the number of events in disjoint time intervals are stochastically independent. We denote this process by NHPP($\lambda(\cdot)$). A nice reference on point process models in reliability is Thompson (1988).

As we shall see in the next section, the RP and NHPP can be viewed as corresponding to the two extreme repair actions, *perfect repair* and *minimal repair*, respectively. In order to describe models for more general repair strategies, we need some more notation from the theory of point processes.

Let $\mathcal{F}_{t-}$ denote the *history* of the failure process up to, but not including, t. Usually we shall think of $\mathcal{F}_{t-}$ as $\{N(s) : 0 \leq s < t\}$ which keeps record of all the failures before time t. It could, however, contain more information, e.g. in the case where a marked point process is used as mentioned earlier. The conditional intensity $\lambda(t|\mathcal{F}_{t-})$ is defined by [Andersen et al. (1992)]

$$\lambda(t|\mathcal{F}_{t-}) = \lim_{\Delta t \to 0} \frac{Pr(\text{failure in } [t, t + \Delta t)|\mathcal{F}_{t-})}{\Delta t}.$$

In other words, $\lambda(t|\mathcal{F}_{t-})\Delta t$ is approximately the probability of failure in the time interval $[t, t + \Delta t)$, conditional on the failure history before time t.

For statistical inference in models defined by the conditional intensity we need to write down the *likelihood function*. Suppose that a single repairable system with failure process $N(t)$ and conditional intensity function $\gamma(t) \equiv \lambda(t|\mathcal{F}_{t-})$ is observed from time 0 to time τ, resulting in failure times $T_1, T_2, \ldots, T_{N(\tau)}$.

The *likelihood* function is then given by [Andersen et al. (1992)]

$$L_\tau = \left\{ \prod_{i=1}^{N(\tau)} \gamma(T_i) \right\} \exp\left(- \int_0^\tau \gamma(u) \mathrm{d}u \right) \tag{1.1}$$

This likelihood function is valid under the assumption that τ is a so-called *stopping time*, which holds for most censoring schemes occurring in practice. In particular it holds for the two most common observation schemes for repairable systems, *time truncation* in which τ is a prespecified point in time, and *failure truncation* where the process is observed until a given number n of failures have occurred.

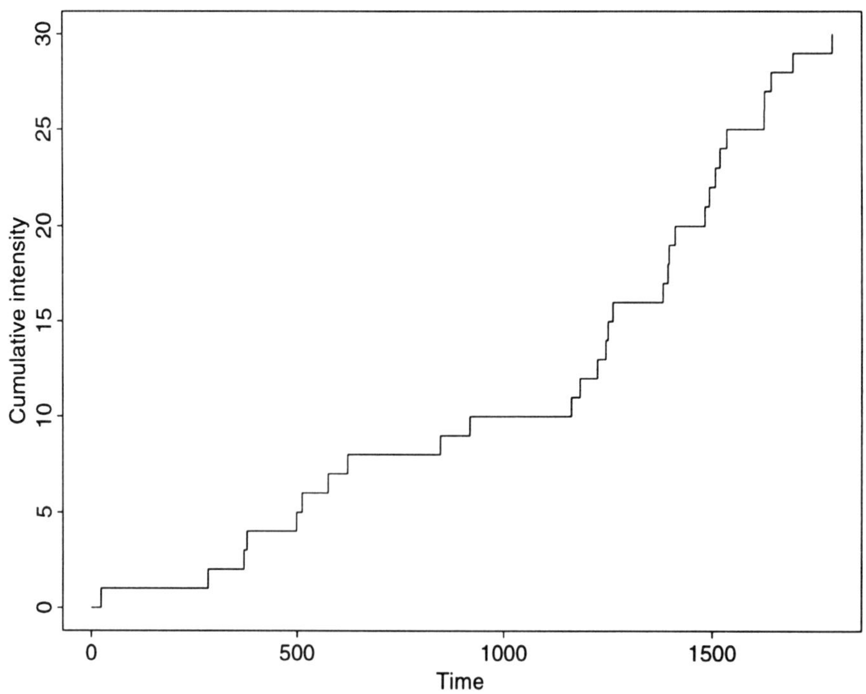

Figure 1.2: Plot of observed cumulative intensity, $N(t)$, for airconditioner failures of plane 7912 in the Proschan (1963) data

For parametric models for $\gamma(t)$ there is a well developed theory for maximum likelihood estimation based on the martingale approach to point processes, see Andersen et al. (1992). Other relevant references for statistical inference in reliability models are Crowder et al. (1991) and Høyland and Rausand (1994).

1.4 Classification of Repair Actions

A common recipe for analysis of a set of data from a repairable system seems to be as follows. First, apply a test for *trend* in the interfailure times X_i (see Section 1.7). If no significant trend is found, then model the X_i by an RP using the well established machinery for analysis of independent and identically distributed (i.i.d.) observations. Otherwise, use an NHPP model which is able to handle various kinds of trends through specification of the intensity function $\lambda(t)$. For example, a *deteriorating* system corresponds to an increasing function $\lambda(t)$, while an *improving* system corresponds to a decreasing $\lambda(t)$. Of course a *homogeneous* Poisson process, HPP(λ), corresponds to a constant intensity $\lambda(t) \equiv \lambda$.

As mentioned in the previous section, the RP and NHPP models correspond to, respectively, perfect and minimal repair. We shall now consider a unifying approach which makes these two models the extremes of a more general model. The unifying approach involves the specification of a *hazard rate* function $z(t)$, which we think of as being the hazard function of the time to first failure of the system. This function will then be used in describing the various models by means of their conditional intensity function $\lambda(t|\mathcal{F}_{t-})$. Recall that if T is a positive, continuous random variable, then the hazard rate is $z(t) = \lim_{\Delta t \to 0} Pr(T \leq t + \Delta t | T > t)/\Delta t$.

For a renewal process, RP(F), it is well known that

$$\lambda(t|\mathcal{F}_{t-}) = z(t - T_{N(t-)})$$

where z is the hazard rate corresponding to F, and $t - T_{N(t-)}$ is the time since the last failure strictly before time t (often called *local time*). Thus, after each failure, the rate of failures is returned back to the state at time 0, i.e. the system is as good as new immediately after each failure. This is the reason for calling the RP a *perfect repair* model.

On the other hand, an NHPP has the conditional intensity

$$\lambda(t|\mathcal{F}_{t-}) = z(t)$$

which means that after a failure, the failure rate is exactly as if no failure had ever occurred. Thus we have a reason for calling this a "minimal repair", since the system is only restored to a state where it is exactly as good (or bad) as it was immediately before the failure. It should also be remarked that for an NHPP the intensity function $\lambda(t)$ equals, for all t, the hazard rate $z(t)$ of the time to first failure.

For many applications it is more reasonable to model the repair action by something in between the two given extremes, so called "better-than-minimal repair" models. One such model is suggested by Brown and Proschan (1983), where at the time of each failure a perfect repair occurs with probability p and a minimal repair occurs with probability $1-p$. More general suggestions were given by Kijima (1989).

In a recent paper, Dorado, Hollander and Sethuraman (1997) suggest a general repair model which contains all the above models as special cases. The definition in their paper is given in terms of conditional distributions of the X_i. An equivalent definition in terms of conditional intensities is

$$\lambda(t|\mathcal{F}_{t-}) = \theta(t)z\left(a(t) + \theta(t)(t - T_{N(t-)})\right)$$

which in addition to the basic hazard rate $z(t)$ is defined in terms of two random processes

$$\begin{aligned} a(t) &= \text{effective age at last failure before } t \text{, i.e. at time } T_{N(t-)} \\ \theta(t) &= \text{repair characteristic of last repair before } t \\ &\quad \text{(called "life supplement")}, 0 < \theta(t) \le 1. \end{aligned}$$

It is easy to see that an RP corresponds to $\theta(t) \equiv 1$, $a(t) \equiv 0$, while an NHPP has $\theta(t) \equiv 1$, $a(t) = T_{N(t-)}$. Kijima's (1989) Model I has $\theta(t) \equiv 1$, while the effective age $a(t)$ increases by the amount $D_j X_j$ at the jth failure time T_j, where $D_1, D_2, \ldots$ is a sequence of random variables independently distributed on $[0,1]$ and independent of other processes. In Kijima's Model II, we have $\theta(t) \equiv 1$, while the effective age $a(t)$ immediately after the jth failure is set to $\sum_{k=1}^{j}(\prod_{i=k}^{j} D_i)X_k$. Moreover, when D_j is 1 with probability p and 0 with probability $1-p$, we obtain the Brown-Proschan model.

Dorado et al. (1997) consider estimation in the general repair model when the effective ages $a(t)$ and life supplements $\theta(t)$ are observed at each failure, in addition to the failure times T_i. Using the general counting process framework they derive estimates and confidence bands for the distribution of time to first failure, under the condition that this has increasing failure rate. Statistical inference in the Brown-Proschan model was first studied by Whitaker and Samaniego (1989).

Berman and Turner (1992) consider estimation in the fairly general parametric model

$$\lambda(t|\mathcal{F}_{t-}) = g\left\{\sum_{k=1}^{p} \theta_k Q_k(t|\mathcal{F}_{t-})\right\} \tag{1.2}$$

where g is a known monotonic continuous function and the $Q_k(t|\mathcal{F}_{t-})$ are known functions of t and the history $\mathcal{F}_{t-}$. The data needed for estimation are just the failure times T_i.

The special case when $g(y) = e^y$ is applied to repairable systems by Lawless and Thiagarajah (1996). In particular they consider the model

$$\lambda(t|\mathcal{F}_{t-}) = e^{\alpha + \beta g_1(t) + \gamma g_2(t - T_{N(t-)})} \tag{1.3}$$

where α, β, γ are parameters and g_1 and g_2 are known functions. This is, in fact, an example of what Cox (1972) called a *modulated renewal process.*

1.5 The Trend-Renewal Process

An interesting feature of the model (1.3) is that it contains both an NHPP and an RP as special cases. This makes it possible to choose between a minimal and perfect repair model by use of statistical (likelihood ratio) tests for the parameters β and γ. We shall present and study an alternative model, which at the outset may seem very similar to (1.3), but which is not even covered by the general formula (1.2).

The idea behind the process, which was termed the *trend-renewal process* by Lindqvist (1993) and Lindqvist et al. (1994), is to generalize the following well known property of the NHPP. First, define the *cumulative* intensity function corresponding to an intensity $\lambda(\cdot)$ by $\Lambda(t) = \int_0^t \lambda(u)\mathrm{d}u$.

The result is that if $T_1, T_2, \ldots$ is a NHPP($\lambda(\cdot)$), then the time-transformed stochastic process $\Lambda(T_1), \Lambda(T_2), \ldots$ is HPP(1).

We define the *trend-renewal process* (TRP) simply by allowing the above HPP(1) to be *any* renewal process RP(F). Thus, in addition to the intensity function $\lambda(t)$, for a TRP we need to specify a distribution function F of the interarrival times of this renewal process. Formally we can define the process TRP($F, \lambda(\cdot)$) as follows:

Let $\lambda(t)$ be a nonnegative function defined for $t \geq 0$, and let $\Lambda(t) = \int_0^t \lambda(u)\mathrm{d}u$. The process $T_1, T_2, \ldots$ is called TRP($F, \lambda(\cdot)$) if the time-transformed process $\Lambda(T_1), \Lambda(T_2), \ldots$ is RP(F), i.e. if the $\Lambda(T_i) - \Lambda(T_{i-1})$; $i = 1, 2, \ldots$ are i.i.d. with distribution function F.

Figure 1.3 illustrates the definition. Note that for an NHPP($\lambda(\cdot)$), the RP(F) is replaced by HPP(1). Thus TRP($1 - e^{-x}, \lambda(\cdot)$) = NHPP($\lambda(\cdot)$). Also, TRP($F$, 1) = RP($F$), which shows that the TRP generalizes both the RP and the NHPP.

Remark, however, that the representation TRP($F, \lambda(\cdot)$) is unique only up to a scale factor in the following manner,

$$\mathrm{TRP}(F(x), \lambda(t)) = \mathrm{TRP}(F(cx), c^{-1}\lambda(t)) \tag{1.4}$$

for any $c > 0$. This means in particular that if the expected value of F exists, it can without loss of generality be assumed to be 1 (which it is, indeed, for the HPP(1)).

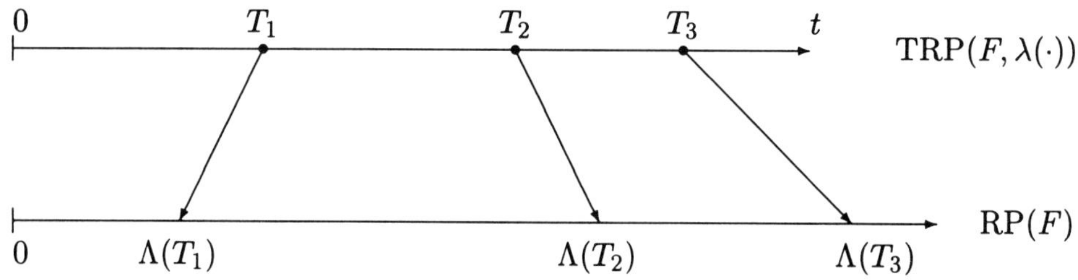

Figure 1.3: The defining property of the trend-renewal process

The *inhomogeneous gamma process* of Berman (1981) is a special case of the TRP, given by TRP$(G, \lambda(\cdot))$ with G corresponding to a gamma-distribution with scale parameter 1. Berman studied in particular what he calls a *modulated gamma process*, with intensity function given by

$$\lambda(t) = \rho \exp\{\sum_{i=1}^{p} \beta_i z_i(t)\} \tag{1.5}$$

where ρ and the β_i are unknown parameters and the $z_i(\cdot)$ are given ("explanatory") functions. Of particular interest is the model for "cyclic trend of fixed frequency", for which

$$\lambda(t) = \rho \exp\{\beta_1 \cos(\omega t) + \beta_2 \sin(\omega t)\}$$

with ω fixed. Testing the null hypothesis $\beta_1 = \beta_2 = 0$ here leads to a test for the null hypothesis of an RP with gamma-distributed interfailure times vs. a *non*-monotonic (cyclic) trend.

We obtain the conditional intensity function for the TRP$(F, \lambda(\cdot))$ from the following informal derivation, where $z(t)$ here denotes the hazard rate corresponding to F. We have

$$\begin{aligned} \lambda(t|\mathcal{F}_{t-})\Delta t &\doteq P(\text{failure in } [t, t+\Delta t)|\mathcal{F}_{t-}) \\ &= P(\text{failure in } RP(F) \text{ in } [\Lambda(t), \Lambda(t+\Delta t))|\mathcal{F}_{t-}) \\ &\doteq P(\text{failure in } RP(F) \text{ in } [\Lambda(t), \Lambda(t) + \lambda(t)\Delta t))|\mathcal{F}_{t-}) \\ &= z(\Lambda(t) - \Lambda(T_{N(t-)}))\lambda(t)\Delta(t) \end{aligned}$$

so that

$$\lambda(t|\mathcal{F}_{t-}) = z(\Lambda(t) - \Lambda(T_{N(t-)}))\lambda(t). \tag{1.6}$$

It is seen that the intensity function of the TRP is a product of one factor, $\lambda(t)$, which depends on the age t of the system and one factor which depends on a transformed time from the last previous failure. In fact, time since last failure is measured on a scale depending on the current intensity of failures, in contrast to the model (1.3) of Lawless and Thiagarajah (1996).

We conclude the present section by showing how the trend-renewal process naturally arises in connection with bootstrapping of NHPP models. The idea of *bootstrapping* in statistics can briefly be explained as follows [Efron and Tibshirani (1993)]:

Suppose that the observed phenomenon obeys the unknown probability law $\mathcal{P}$, and let $\hat{\mathcal{P}}$ be an estimate of $\mathcal{P}$ obtained from the observations. Bootstrapping means to estimate a desired characteristic of $\mathcal{P}$ by deriving the corresponding characteristic for $\hat{\mathcal{P}}$. In practice, these characteristics are usually found by simulating a large number of realizations from the estimated law $\hat{\mathcal{P}}$.

Now consider an NHPP($\lambda(\cdot)$), which we for simplicity assume is observed until the nth failure, for a fixed number n (failure truncation). Let $T_1, T_2, \ldots, T_n$ be the observed failure times, and let $\mathcal{P}$ be the unknown probability law of these observations.

Suppose we have found an estimate $\hat{\lambda}(\cdot)$ of $\lambda(\cdot)$ based on the observations $T_1, \ldots, T_n$. Then a natural estimate $\hat{\mathcal{P}}$ of $\mathcal{P}$ is NHPP($\hat{\lambda}(\cdot)$), from which it would be easy to simulate new realizations, so called bootstrap samples.

Another way of constructing an estimated model, $\tilde{\mathcal{P}}$, is the following. Again, let $\hat{\lambda}(\cdot)$ be an estimate of $\lambda(\cdot)$. Then we can compute *estimated* residuals by

$$\hat{R}_i = \hat{\Lambda}(T_i) - \hat{\Lambda}(T_{i-1}) \; ; \; i = 1, \ldots, n.$$

Let $\hat{F}$ denote the empirical distribution function corresponding to the $\hat{R}_i$, i.e. the distribution putting mass $1/n$ to each of $\hat{R}_1, \hat{R}_2, \ldots, \hat{R}_n$. Then we shall define $\tilde{\mathcal{P}}$ as TRP($\hat{F}, \hat{\lambda}(\cdot)$), and may instead simulate bootstrap samples from this process. The motivation for this method is that the theoretical residuals of the NHPP are i.i.d. exponential with parameter 1.

To compare the two ways of bootstrapping, the $\tilde{\mathcal{P}}$ has the advantage of being more robust against departures from the NHPP assumption in the direction of TRP. This is quite similar to bootstrapping in ordinary regression models, where bootstrapping of residuals is more model-robust than a parametric bootstrapping with normally distributed residuals.

Bootstrapping in NHPP and TRP models is studied in more detail in Elvebakk (1998).

1.6 Statistical Inference in Trend-Renewal Processes

We shall consider statistical inference in cases where failure times $T_1, T_2, \ldots$ are observed for one or more systems of the same kind, and the failure process of each system is modeled by a TRP as defined in the previous section.

Recall that a TRP model is specified in terms of a distribution function F and an intensity function $\lambda(\cdot)$. A parametric model can thus conveniently be described in terms of parametric representations for each of F and $\lambda(\cdot)$. Semiparametric models may involve either a parametric F and nonparametric λ, or vice versa. Completely nonparametric models may also be considered.

An S-Plus library, called REIDAR (=REpaIrable systems Data Analysis Routines), containing routines for analysis of repairable systems, has been developed at the Norwegian University of Science and Technology as part of Elvebakk (1998), (for information, contact this author). In particular, the program has routines for graphical analyses, parametric estimation in TRP models, estimation of heterogeneity, trend testing and model checking.

Parametric estimation in TRP models using REIDAR is possible with F either a Weibull-, gamma- or exponential distribution, and $\lambda(\cdot)$ being either a constant, a *power law* intensity $\lambda(t) = abt^{b-1}$, or a *log linear* intensity $\lambda(t) = \exp\{a + bt\}$.

As an example we shall describe what we call the *Weibull-trend-renewal process*, WTRP(a, b, β), which in the notation of the previous section is given by TRP$(F_W(\cdot|1, \beta), \lambda_W(\cdot|a, b))$, where $F_W(t|\alpha, \beta) = 1 - \exp(-(\alpha t)^\beta)$ is the Weibull distribution with parameters α, β, and λ_W is the power law intensity given above. Note that the scale parameter α in the Weibull distribution is set to 1, since by (1.4) the scale is taken care of by the parameter a in λ_W.

Three special cases are easily identified: WTRP$(a, 1, \beta)$ = RP$(F_W(\cdot|a, \beta))$, WTRP$(a, b, 1)$= NHPP$(\lambda_W(\cdot|a, b))$, WTRP$(a, 1, 1)$= HPP(a).

Using (1.6) we find the following expression for the intensity function of the WTRP,

$$\gamma(t) = \beta a^\beta b t^{b-1}[t^b - T^b_{N(t-)}]^{\beta-1}$$

which when substituted into (1.1) leads to the *log* likelihood

$$\begin{aligned} l(a, b, \beta) &= N(\tau)\{\log\beta + \log b + \beta\log a\} \\ &+ \sum_{i=1}^{N(\tau)} \left\{(b-1)\log T_i + (\beta - 1)\log[T_i^b - T_{i-1}^b] - a^\beta[T_i^b - T_{i-1}^b]^\beta\right\} \\ &- a^\beta[\tau^b - T^b_{N(\tau)}]^\beta. \end{aligned} \tag{1.7}$$

The unknown parameters are estimated by maximum likelihood.

If *several* systems of the same kind are under observation, and if the corresponding failure processes are taken as independent and identically distributed, then the full log likelihood consists of a sum of terms (1.7) for each system.

However, in the case of several systems, even being of same type, they may well exhibit different failure rates due to various causes. For example, systems may be used under different environmental or operational conditions, like compressors or pumps installed on different oil platforms situated in different fields of the North Sea.

If the differing conditions can be quantified, then they may be included in the model as *observable* covariates, and treated with models like (1.2) and (1.5). For estimation in NHPP models with covariates, see e.g. Lawless (1987).

Unobserved heterogeneity is more difficult to handle, but can be incorporated into the WTRP process by allowing, e.g., the scale parameter a to vary from system to system. The common approach [Andersen et al. (1992)] is to model a by a gamma-distributed random variable with shape parameter γ and scale parameter δ, say. Thus what we shall call the GWTRP has four parameters, b, β, γ, δ, where b, β have the same meaning as for the WTRP. The likelihood function for each single system is now obtained by taking the expected value of the likelihood function (exponential of (1.7)), with respect to the gamma distributed a. It is not possible to do the integration explicitly, and the program REIDAR uses numerical integration here and computes the maximum likelihood estimates by a simplex method. Heggland (1994) studied heterogeneity in the inhomogeneous gamma process of Berman (1981).

As an example of applying the GWTRP model, we consider the airconditioner data studied by Proschan (1963) (see Section 1.2). In order to determine the "best" model for the data, we have used REIDAR to compute maximum values of the log likelihood for the GWTRP and various submodels. For notation, note that a G in front of the earlier notation for NHPP etc., means a model where the scale parameter a is replaced by a gamma-distributed random variable with parameters γ and δ. The results are given in Table 1.1.

Table 1.1: Maximum log likelihood values for various models for the Proschan (1963) airconditioner data

Model	Parameters in model	Log likelihood
GWTRP	b, β, γ, δ	-1172.1
GNHPP	$b, 1, \gamma, \delta$	-1172.3
GHPP	$1, 1, \gamma, \delta$	-1174.9
GRP	$1, \beta, \gamma, \delta$	-1174.8
HPP	$b = 1, \beta = 1, a =$ constant	-1178.9

For testing hypotheses of the parameters we can use the general result that *two times differences in log likelihood* are approximately chi-square distributed, with degrees of freedom (df) equal to the difference in number of parameters in the largest model and the null hypothesis model, when the null hypothesis holds.

Recall from Section 1.2 that Proschan (1963) tested the null hypothesis of a common HPP model for all aircrafts, against the alternative of HPPs with intensities differing from plane to plane. In the WTRP model this translates to testing HPP vs. GHPP. Using Table 1.1 we get that twice the difference in log likelihood is 7.84, which with df $= 1$ gives a P-value of 0.005. This is not much different from what Proschan got (0.007). However, from Table 1.1 it seems that the best model is an NHPP with heterogeneity, since testing GHPP vs. GNHPP gives a P-value of 0.022. Follmann and Goldberg (1988) studied Weibull renewal processes with gamma heterogeneity (i.e. GRP) and concluded for the Proschan data that a GHPP would be the best model, since the log likelihood increases by 0.1 only when going from GHPP to GRP.

1.7 Trend Testing

For the proper operation of repairable systems it is important to detect possible changes in the pattern of failures occurring over time. For example, reliability growth corresponds to times between failures becoming longer as time goes (improving system), whereas aging effects often lead to decreasing interfailure times (deteriorating system). Various types of nonmonotonic trend can also be present, e.g. *cyclic* trend or *bathtub shaped* trend.

A simple but informative way of checking for a possible trend in the pattern of failures is to study plots like Figure 1.2. Departures from a straight line may indicate some kind of trend. In particular, a convex shape will be indicative of a deteriorating system, while a concave plot may indicate an improving system. In Figure 1.2 there seems to be an increase in the failure intensity from about 1200 hrs.

By a *trend test* we shall mean a statistical test for the null hypothesis that the failure process is stationary, in some sense to be made precise, versus alternatives depending on the kind of trend one would like to detect. Here we shall consider the null hypotheses either "the process is an HPP" or "the process is an RP", whereas the alternative will for brevity of exposition be restricted to "monotone trend". Note that trend tests of "RP" versus "cyclic trend" already are mentioned in connection with Berman's inhomogeneous gamma process of Section 1.5. Trend tests versus bathtub shaped trend are considered for the NHPP case in Kvaløy and Lindqvist (1998).

In parametric modeling of failure processes one may often represent trend by means of one or more of the parameters. For example in the WTRP(a, b, β) model, the parameter b represents trend and equals 1 for the case of no trend. Thus a likelihood ratio test of the null hypothesis $b = 1$ is in fact a trend test, with null hypothesis corresponding to a Weibull renewal process. An example of such a test is the likelihood ratio test for H_0 : GHPP vs. H_1 : GNHPP that we performed for the Proschan data in Table 1.1.

Two classical trend tests will be given much attention in the following, For simplicity we shall always assume that a single system is under observation, with failure truncation after n failures.

The Laplace test

This is a test for the null hypothesis of "HPP" vs. the alternative of a monotonic trend. The test statistic is

$$U = \frac{\frac{1}{n-1}\sum_{i=1}^{n-1} T_i - \frac{1}{2}T_n}{T_n\sqrt{\frac{1}{12(n-1)}}}$$

which under the null hypothesis is approximately standard normally distributed. The null hypothesis is rejected for too small or too large values of U. Moreover, $U > 0$ (resp. $U < 0$) indicate deteriorating (respectively improving) system.

The Military Handbook test

The null hypothesis is "HPP" and the test statistic is $Z = 2\sum_{i=1}^{n-1}\log(T_n/T_i)$, which under the null hypothesis is approximately χ^2-distributed with $2(n-1)$ degrees of freedom. The null hypotheses is rejected for small or large values of Z, low values indicating deteriorating system and large values indicating improving system.

Many users seem to forget the fact that the Laplace- and Military Handbook tests really are tests for the null hypothesis that the data come from an HPP. Thus *rejection* of the null hypothesis in strict terms means just that the process is *not an HPP* – but it could still in principle be an RP and thus still have no trend. That such false rejections may occur with considerable probability will be seen from simulations reported below.

There exist, however, tests which are constructed to test the null hypothesis of an RP vs. a monotone trend, e.g. the *Mann test* and the *Lewis-Robinson test* [Ascher and Feingold (1984)]. The latter test is defined and studied below.

The Lewis-Robinson test

This is a modification of the Laplace test, with test statistic $U_{LR} = U/\widehat{CV}(X)$. Here U is the Laplace test statistic, while $\widehat{CV}(X)$ is an estimate of the *coefficient of variation* of the X_i. Note that the coefficient of variation of the exponential distribution is 1. The reason for introducing the denominator in U_{LR} is to account for non-exponential distributions of the interfailure times. In fact, the denominator assures that U_{LR} is approximately standard normally distributed whenever the data come from an RP.

Table 1.2 shows the results of simulations reported in Lindqvist et al. (1994), performed in order to show the deficiency of the Laplace- and Military Handbook tests as tests for the null hypothesis of RP. An RPs with Weibull$(1,\beta)$-distribution and varying shape parameter β was simulated. The number n of failures per case was set to 20, and the nominal significance level was $\alpha = 0.05$.

Each estimated value is based on 2000 simulations. Table 1.2 shows the effective significance level (i.e. relative number of rejections of the null hypothesis made in the 2000 simulations).

Table 1.2: Simulated rejection probabilities for $n = 20$ failures of an RP with Weibull$(1, \beta)$ interfailure times

β	Laplace	MilHbk	Lew-Rob
0.2	.591	.738	.018
0.5	.289	.333	.043
0.8	.120	.110	.051
1.0	.046	.049	.046
1.2	.022	.023	.053
1.5	.003	.003	.055
2.0	.000	.000	.053

The effective significance level of the Lewis Robinson test correspond very well to the nominal level, which should be expected. For the two classical tests, however, the effective level is 0.05 only when $\beta = 1$, which corresponds to an HPP. The tests reject much too often for $\beta < 1$ (improving system), but are surprisingly conservative, i.e. have too low effective significance level, when $\beta > 1$ (deteriorating system).

Lindqvist et al. (1994) also compare the power of the tests against alternatives with monotonic trend. As is well known [Ascher and Feingold (1984)] the Laplace test and the Military Handbook test are optimal tests for alternatives of a NHPP with log linear intensity and power law intensity functions, respectively. The general conclusion regarding testing vs. NHPP trend is that the two classical tests out perform the Lewis-Robinson test for NHPP alternatives with log linear and power law intensities.

In a study of the properties of the three tests with respect to rejection power against non-NHPP cases with trend, it is concluded, however, that the Laplace and Military Handbook tests should be used only with extreme care. Tests like the Lewis Robinson test and the Mann test are to be preferred here, due to the fact that they behave correctly under the null hypothesis of an RP. Alternative tests based on Monte Carlo simulation are suggested in the next section.

1.8 Monte Carlo Trend Tests

As mentioned above, the Lewis-Robinson test is essentially a modified Laplace test, with the aim of attaining the correct significance level under the null hypothesis of an RP. We shall in this section see that corresponding modifications can be done with any trend test by the use of Monte Carlo simulations. The work described in this section will appear in more detail in Elvebakk (1998). A general reference on Monte Carlo tests is Ripley (1987).

Consider again for simplicity a single failure truncated system, observed until n failures have occurred. Thus the observations are the failure times $T_1, T_2, \ldots, T_n$, or equivalently the interfailure times $X_1, X_2, \ldots, X_n$. The null hypothesis H_0 is that the observations come from an RP.

Suppose now that $W \equiv W(X_1, \ldots, X_n)$ is a test statistic, for which H_0 is rejected for large values. We may, for example, think of W as $|U|$, where U is the Laplace test statistic. The problem in practice is to find the critical value for a given significance level. We shall describe two types of Monte Carlo tests for this problem.

Permutation test

Let F be the cumulative distribution function of the interfailure times under the null hypothesis of RP. It is well known that the order statistic $S \equiv (X_{(1)}, X_{(2)}, \ldots, X_{(n)})$ is statistically sufficient for F. The permutation test with given significance level α is defined as a conditional test given S, which rejects H_0 if

$$Pr(W \geq w_{obs}|S = s) \leq \alpha \tag{1.8}$$

where w_{obs} is the observed value of W and s is the observed value of S. By sufficiency, the distribution of W given $S = s$ is independent of the unknown distribution F, and hence the left hand side of (1.8) can in principle be found. In practice the permutation test can be performed as follows:

1. Record interfailure times $x_1, x_2, \ldots, x_n$.

2. Compute $w_{obs} = W(x_1, x_2, \ldots, x_n)$

3. Compute $W(x_{\pi(1)}, x_{\pi(2)}, \ldots, x_{\pi(n)})$ for a large number K of randomly chosen permutations π of $x_1, x_2, \ldots, x_n$. Order the computed values of W as $w_1 \leq w_2 \leq \ldots \leq w_K$.

4. Reject H_0 if $w_{obs} \geq w_{[(1-\alpha)(K+1)]}$ (where $[a]$ denotes the integer part of a).

Bootstrap test

As above, let F be the distribution of the RP under H_0. If F were known, then we would reject H_0 with significance level α if $P_F(W \geq w_{obs}) \leq \alpha$. In practice, however, the distribution F is unknown. The idea of the bootstrap test is to replace F by the empirical distribution $\hat{F}$ giving mass $1/n$ to each of the observations $x_1, \ldots, x_n$, and then reject H_0 if $P_{\hat{F}}(W \geq w_{obs}) \leq \alpha$. It can be shown that, when certain conditions hold (which will not be discussed here) we have an asymptotic α level test. The practical recipe of the bootstrap test turns out to be as for the permutation test, but with step 3 replaced by 3*: Compute $W(x_1^*, x_2^*, \ldots, x_n^*)$ for a large number K of random, ordered samples $(x_1^*, x_2^*, \ldots, x_n^*)$ drawn *with replacement* from $x_1, x_2, \ldots, x_n$. Order the computed values of W as $w_1 \leq w_2 \leq \ldots \leq w_K$.

It is interesting to note that the only difference between the permutation test and the bootstrap test is that the bootstrap test draws *with* replacement, while the permutation test draws *without* replacement from the observed set of interfailure times.

Note also that the permutation tests are exact conditional tests, which in principle achieve the correct significance level, while the bootstrap tests only have approximately correct level, due to the approximation of F by $\hat{F}$. In practice, however, the level of the permutation test also becomes approximate, due to the limited number K of permutations considered.

Comprehensive simulation studies of the Monte Carlo tests described above are found in Kvaløy (1994) and Elvebakk (1998). We merely extract a few numbers from their studies in order to illustrate the main features of the Monte Carlo tests.

The first question one may ask is whether much is lost in power for NHPP alternatives when going from the classical Laplace- and Military Handbook tests to the corresponding Monte Carlo tests. Table 1.3 gives estimated rejection probabilities when data are simulated from an NHPP with $\lambda(t) = bt^{b-1}$, with varying n and b. The tests compared are the ordinary Military Handbook test, its two Monte Carlo versions, and the Lewis Robinson test.

Table 1.3: Simulated rejection probabilities for NHPP alternatives with $\lambda(t) = bt^{b-1}$

n	b	MilHbk	MilH Boot	MilH Permut	Lew-Rob
	0.50	0.976	0.967	0.967	0.867
	0.75	0.409	0.393	0.398	0.281
35	1.00	0.045	0.050	0.049	0.049
	1.50	0.637	0.602	0.576	0.468
	2.00	0.983	0.973	0.962	0.880
15	0.50	0.750	0.677	0.693	0.459
	1.50	0.310	0.268	0.236	0.228
5	2.50	0.396	0.121	0.138	0.083

We conclude that rather little power seems to be lost in this case by performing Monte Carlo tests, but that more is lost for small sample sizes. Moreover, there is only a small difference between the power of the bootstrap- and the permutation tests.

It is furthermore seen that the Monte Carlo tests are always better than the Lewis-Robinson test. But this can be explained by the fact that the Military Handbook test is optimal for the NHPP alternative considered here. In fact, simulations of Elvebakk (1998) show small differences between the power of the Lewis-Robinson test and the Monte Carlo tests based on the Laplace test, but with the Monte Carlo tests always slightly better.

Recall from the previous section that the ordinary Laplace and Military Handbook tests may behave badly in non-Poisson cases. Table 1.4 shows estimated rejection probabilities for alternatives given by WTRP$(1, b, \beta)$, with $\beta = 1.5$ and 0.75, and varying b. As before we can conclude that the Military Handbook test does not achieve the correct significance level (see $b = 1.0$). On the other hand, the results for the Monte Carlo tests based on the Military Handbook are satisfactory.

Table 1.4: Simulated rejection probabilities for $n = 15$ failures of a WTRP$(1, b, \beta)$

β	b	MilHbk	MilH Boot	MilH Permut	Lew-Rob
	0.5	0.780	0.908	0.912	0.766
1.5	1.0	0.006	0.051	0.049	0.047
	1.5	0.299	0.496	0.450	0.375
	2.0	0.842	0.899	0.851	0.721
	0.5	0.749	0.505	0.528	0.266
	1.0	0.136	0.053	0.052	0.047
0.75	1.5	0.299	0.172	0.145	0.125
	2.0	0.621	0.415	0.355	0.291
	2.5	0.844	0.634	0.551	0.443

The general conclusion of Elvebakk (1998) is that Monte Carlo trend tests are favorable to the Lewis-Robinson test and the Mann test in any case considered. As regards the Laplace test and the Military Handbook test, the corresponding Monte Carlo tests have lower power than the original tests for NHPP alternatives, but the difference is small for large sample sizes. This fact, together with the fact that the Laplace and Military Handbook tests are invalid in non-Poisson cases, leads to a recommendation of the Monte Carlo tests as an overall conclusion.

1.9 Concluding Remarks and Topics for Further Study

Goodness of fit tests for NHPP models

The literature contains several suggestions of how to test the adequacy of parametric NHPP models. Particularly much studied are NHPP with power law intensity. Recent references are Baker (1996) and Gaudoin (1997). Less studied is the case of NHPP with log linear intensity, apparently due to mathematical intractability. Lindqvist and Elvebakk (1998) study a general setup for using conditional tests given the minimal sufficient statistic under the null hypothesis.

Nonparametric and semiparametric inference

Boswell (1966) considered nonparametric estimation in NHPP($\lambda(\cdot)$) under the assumption of a monotone $\lambda(\cdot)$. Heggland (1994) considered the semiparametric case of estimating a monotone $\lambda(\cdot)$ nonparametrically in a TRP($F, \lambda(\cdot)$), with F being a parametric Weibull($1, \beta$) distribution. As mentioned in Section 1.6, there are several possibilities for non- and semiparametric inference in the TRP model. Kernel estimators and other smoothing techniques applied to the $\lambda(\cdot)$ or the F are of particular interest.

Complex modeling for repairable systems

The TRP model and in particular the NHPP model can in principle easily be extended by assuming that the intensity function $\lambda(\cdot)$ itself is stochastic. For the NHPP, this is the well known *doubly stochastic* Poisson process.

For repairable systems one could also imagine using multiple time scales, e.g. *time* and *number of cycles*. In practice there may also be several competing failure modes. It would be nice to see how this can be incorporated in the common models for repairable systems.

For many applications it is natural to model the interfailure times in more detail. For example this could be done by Markov models, which would lead to phase type distributions for times between failures. Hokstad and Frøvig (1996) suggest simple models of this kind.

References

1. Aalen, O. O. (1975). Statistical inference for a family of counting processes, *Ph.D. Thesis*, University of California, Berkeley.

2. Andersen, P., Borgan, O., Gill, R. and Keiding, N. (1992). *Statistical Models Based on Counting Processes*, New York: Springer-Verlag.

3. Ascher, H. and Feingold, H. (1984). *Repairable Systems—Modeling, Inference, Misconceptions and Their Causes*, New York: Marcel Dekker.

4. Baker, R. D. (1996). Some new tests for the power law process, *Technometrics*, **38**, 256–265.

5. Barlow, R. E., Marshall, A. W. and Proschan, F. (1963). Properties of probability distributions with monotone hazard rate, *Annals of Mathematical Statistics*, **34**, 375–389.

6. Berman, M. (1981). Inhomogeneous and modulated gamma processes, *Biometrika*, **68**, 143–152.

7. Berman, M. and Turner, T. R. (1992). Approximate point process likelihoods with GLIM, *Applied Statistics*, **41**, 31–38.

8. Boswell, M. T. (1966). Estimating and testing trend in a stochastic process of Poisson type, *Annals of Mathematical Statistics*, **37**, 1564–1573.

9. Brown, M. and Proschan, F. (1983), Imperfect repair, *Journal of Applied Probability*, **20**, 851–859.

10. Cox, D. R. (1955). Some statistical methods connected with series of events (with discussion), *Journal of the Royal Statistical Society, Series B*, **17**, 129–164.

11. Cox, D. R. (1972). The statistical analysis of dependencies in point processes, In *Stochastic Point Processes* (Ed., P. A. Lewis), pp. 55–66, New York: John Wiley & Sons.

12. Cox, D. R. and Lewis, P. W. (1966). *The Statistical Analysis of Series of Events*, London: Methuen.

13. Crowder, M. J., Kimber, A. C., Smith, R. L. and Sweeting, T. J. (1991). *Statistical Analysis of Reliability Data*, Great Britain: Chapman & Hall.

14. Dorado, C., Hollander, M. and Sethuraman, J. (1997). Nonparametric estimation for a general repair model, *Annals of Statistics*, **25**, 1140–1160.

15. Efron, B. and Tibshirani, R. J. (1993). *An Introduction to the Bootstrap*, London: Chapman and Hall.

16. Elvebakk, G. (1998). Resampling methods for analysis of repairable systems, *Thesis work*, Norwegian University of Science and Technology (in preparation).

17. Follmann, D. A. and Goldberg, M. S. (1988). Distinguishing heterogeneity from decreasing hazard rates, *Technometrics*, **30**, 389–396.

18. Gaudoin, O. (1997). CPIT goodness-of-fit tests for the power-law process, *Communications in Statistics—Theory and Methods* (to appear).

19. Heggland, K. (1994). Analysis of repairable systems: Modeling of heterogeneity, especially in trend-renewal processes, *Project work*, Norwegian Institute of Technology (in Norwegian).

20. Heggland, K. (1994). Non-parametric estimation in trend-renewal processes, *Diploma Thesis*, Norwegian Institute of Technology.

21. Hokstad, P. and Frøvig, A. T. (1996). The modelling of degraded and critical failures for components with dormant failures, *Reliability Engineering and System Safety*, **51**, 189–199.

22. Høyland, A. and Rausand, M. (1994). *System Reliability Theory. Models and Statistical Methods*, New York: John Wiley & Sons.

23. Kijima, M. (1989). Some results for repairable systems with general repair, *Journal of Applied Probability*, **26**, 89–102.

24. Kvaløy, J. T. (1994). Analysis of repairable systems: Trend testing, *Diploma Thesis*, Norwegian Institute of Technology.

25. Kvaløy, J. T. and Lindqvist, B. H. (1998). TTT-based tests for trend in repairable systems data, *Reliability Engineering and System Safety* (to appear).

26. Lawless, J. F. (1987). Regression methods for Poisson process data, *Journal of American Statistical Association*, **82**, 808–815.

27. Lawless, J. F. and Thiagarajah, K. (1996). A point-process model incorporating renewals and time trends, with application to repairable systems, *Technometrics*, **38**, 131–138.

28. Lindqvist, B. H. (1993). The trend-renewal process, a useful model for repairable systems, Society of Reliability Engineers, Scandinavian Chapter, Annual Conference, Malmö, Sweden.

29. Lindqvist, B. H., Kjønstad, G. A. and Meland, N. (1994). Testing for trend in repairable systems data, *Proceedings of ESREL'94*, La Baule, France.

30. Lindqvist, B. H. and Elvebakk, G. (1998). Exact tests for deviation from nonhomogeneous Poisson processes (in preparation).

31. Proschan, F. (1963). Theoretical explanation of observed decreasing failure rates, *Technometrics*, **5**, 375–383.

32. Ripley, B. D. (1997). *Stochastic Simulation*, New York: John Wiley & Sons.

33. Thompson, W. A. Jr. (1988). *Point Process Models with Applications to Safety and Reliability*, London: Chapman and Hall.

34. Whitaker, L. R. and Samaniego, F. J. (1989). Estimating the reliability of systems subject to imperfect repair, *Journal of American Statistical Association*, **84**, 301–309.

2

CPIT Goodness-Of-Fit Tests for Reliability Growth Models

Olivier Gaudoin

Laboratoire IMAG-LMC, Grenoble, France

Abstract: Assume that the n first failure times of a repairable system are observed. In order to choose an appropriate stochastic model for these data, goodness-of-fit tests have to be performed. The aim of this work is to study the goodness-of-fit tests based on the Conditional Probability Integral Transformation of O'Reilly-Quesenberry. The general CPIT methodology is described. Then the CPIT tests are derived for the Homogeneous Poisson Process, Jelinski-Moranda model, Goel-Okumoto model and the Power-Law Process. The power of these tests is assessed through simulations and finally, they are applied to real software reliability data.

Keywords and phrases: Reliability growth, software reliability, goodness-of-fit tests, conditional probability integral transformation, non homogeneous Poisson processes

2.1 Introduction

Let $T_1 < \ldots < T_n$ be the n first failure times of a repairable system, starting from $T_0 = 0$. Let $X_i = T_i - T_{i-1}$ be the successive times between failures and N_t be the number of failures occurred between 0 and t. We consider here systems which exhibit reliability growth. Then, the times between failures are growing, so the X_i's are a sequence of stochastically increasing random variables. Lots of stochastic models have been proposed for this situation, especially for the case where the system is a software. The reliability growth of software systems is due to the debugging of design faults. A detailed review of software reliability models can be found in Xie (1991) or Lyu (1996).

There are two main classes of reliability growth models. In the first class, the times between failures are independent but not identically distributed. The most famous of these models are:

- The Jelinski-Moranda model (1972), denoted JM: the X_i's are independent and exponentially distributed with respective parameters $\phi(N-i+1)$, $i = 1, \ldots, n$. N is the number of faults initially in the system, and ϕ is a positive real number which represents the amount of improvement of the system at each debugging.

- The Moranda Geometric model (1979), denoted MG: the X_i's are independent and exponentially distributed with respective parameters $\lambda e^{-(i-1)\theta}$, $i = 1, \ldots, n$, $\lambda > 0$, $\theta \in \mathbb{R}$.

In the second class, the process $\{N_t\}_{t \geq 0}$ is a Non Homogeneous Poisson Process (NHPP). The most famous of these models are:

- The Power-Law Process [Crow (1974)] or Duane model, denoted PLP: a NHPP with intensity $\lambda(t) = \alpha\beta t^{\beta-1}$, $\alpha > 0$, $\beta > 0$.

- The Goel-Okumoto model (1979), denoted GO: a NHPP with intensity $\lambda(t) = \lambda e^{-\phi t}$, $\lambda > 0$, $\phi \in \mathbb{R}$.

- The delayed S-shaped model [Yamada, Ohba and Osaki (1984)], denoted S: a NHPP with intensity $\lambda(t) = \alpha\beta^2 t e^{-\beta t}$, $\alpha > 0$, $\beta \in \mathbb{R}$.

Since a lot of reliability growth models exist, the problem of choosing an appropriate model for a particular data set is very important. When the random variables concerned are independent and identically distributed (iid), well-known methods exist for choosing a model: graphical methods, such as the probability plots or Q-Q plots, and statistical methods, the Goodness-of-Fit (GOF) tests. The most usual GOF tests are the Chi-Square tests and the tests based on the Empirical Distribution Function of the data: Kolmogorov-Smirnov, Cramer-von Mises and Anderson-Darling. A detailed review of these techniques is in D'Agostino and Stephens (1986). For example, it has been proved that the Anderson-Darling test is one of the most powerful tests for a large range of alternatives.

Here, the observations are not iid. So a simple idea to make a GOF test is to find a function which transforms the failure times into iid random variables. Such transformations exist for some models. For example, for the Power-Law Process, the random vector $(Z_1, \ldots, Z_{n-1})$, where $Z_i = \ln T_n / T_{n-i}$, is distributed like the order statistics of a $n-1$ sample of the exponential distribution with parameter β. So it is possible to test the fit of the PLP by testing the fit of the Z_i's to the exponential distribution [Rigdon (1989)]. For all the other models mentioned here, it is not easy to find such a transformation.

The aim of this work is to study the use of the so-called Conditional Probability Integral Transformation (CPIT), in order to test the Goodness-of-Fit of several reliability growth models.

2.2 The Conditional Probability Integral Transformation

The Conditional Probability Integral Transformation (CPIT) has been proposed by Rosenblatt (1952) and used by O'Reilly and Quesenberry (1973) to build GOF tests.

Assume that $(T_1, \ldots, T_n)$ is a random vector with absolutely continuous multivariate cumulative distribution function (cdf), depending on an unknown parameter θ. Assume that there exists a sufficient statistic $S_n(T_1, \ldots, T_n)$ for θ. Let $\tilde{F}_n(t_1, \ldots, t_n)$ be the conditional cdf of $(T_1, \ldots, T_n)$ given S_n, and $\tilde{F}_n(t_i|t_1, \ldots, t_{i-1})$, $i = 2, \ldots, n$, be the conditional cdf of T_i given $T_1, \ldots, T_{i-1}$ and S_n. Let $V_1 = \tilde{F}_n(T_1)$ and $V_i = \tilde{F}_n(T_i|T_1, \ldots, T_{i-1})$, $i = 2, \ldots, n$.

The rank of absolute continuity k is the greatest value of i such that $\tilde{F}_i$ is absolutely continuous. Usually, $k = n - p$, where p is the dimension of S_n. Then, the result of O'Reilly-Quesenberry is that $V_1, \ldots, V_k$ are independent and uniformly distributed over [0,1].

The conditioning by the sufficient statistic ensures that the V_i's do not depend on the unknown parameter. So goodness-of-fit tests for uniformity can be performed on the V_i's by the usual methods mentioned in Section 2.1.

Note that, for most of the reliability growth models, $\{T_i\}_{i \geq 1}$ is a Markov process, so the expression of V_i reduces to $V_i = \tilde{F}_n(T_i|T_{i-1})$, $i = 2, \ldots, n$.

A CPIT test is entirely determined by the V_i's. So in the following, only the expressions of the V_i's are given for several reliability models. As a first example, the case of the Homogeneous Poisson Process is presented in Section 2.3. Then, two reliability growth models are tested, the Jelinski-Moranda model (1972) in Section 2.4 and the Power-Law process (1974) in Section 2.5. Finally, the power of these tests is assessed through simulations in Section 2.6 , then the tests are applied to real software reliability data.

2.3 CPIT GOF Tests for the Homogeneous Poisson Process

When $\{N_t\}_{t \geq 0}$ is an Homogeneous Poisson Process (HPP), the times between failures X_i are independent and exponentially distributed with parameter λ.

Of course, this is not a reliability growth model, but it is interesting to study how the CPIT methodology works in the most simple case.

The likelihood function associated to the observation of the n first failure times is:

$$\mathcal{L}(\lambda; t_1, \ldots, t_n) = \lambda^n e^{-\lambda t_n}, \qquad 0 \le t_1 \le \ldots \le t_n.$$

So T_n is a sufficient statistic for λ. Then the calculation of the V_i's need to know the distribution of T_i given T_{i-1} and T_n. Now it is well known that, conditionally to $[T_n = t_n]$, the vector $(T_1, \ldots, T_{n-1})$ is distributed like the order statistics of a $n-1$ sample of the uniform distribution over $[0, t_n]$. So conditionally to $[T_{i-1} = t_{i-1}]$ and $[T_n = t_n]$, T_i is distributed like the first order statistics of a $n-i$ sample of the uniform distribution over $[t_{i-1}, t_n]$. Then a straightforward calculation leads to:

$$V_i = 1 - \left(\frac{T_n - T_i}{T_n - T_{i-1}} \right)^{n-i}, \qquad i = 1, \ldots, n-1.$$

These tests have been studied by O'Reilly and Stephens (1982), and compared with the usual GOF tests to the exponential distribution, by ways of simulations. The conclusion of their study is that the CPIT tests are not among the most powerful tests for exponentiality, but they perform correctly under a reasonably large range of alternatives.

2.4 CPIT GOF Tests for the Jelinski-Moranda and Goel-Okumoto Models

The first application of the CPIT tests to reliability growth models has been done by Lee and Finelli (1989), for the Jelinski-Moranda model. The authors used a modified version of the O'Reilly-Quesenberry theorem, so they obtained for the V_i's a slightly different expression from that presented below.

The likelihood function associated to the observation of the n first failure times in a JM model is:

$$\mathcal{L}(N, \phi; t_1, \ldots, t_n) = \frac{\phi^n N!}{(N-n)!} \exp\left[-\phi(N-n+1)t_n - \phi \sum_{i=1}^{n} t_i \right], \; 0 \le t_1 \le \ldots \le t_n.$$

So $\left(T_n, \sum_{i=1}^{n-1} T_i\right)$ is a sufficient statistic for (N, ϕ). Then the calculation of the V_i's need to know the distribution of T_i given T_{i-1},$\sum_{j=i}^{n-1} T_j$ and T_n. It is easy to show that, conditionally to $[T_n = t_n]$, the vector $(T_1, \ldots, T_{n-1})$ is distributed like the order statistics of a $n-1$ sample of the distribution with density:

$$g(t) = \frac{\phi e^{-\phi t}}{1 - e^{-\phi t_n}}, \qquad 0 \le t \le t_n$$

A very important point is that the conditioning by the sufficient statistic proves that V_i does not depend on ϕ, so it is possible to set $\phi = 0$. In that case, the above distribution is the uniform distribution over $[0, t_n]$. Then, the calculation can be done as if the true process was an HPP. Knowing the distribution of the sum of independent uniform variables, we obtain finally:

$$V_i = 1 - \frac{\displaystyle\sum_{k=0}^{n-i} C_{n-i}^k (-1)^k \left[\sup\left(0, \sum_{j=i}^{n-1} W_j - k - (n-i-k)W_i\right)\right]^{n-i-1}}{\displaystyle\sum_{k=0}^{n-i} C_{n-i}^k (-1)^k \left[\sup\left(0, \sum_{j=i}^{n-1} W_j - k - (n-i-k)W_{i-1}\right)\right]^{n-i-1}}$$

$i = 1, \ldots, n-2$, where $W_i = \dfrac{T_i}{T_n}$.

Lee-Finelli showed that the V_i's can have distributions which are very far from the uniform under a family of alternative models based on the gamma distribution. But they did not provide simulations in order to assess the power of the tests.

An important remark is that the conditional distribution of $(T_1, \ldots, T_{n-1})$ given $[T_n = t_n]$ is exactly the same for the Jelinski-Moranda and Goel-Okumoto models. Then, since the calculations are always done conditionally to T_n, the CPIT tests are identical for these two models. This means that, as noted by Miller (1986), the JM and GO models are indistinguishable by this method. However, the predictions made by these two models on the same data can be significantly different. So it seems that the CPIT tests are probably not very powerful. But they remain very useful since they are the only GOF tests available for these two models.

2.5 CPIT GOF Tests for the Power-Law Process

The results in this section are detailed in Gaudoin (1998).

The likelihood function associated to the observation of the n first failure times of a Power-Law Process is:

$$\mathcal{L}(\alpha, \beta; t_1, \ldots, t_n) = \alpha^n \beta^n \exp\left[(\beta - 1)\sum_{i=1}^{n} \ln t_i - \alpha t_n^\beta\right], \qquad 0 \le t_1 \le \ldots \le t_n.$$

So $\left(T_n, \sum_{i=1}^{n-1} \ln T_i\right)$ is a sufficient statistic for (α, β). Then the calculation of the V_i's need to know the distribution of T_i given T_{i-1},$\sum_{j=i}^{n-1} \ln T_j$ and T_n. For the same reasons as before, the calculations can be done setting $\alpha = \beta = 1$, so we can do as if the true process was an HPP with intensity 1.

Then, conditionally to $[T_{i-1} = t_{i-1}]$ and $[T_n = t_n]$, $-\sum_{j=i}^{n-1} \ln T_j/T_n$ is distributed like the sum of $n-i$ independent random variables with the same distribution as $X_i = -\ln U_i$, where U_i has the uniform distribution over $[t_{i-1}/t_n, 1]$. The derivation of this distribution can be done, and we obtain finally:

$$V_i = 1 - \frac{\sum_{k=0}^{n-i} C_{n-i}^k (-1)^k \left[\sup\left(0, \sum_{j=1}^{n-i} Z_j - kZ_{n-i}\right)\right]^{n-i-1}}{\sum_{k=0}^{n-i} C_{n-i}^k (-1)^k \left[\sup\left(0, \sum_{j=1}^{n-i} Z_j - kZ_{n-i+1}\right)\right]^{n-i-1}}$$

$i = 1, \ldots, n-2$, where $Z_i = \ln \dfrac{T_n}{T_{n-i}}$.

There is obviously a great similarity between the expressions of the V_i's for the JM, GO and PLP models. This is due to the similarity between the sufficient statistics and to the fact that the calculations are always done under the HPP assumption. However, there does not exist a simple link between those models, so we can not use this similarity to establish a general formula of the CPIT tests for a large class of models.

2.6 Experimental Results

In this section, we want to assess the power of the tests presented above. For the Power-Law Process, the CPIT tests can be compared to the Rigdon tests for exponentiality of the Z_i's. But for the JM-GO models, no comparison can be done since no other GOF tests exist for these models.

Then, we have three classes of tests: the Rigdon and CPIT tests for the PLP, and the CPIT tests for JM-GO. In each class, the usual GOF tests based on the Empirical Distribution Function are applied: Kolmogorov-Smirnov (KS), Cramer-von Mises (CM), Anderson-Darling (AD).

In order to estimate the power of these tests, they are applied on data simulated according to the five reliability growth models defined in Section 2.1. For each model, two sets of parameters are proposed, detailed below:

JM1: $N = 200, \phi = 0.0001$,	JM2: $N = 100, \phi = 0.00009$
MG1: $\lambda = 0.01, \theta = 0.03$,	MG2: $\lambda = 0.01, \theta = 0.07$
PLP1: $\alpha = 0.3, \beta = 0.5$,	PLP2: $\alpha = 2.0, \beta = 0.1$
GO1: $\lambda = 0.005, \phi = 0.00003$,	GO2: $\lambda = 0.0004, \phi = 0.000004$
S1: $\alpha = 200, \beta = 0.0001$,	S2: $\alpha = 120, \beta = 0.0001$

The first set corresponds to models which are close to a PLP, and the second to models which are far from the PLP (except for PLP2, of course). The PLP-GOF tests are expected to be more powerful on sets 2 than on sets 1.

For each alternative, 2000 simulations have been done. The sample sizes of simulated data are n=20, 60 and 100. The KS, CM and AD tests are applied with the usual modifications for small samples [see D'Agostino and Stephens (1986)]. The significance level for all the tests is 5%. Tables 2.1 to 2.5 give the fraction of the 2000 replicates for which the null hypothesis (PLP or JM-GO) has been rejected.

A more detailed experimental study can be found in Cretois and Gaudoin (1997).

Table 2.1: Power of the tests for the Jelinski-Moranda alternative

		Rigdon PLP			CPIT PLP			CPIT JM-GO		
n		KS	CM	AD	KS	CM	AD	KS	CM	AD
20	JM1	4.6	4.7	4.8	5.1	5.5	5.5	5.2	5.3	6.5
	JM2	4.5	4.8	4.7	5.1	5.5	5.4	5.0	5.5	6.3
60	JM1	6.5	7.0	6.2	4.9	5.5	5.3	5.5	5.7	6.2
	JM2	14.6	18.2	16.2	7.7	8.3	8.3	5.1	5.8	6.0
100	JM1	13.8	16.4	14.0	9.1	8.9	9.7	5.3	5.2	5.9
	JM2	100	100	100	99.0	99.0	99.6	5.3	4.9	5.2

Table 2.2: Power of the tests for the Moranda Geometric alternative

		Rigdon PLP			CPIT PLP			CPIT JM-GO		
n		KS	CM	AD	KS	CM	AD	KS	CM	AD
20	MG1	5.1	5.6	5.1	4.9	5.4	5.4	4.9	5.1	8.1
	MG2	7.7	9.1	7.0	6.2	6.5	6.7	4.8	4.1	5.7
60	MG1	26.8	34.3	30.8	9.0	9.8	10.2	6.4	6.4	6.3
	MG2	71.1	85.8	82.4	12.5	13.4	17.7	71.2	77.8	79.1
100	MG1	78.0	88.8	86.8	21.0	22.3	27.5	38.1	42.8	43.0
	MG2	99.8	100	100	50.4	47.2	60.2	100	100	100

Table 2.3: Power of the tests for the Power-Law Process alternative

		Rigdon PLP			CPIT PLP			CPIT JM-GO		
n		KS	CM	AD	KS	CM	AD	KS	CM	AD
20	PLP1	4.3	4.7	4.8	5.1	5.8	5.5	33.8	39.0	42.0
	PLP2	4.3	4.7	4.8	5.1	5.8	5.5	100	100	100
60	PLP1	5.6	5.5	5.3	5.1	5.4	5.4	81.2	85.9	85.8
	PLP2	5.6	5.5	5.3	5.1	5.4	5.4	100	100	100
100	PLP1	4.6	5.4	5.2	5.4	5.7	5.5	96.5	98.0	98.2
	PLP2	4.6	5.4	5.2	5.4	5.7	5.5	100	100	100

Table 2.4: Power of the tests for the Goel-Okumoto alternative

		Rigdon PLP			CPIT PLP			CPIT JM-GO		
n		KS	CM	AD	KS	CM	AD	KS	CM	AD
20	GO1	4.5	4.8	4.7	5.0	5.4	5.3	5.2	5.3	6.4
	GO2	4.6	4.8	4.6	5.0	5.6	5.4	5.2	5.3	6.3
60	GO1	7.1	7.9	7.1	5.0	5.4	5.4	5.5	5.8	6.2
	GO2	15.4	19.0	17.2	7.8	8.3	8.9	5.4	5.8	6.1
100	GO1	22.8	27.2	24.7	11.2	11.5	10.6	5.1	5.5	5.9
	GO2	90.7	94.4	93.7	68.3	70.1	75.3	6.0	5.9	5.7

Table 2.5: Power of the tests for the S-shaped alternative

		Rigdon PLP			CPIT PLP			CPIT JM-GO		
n		KS	CM	AD	KS	CM	AD	KS	CM	AD
20	S1	4.7	5.0	4.6	5.1	5.5	5.4	12.8	14.4	34.9
	S2	4.8	5.0	5.1	5.0	5.6	5.5	13.6	15.2	35.2
60	S1	8.3	10.9	9.5	5.7	6.0	6.1	27.3	32.4	46.2
	S2	17.7	22.0	19.7	7.4	8.2	8.6	34.0	40.4	56.7
100	S1	28.6	32.8	30.5	11.0	11.7	13.0	42.1	49.2	59.3
	S2	82.5	88.1	87.6	46.7	49.0	55.0	52.6	62.2	73.0

Tables 2.1 to 2.5 show that, for the PLP, the CPIT tests are much less powerful than the Rigdon tests. The power of all tests is very low for small samples and grows when the sample size grows. The Anderson-Darling is quite always slightly the most powerful test. As expected, the PLP is more easily rejected for sets 2 than for sets 1.

The JM-GO models are more rejected than the PLP for the MG and S alternatives. But it is not meaningful to compare the powers of these tests since the null hypothesis is not the same in both cases.

Finally, the tests are applied to software reliability data collected by Musa (1979). These data consist in the times between failures of nine control-command software (denoted M1, M2, M3, M4, M6, M14C, M17, M27, M40) in the test period and operational life.

The results of the tests are presented in Table 2.6 in the following form:

RR: strong reject (p-value $<$ 1%)
R: moderate reject (1% $<$ p-value $<$ 5%)
A: weak acceptance (5% $<$ p-value $<$ 10%)
AA: moderate acceptance (10% $<$ p-value $<$ 25%)
AAA: strong acceptance (p-value $<$ 25%)

Table 2.6: Application of the tests to real software reliability data

		Rigdon PLP			CPIT PLP			CPIT JM-GO		
n		KS	CM	AD	KS	CM	AD	KS	CM	AD
136	M1	R	RR	RR	AA	AA	A	AAA	AAA	AA
54	M2	A	A	A	AA	A	A	AA	AA	A
38	M3	R	R	R	AAA	AAA	AAA	AA	R	R
53	M4	RR	RR	RR	AA	AA	RR	AA	A	R
73	M6	R	R	R	RR	RR	RR	RR	RR	RR
36	M14C	RR	RR	RR	A	A	RR	A	R	RR
38	M17	RR	RR	RR	AA	AAA	A	AAA	AAA	AA
41	M27	RR	RR	RR	AAA	AA	R	R	R	R
101	M40	RR	RR	RR	RR	RR	RR	RR	RR	RR

From Table 2.6, it is obvious that the behavior of the tests is not the same in the three classes. The Rigdon tests tend to reject the PLP while the CPIT tests tend to accept it, except for M6 and M40. Since the Rigdon tests are more powerful, we will conclude that the PLP is strongly rejected for all Musa's data, except M2. The JM-GO models are slightly more rejected than the PLP by the CPIT tests. For example, they are rejected by the AD CPIT test for all data sets, except M1, M2 and M17. So it seems that, except for these cases, the JM and GO models fit Musa's data worse than the PLP. Then a reliability engineer should apply the PLP for M2, JM or GO for M1 and M17, and none of these models for the other data sets.

2.7 Conclusion

The experimental study exposed here has the same results as the study of O'Reilly-Stephens for the Homogeneous Poisson Process: when other Goodness-of-fit tests exist for a particular model, the CPIT tests are less powerful, so they should not be used. But they become very interesting for testing the fit of models for which no GOF test is available at the present time. This is the case for the Jelinski-Moranda and Goel-Okumoto models, and this will be also the case for several other reliability growth models. The only property that these models must have in order to build CPIT tests is the existence of a non trivial sufficient statistic. For example, a sufficient statistic for the S-shaped model is $\left(T_n, \sum_{i=1}^{n-1} T_i, \sum_{i=1}^{n-1} \ln T_i\right)$. So the CPIT methodology will allow us to test the fit of a large class of reliability growth models. The major drawback of the method is that the derivation of the V_i's can be long and difficult. Finally, the method can not be applied to the MG model, since a non trivial sufficient statistic does not exist.

Acknowledgement. The author is grateful to Emmanuelle Cretois for her help in the experimental part of this work.

References

1. Cretois E. and Gaudoin O. (1997). New results on goodness-of-fit tests for the power-law process and applications to software reliability, *Proceedings of the 3rd ISSAT International Conference on Reliability and Quality in Design*, Anaheim, 111–115.

2. Crow, L. H. (1974). Reliability analysis for complex repairable systems, In *Reliability and Biometry - Statistical Analysis of Lifelength*, SIAM Philadelphia, pp. 379–410.

3. D'Agostino R. B. and Stephens M. A. (1986). *Goodness-of-fit Techniques*, New York: Marcel Dekker.

4. Gaudoin O. (1998). CPIT Goodness-of-fit tests for the power law process, *Communications in Statistics—Theory and Methods*, **27**, 1.

5. Goel, A. L. and Okumoto, K. (1979). Time dependent error detection rate model for software reliability and other performance measures, *IEEE Transactions on Reliability*, **28**, 1, 206–211.

6. Jelinski, Z. and Moranda, P. B. (1972). Statistical computer performance evaluation, In *Software Reliability Research* (Ed., W. Freiberger), pp. 465–497, New York: Academic Press.

7. Lee, L. and Finelli, G. B. (1989). A transformation for testing the fit of an exponential order statistics model, *Stochastic Processes and Their Applications*, **33**, 299–307.

8. Lyu, M. R. (Ed.) (1996). *Handbook of Software Reliability Engineering*, IEEE Computer Society Press and McGraw-Hill Book Company.

9. Miller, D. R. (1986). Exponential order statistics models of software reliability growth, *IEEE Transactions on Software Engineering*, **12**, 12–24.

10. Moranda, P. B.(1979). Event altered rate models for general reliability analysis, *IEEE Transactions on Reliability*, **28**, 5, 376–381.

11. Musa J. D. (1979). Software reliability data, *Technical Report*, Rome Air Development Center.

12. O'Reilly, F. J. and Quesenberry, C. P. (1973). The conditional probability integral transform and applications to obtain composite chi-square goodness-of-fit tests, *Annals of Statistics*, **1**, 74–83.

13. O'Reilly, F. J. and Stephens, M. A. (1982). Characterizations and goodness-of-fit tests, *Journal of the Royal Statistical Society, Series B*, **44**, 353–360.

14. Rigdon S. E. (1989). Testing goodness-of-fit for the power-law process, *Communications in Statistics—Theory and Methods*, **18**, 4665–4676.

15. Rosenblatt, M. (1952). Remarks on a multivariate transformation, *Annals of Mathematical Statistics*, **23**, 470–472.

16. Xie, M. (1991). *Software reliability modelling*, Singapore: World Scientific.

17. Yamada, S., Ohba, M. and Osaki, S. (1983). S-shaped reliability growth modelling for software error detection, *IEEE Transactions on Reliability*, **35**, 475–478.

3

On the Use of Minimally Informative Copulae in Competing Risk Problems

Tim Bedford

Delft University of Technology, Delft, The Netherlands

Abstract: The competing risks problem in reliability estimation is described. Various approaches to this problem are briefly discussed, including the copula-graphical estimator of Zheng and Klein. This estimator requires the specification of a copula to make the underlying distribution functions identifiable. In this paper we propose use of the minimally informative copula with given rank correlation coefficient. This generalizes the usual approach in the literature as the Kaplan-Meier estimator is obtained when the rank correlation is set to zero. Two examples are discussed. In these cases the estimated distribution function is highly sensitive to rank correlation, and increasing the correlation reduces the mean lifetime.

Keywords and phrases: Competing risk, copula, entropy, information, rank correlation

3.1 Competing Risk

A valid statistical treatment of reliability data is extremely difficult. Various authors [Ascher and Feingold (1984), Paulsen (1996)] have described methods for analysing data in which one begins by considering stationarity, and progresses through various levels of generality of models. The analyst may often conclude that the data can be modelled with a renewal system. We assume here that this situation has been reached, and that the reliability data consists of a sequence of times

$$\ldots x_0, x_1, x_2, \ldots$$

which are the times between events in the system. These times will be modelled as an i.i.d. sequence

$$\ldots X_0, X_1, X_2, \ldots \; .$$

In many situations the event times X_i are not failure times, but are times at which the system was restored to "as good as new". Often X_i is actually either the time of failure or of preventive maintenance. In this case it is reasonable to model X_i by an i.i.d. sequence of random variables (Y_i, Z_i) in the following way. The r.v. Y_i is the failure time of the component and Z_i is the time of preventative maintenance. The observable data consists of $X_i = \min(Y_i, Z_i)$ and the label I_i (telling us whether a failure occurred or a preventive maintenance). The variables Y_i and Z_i are not directly observed. This situation is known as *competing risk*. For many applications, for example in maintenance planning, we wish to estimate the distribution function of Y.

A general result of Tsiatis (1975), showed that the failure data arising in such a situation determine a unique independent model, so that we can in principle, *under the assumption of independence*, determine the distribution for Y_i and Z_i from the data. This result is easily extended to the case of more than two competing risks. The Kaplan-Meier estimator [Kaplan and Meier (1958)] gives non-parametric maximum likelihood estimators of the distributions of Y and Z given data of the form described above.

Testing the assumption of independence is however impossible due to unidentifiability: With the observable data we can only estimate the sub-distribution functions

$$F_Y^*(t) = P(Y \le t, Y < Z)$$

and

$$F_Z^*(t) = P(Z \le t, Z < Y).$$

There are many possible joint distributions (with many different marginal distributions) which possess the same sub-distribution functions.

We illustrate this with an example. Suppose we have failure data as follows:

Observed failure times, Y	2, 4.5, 5, 7.5, 7.8
Prev. maintenance times, Z	1, 1.5, 3.1, 4.2, 6.1.

The data may be used to produce (empirical) sub-distribution functions as shown in Figure 3.1.

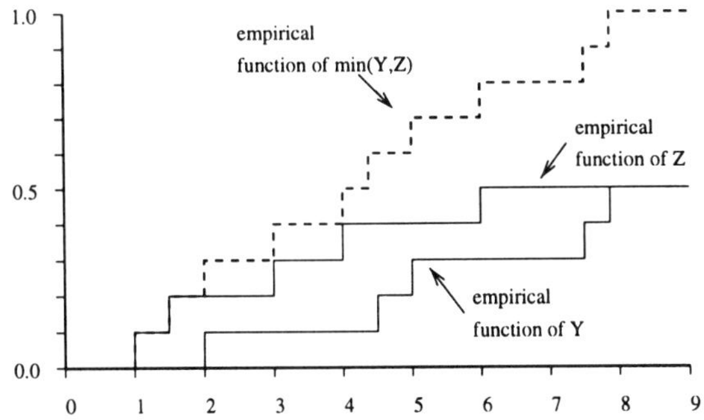

Figure 3.1: Empirical sub-distribution functions

Also shown on the figure is the empirical distribution function of $\min(Y, Z)$. We want to know what the "true" empirical distribution function of Y would have been if we had been able to observe it. Clearly the only thing we can say with any certainty is that each unobserved failure occurred *after* the corresponding censoring time. This implies that the (unseen) empirical distribution function

1. lies between the sub-distribution function for Y and the distribution function of the minimum, and

2. has a jump wherever the sub-distribution function for Y has a jump.

This gives à wide range of possibilities for the marginal of Y. In Figure 3.2 we show the sub-distribution function for Y, the distribution function of $\min(Y, Z)$, and two possible distribution functions for Y (one with high correlation between Y and the censoring variable Z, and one which assumes independence between Y and Z).

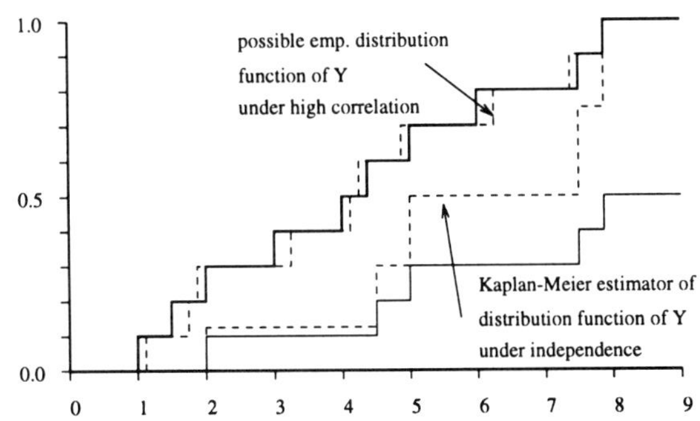

Figure 3.2: Two possible marginal distribution functions

When we take the example of Y being failure time and Z being preventive maintenance time, it is evident that they *cannot* be assumed independent. The assumption of independence would imply that maintenance engineers take no account of the state of a component when taking the decision to preventively maintain. It is more reasonable to make an *a-priori* assumption of positive dependence between Y and Z.

3.2 Bounds Without Assumptions on a Dependence Structure

As we have seen above, the marginal distribution of Y is unidentifiable from the (observable) subsurvival functions. This suggests that we could try to give bounds on the marginal distribution without making assumptions on the form of dependence between Y and Z.

3.2.1 Peterson bounds

The Peterson bounds are pointwise upper and lower bounds on the value of the marginal distribution function. We state the bounds in terms of the subdistribution functions, and suppose (for simplicity) that $P(Y = Z) = 0$. Note that if the subdistribution functions are given then the distribution function of $\min(Y, Z)$ is also known, indeed $F_{\min(Y,Z)}(t) = F_Y^*(t) + F_Z^*(t)$.

Theorem 3.2.1 *[Peterson (1976)]*

1. *For any $t \geq 0$ we have $F_Y^*(t) \leq F_Y(t) \leq F_{\min(Y,Z)}(t)$.*
2. *For any $t \geq 0$ and any u satisfying $F_Y^*(t) < u < F_{\min(Y,Z)}(t)$, there is a joint distribution on (Y, Z) with the given subdistribution functions such that $u = F_Y(t)$.*

The Peterson bounds are *pointwise* bounds, and are "too wide to be useful" according to the entry on *competing risks* in [Kotz et al. (1983)].

3.2.2 Crowder-Bedford-Meilijson bounds

Functional bounds were found by Crowder (1991) and improved by Bedford and Meilijson (1997) to give a complete characterisation of the marginals consistent with given subdistribution functions. Bedford and Meilijson also used the characterisation to produce a statistical test based on the Kolmogorov-Smirnov statistic in which a hypothesised marginal distribution can be tested against available data.

The intuition behind the functional bounds is easy to understand. Firstly it is obvious that not all functions K satisfying $F_Y^*(t) \leq K(t) \leq F_{\min(Y,Z)}(t)$

for all t are distribution functions. It is also quite easy to see that not all distribution functions satisfying this inequality are allowable marginals. Indeed, if we consider the gap between each marginal distribution function and its lower bound,

$$\begin{aligned} F_Y(t) - F_Y^*(t) &= P(Y \leq t, Y > Z) \\ F_Z(t) - F_Y^*(t) &= P(Z \leq t, Z > Y) \end{aligned} \tag{3.1}$$

then, as observed by Crowder (1991), these gaps must be nonnegative *and nondecreasing* as a function of t. In other words, feasible marginal distributions F_i are *co-monotone* with their lower Peterson bound F_i^*.

These bounds allow one to narrow the bounds given by the Peterson bounds, but, as noted by Bedford and Meilijson (1995) confidence intervals based on their test narrow very slowly as a function of the number of data points.

Bedford and Meilijson (1995) use the bounds to demonstrate the alarming over-optimism of the assumption of independence for the joint distribution of (Y, Z). Further evidence for the extent of this unsafe feature of the independence model is provided by Zheng and Klein (1995), whose work will be described in the next section:

> "Note that as the assumed strength of association increases the estimated survival function becomes smaller, so that an assumption of independence gives us overly optimistic estimates of the survival function."

3.3 Estimators Using Dependence Assumptions

Recently, Zheng and Klein (1995) generalised the Kaplan-Meier estimator so that the modeller is able make assumptions about the dependence structure between Y and Z. Specifically, they allow the modeller to specify the *copula* between Y and Z. Recall that for a continuous random variable Y with c.d.f. F_Y, the random variable $F_Y(Y)$ is always uniformly distributed on $[0, 1]$.

Definition 3.3.1 The *copula* of two random variables Y and Z is the distribution C on the unit square $[0, 1]^2$ of the pair $(F_Y(Y), F_Z(Z))$.

The functional form of $C : [0, 1]^2 \rightarrow \mathbb{R}$ is

$$C(u, v) = H(F_Y^{-1}(u), F_Z^{-1}(v))$$

where H is the joint distribution function of (Y, Z).

If Y and Z are independent random variables then it is easy to see that their copula is the uniform distribution on the unit square, $C(y,z) = yz$. A well known family is the Morgenstern family, given by $C(y,z) = yz(1+\alpha(1-y)(1-z))$, where $-1 \leq \alpha \leq 1$.

Definition 3.3.2 The *rank correlation* $\tau(Y,Z)$ of (Y,Z) is the product-moment correlation coefficient of the pair $(F_Y(Y), F_Z(Z))$.

The rank correlation is a non-parametric measure of association between Y and Z. By choosing a copula with a high rank correlation we can model a high degree of association between Y and Z.

There are various definitions of rank correlation. The one given above is due to Spearman, but also well known is Kendall's tau which is defined as

$$\tau(X,Y) = P((Z-Z^*)(Y-Y^*) > 0) - P((Z-Z^*)(Y-Y^*) < 0)$$

where (Y,Z) and (Y^*,Z^*) are independent and identically distributed. See [Genest and Mackay (1986)] for a discussion of various equivalent definitions of Kendall's tau. The differences in values taken between the various rank correlations are not large, but the definitions are not equivalent. Kendall's tau has the advantage of a definition which can be explained to a non-specialist, but the Spearman rank correlation is easier to work with from a mathematical point of view. In particular we have an algorithm to calculate the minimally informative copula with a given (Spearman) rank correlation, which will be discussed in the next section. When we refer to the rank correlation later in this paper we shall always mean the Spearman correlation.

The assumption of a particular copula implies nothing about the marginal distributions of Y and Z. Extra information is required to determine these distributions. A special case is the result of Tsiatis (1975) where the assumption of the independence copula plus the subsurvival functions determines the marginals. The following result of Zheng and Klein says that the copula plus the subsurvival functions together determine the marginals, that is, by assuming a copula we have identifiability in the competing risks problem.

Theorem 3.3.1 *[Zheng and Klein (1995)] Suppose the marginal distribution functions of (Y,Z) are continuous and strictly increasing in $(0,\infty)$. Suppose the copula C of (Y,Z) is known, and gives positive probability to any open set of the unit square. Then F_Y and F_Z, the marginal distribution functions of Y and Z, are uniquely determined by the subdistribution functions.*

When Y and Z have a finite upper bound, say F_Y is strictly increasing on $(0,t_1)$ and F_Z is strictly increasing on $(0,t_2)$ with $F_Y(t_1) = 1$ and $F_Z(t_2) = 1$, then F_Y and F_Z can be uniquely determined from the subsurvival functions on the interval $(0, \min(t_1,t_2))$.

3.3.1 The copula-graphic estimator

Zheng and Klein have also produced an estimator which reduces to the Kaplan-Meier estimator in the case that the independence copula is used. This estimator uses the identity

$$F_Y(t) + F_Z(s) = C(F_Y(t), F_Z(s)) + F_Y^*(t) + F_Z^*(s),$$

and can be numerically implemented as follows.

Suppose our data consists of points (t_1, I_1), (t_2, I_2), ..., (t_n, I_n), and define $t_0 = 0$. Our estimators $\hat{F}_Y$ and $\hat{F}_Z$ are step functions such that $\hat{F}_Y(t_0) = \hat{F}_Z(t_0) = 0$, and the functions are constant except possibly at the times t_i. $\hat{F}_Y$ jumps at those times for which a Y was observed, and $\hat{F}_Z$ jumps at those times for which a Z was observed. The heights of the jumps are determined iteratively for $i = 1, \ldots, n$ by numerically solving the equation

$$\hat{F}_Y(t_i) + \hat{F}_Z(t_i) = C(\hat{F}_Y(t_i), \hat{F}_Z(t_i)) + \hat{F}_Y^*(t_i) + \hat{F}_Z^*(t_i),$$

for either $\hat{F}_Y(t_i)$ or $\hat{F}_Z(t_i)$ (depending on which function jumps at t_i). The algorithm recursively determines the points $(\hat{F}_Y(t_i), \hat{F}_Z(t_i))$, $(i = 1 \ldots n)$.

Figure 3.3 shows the operation of the algorithm for the example data given above, for the independent copula, using the data we had before. The smallest observation is of a Z, or more formally of the event $Z = 1, Y > 1$. Starting from $(0, 0)$ we move up to find the value of α so that $\frac{1}{10}$ of the mass on the square is in the rectangle with corners $(0,0)$, $(0, \alpha)$, $(1, 0)$, and $(1, \alpha)$. This gives $\alpha = \frac{1}{10}$, our estimate of $\hat{F}_Z(1)$. The current point on the square is now $(0, \frac{1}{10})$. The second observation is also of a Z, and starting from the current point we move up until $\frac{1}{10}$ of the mass is in a rectangle. This gives a new current point of $(0, \frac{2}{10})$, and estimates $\hat{F}_Z(1.5)$ as $\frac{2}{10}$. The next observation is of a Y. Moving from the current point we now find a vertically oriented rectangle with mass $\frac{1}{10}$. This is the rectangle with corners $(0, 0.2)$, $(0.125, 0.2)$, $(0, 1)$, and $(0.125, 1)$. The new current point is $(0.125, 0.2)$ and our estimate of $\hat{F}_Y(2)$ is 0.125. The algorithm goes on in this way until all the estimates have been made...

Because the rectangles have equal mass, one can view this method as a non-parametric maximum likelihood estimator.

The choice of the independent copula is arbitrary here. We could have substituted any other copula whose support is the whole unit square. By using a copula with a high rank correlation we obtain non-parametric estimates of the marginals of Y and Z under the assumption of a high degree of association.

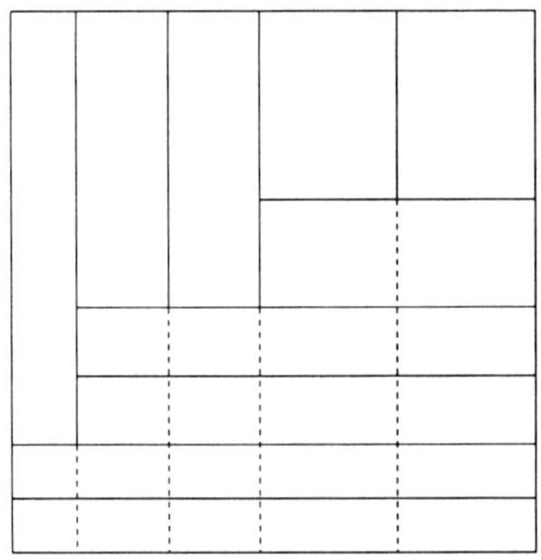

Figure 3.3: Copula-Graphical estimator example

3.4 Minimally Informative Copulae

Given the experiences of Zheng and Klein, it seems reasonable to try to use a family of copulae which allow one to model all possible rank correlations in a simple way. Since the rank correlation is to be treated as a primary parameter, it is necessary to choose a family of copulae which are as "smooth" as possible. One way to do this is to take the family of copulae which minimize the information function (taken with respect to the uniform distribution), $\int\int c(x,y)\log c(x,y)\,dxdy$ (where c is the copula density).

In [Bedford and Meeuwissen (1997)] we showed how the minimally informative copula with given correlation coefficient could be determined. This copula has density of the form

$$\kappa(y,\theta)\kappa(z,\theta)e^{\theta(y-0.5)(z-0.5)}$$

where $\theta = \theta(\tau)$ is a certain monotone increasing function of the correlation coefficient τ, and the function $\kappa(\cdot,\theta)$ is determined as the solution to an integral equation

$$\kappa(y,\theta) = \left[\int_0^1 \kappa(z,\theta)e^{\theta(y-0.5)(z-0.5)}\,dz\right]^{-1}.$$

Although no closed form solution seems possible for θ, it is easy to determine it numerically [see Bedford and Meeuwissen (1997)]. When $\theta = 0$ we obtain the independent copula.

A discrete approximation to the minimally informative copula is easily calculated. We seek a discrete distribution of the form

$$P((x_i + \frac{1}{2}, x_j + \frac{1}{2})) = \frac{1}{n^2}\kappa_i(\theta)\kappa_j(\theta)e^{\theta x_i x_j},$$

where $x_i = \frac{2i-1-n}{n}$ for $i = 1, \ldots, n$. Given θ, we calculate the vector κ by applying a DAD algorithm [Nussbaum (1989)]. The classical *DAD* problem is, given a square symmetric non-negative matrix A, to find a diagonal matrix D with positive diagonal entries such that DAD is doubly stochastic. In our case $A = (e^{\theta x_i x_j})$, and $D = \frac{1}{n}\text{diag}(\kappa_1, \ldots, \kappa_n)$. A simple numerical scheme for determining the values of κ_i is iteration of a function which is defined as follows. Let $\overset{o}{C}$ be the positive cone in $\mathbb{R}^n$ and define $\Psi : \overset{o}{C} \to \overset{o}{C}$ by the composition of normalization,

$$y_i \mapsto \frac{y_i n}{\sqrt{\sum_{j,k} y_j y_k e^{\theta x_j x_k}}}$$

followed by the non-linear map

$$y_i \mapsto \frac{n}{\sum_j y_j e^{\theta x_i x_j}}.$$

This iteration converges geometrically to the fixed point which is the vector κ we required. This enables us to calculate the minimally informative distribution with a given θ, and since we know the distribution we can calculate its τ. This enables us to calculate τ as a function of θ, and as τ is a strict monotone increasing function of θ we can easily numerically invert to solve for the right θ value for any given τ. Even the current author was easily able to program this in Matlab.

This copula was originally introduced to knowledge dependence in uncertainty analysis (see [Cooke (1995)]). In the sense of minimizing information, it is the simplest copula with a given correlation coefficient. Note that information is not changed by coordinate transformations. Hence, if we take the minimum information copula with a given rank correlation ρ, and choose two marginal distributions F and G to form the joint distribution with density

$$f(y, z) = c(F(y), G(z)),$$

this joint distribution will be the unique distribution with minimum information with respect to the independent distribution with marginals F and G, and with rank correlation ρ.

3.5 Examples

3.5.1 Example 1

The data here is that used above, with the empirical sub-distribution functions shown in Figure 3.1. We have applied the algorithm with the minimally informative copula with three values of τ. Figure 3.4(a) shows the estimated distribution function of Y with $\tau = 0$ (i.e. with the independent copula). Figure 3.4(b) shows the estimated distribution function with $\tau = 0.5$, Figure 3.4(c) uses $\tau = 0.9$. Finally, Figure 3.4(d) shows the three graphs superimposed on each other.

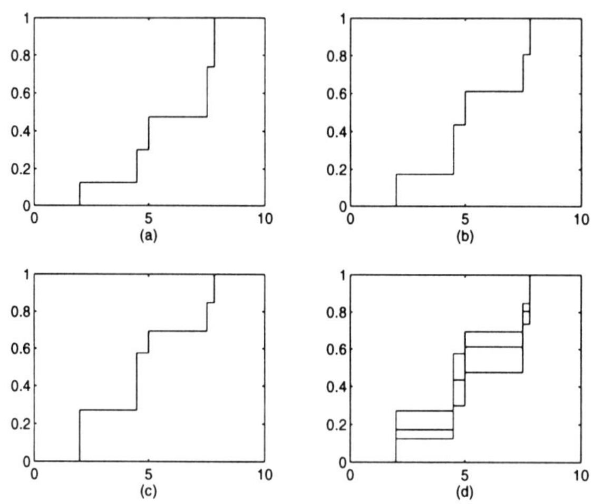

Figure 3.4: Estimated distribution function: Example 1

3.5.2 Example 2

We have assumed that Y and Z have the same sub-distribution function, $t \mapsto \frac{1}{2} - \frac{1}{2}\exp(-t)$, and have generated 50 Y and 50 Z samples. This sub-distribution would have arisen, for example, if the *joint* distribution of Y and Z was independent and Y and Z were exponentially distributed with mean 2. We wish however to explore the consequences of assuming dependence between Y and Z.

We have again applied the algorithm with the minimally informative copula with three values of τ. Figure 3.5(a) shows the estimated distribution function of Y with $\tau = 0$ (i.e. with the independent copula), Figure 3.5(b) with $\tau = 0.5$, Figure 3.5(c) uses $\tau = 0.9$, and Figure 3.5(d) shows the three graphs superimposed on each other.

The estimates of the mean (using the estimated distribution function) are:

τ	0	0.5	0.8	0.9
mean	1.74	1.43	1.25	1.17 .

This shows that making the assumption of a fair degree of correlation makes a substantial difference in estimation as compared to the independent case.

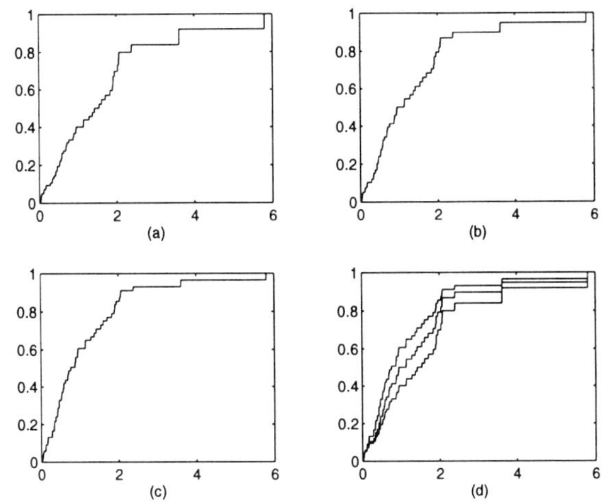

Figure 3.5: Estimated distribution function: Example 2

3.6 Conclusions

We have shown that the minimally informative copula with given rank correlation is appropriate for use in competing risk problems, and in conjunction with the copula-graphical estimator of Zheng and Klein, can be successfully applied to estimate failure distributions.

References

1. Ascher, H. and Feingold, H. (1984). *Repairable Systems Reliability*, Lecture Notes in Statistics Vol. 7, New York: Marcel Dekker.

2. Bedford, T. and Meilijson, I. (1997). A characterisation of marginal distributions of (possibly dependent) lifetime variables which right censor each other, *Annals of Statistics*, **25**, 1622–1645.

3. Bedford, T. and Meilijson, I. (1995). The marginal distributions of lifetime variables which right censor each other, In *Analysis of Censored Data*

(Eds., H. L. Koul and J. V. Deshpande), IMS Lecture Notes Monograph Series, Volume 27, Hayward: Institute of Mathematical Statistics.

4. Bedford, T. and Meeuwissen, A. M. H. (1997). Minimally informative distributions with given rank correlation for use in uncertainty analysis, *Journal of Statistical Computation and Simulation*, **57**, 143–174.

5. Cooke, R. M. (1995). *UNICORN: methods and code for uncertainty analysis*, Published by AEA Technology for ESRA.

6. Crowder, M. (1991). On the identifiability crisis in competing risks analysis, *Scandinavian Journal of Statistics*, **18**, 223–233.

7. Genest, C. and Mackay, J. (1986). The joy of copulas: Bivariate distributions with uniform marginals, *The American Statistician*, **40**, 280–283.

8. Kaplan, E. L. and Meier, P. (1958). Nonparametric estimation from incomplete observations, *Journal of the American Statistical Association*, **53**, 457–481.

9. Kotz, S. ,Johnson, N. L. and Read, C. B. (Eds.) (1983). *Encyclopedia of Statistical Sciences*, New York: John Wiley & Sons.

10. Nussbaum, R. D. (1989). *Hilbert's projective metric and iterated nonlinear maps, II*, Memoirs of the American Mathematical Society, Vol. 79, Number 401.

11. Paulsen, J., Dorrepaal, J., Cooke, R. M. and Hokstadt, P. (1996). The design and use of reliability data base with analysis tool., Riso National Laboratory, Denmark, Riso-R-896(EN) NKS/RAK-1(96)R6.

12. Peterson, A. V. (1976). Bounds for a joint distribution function with fixed subdistribution functions: Application to competing risks, *Proceedings of the National Academy of Sciences, USA*, **73**, 11–13.

13. Tsiatis, A. (1975). A nonidentifiability aspect in the problem of competing risks, *Proceedings of the National Academy of Sciences, USA*, **72**, 20–22.

14. Zheng, M. and Klein, J. P. (1995). Estimates of marginal survival for dependent competing risks based on an assumed copula, *Biometrika*, **82**, 127–138.

4

Model Building in Accelerated Experiments

V. Bagdonavičius and M. S. Nikulin

University of Vilnius, Lithuania,
Université de Bordeaux, Bordeaux, France &
Steklov Mathematical Institute, St. Petersburg, Russia

Abstract: A review of main known and new models used in accelerated life testing is done. An unifying approach of model formulation is used. Only models for which estimation procedures are known are considered.

Keywords and phrases: Accelerated experiments, additive accumulation of damages, generalized multiplicative models, model building, stress

4.1 Introduction

The purpose of this paper is to give models which are used or can be used in accelerated life testing. We are interested only in models for which parametric, semiparametric or nonparametric procedures of estimation (non-Bayes) are given. The most of them will be formulated using the notion of the resource [Bagdonavičius (1990), Bagdonavičius and Nikulin (1994)].

Suppose that the time-to-failure $T_{x(\cdot)}$ is a non-negative random variable with the survival function $S_{x(\cdot)}(t) = \mathbf{P}\{T_{x(\cdot)} > t\}$ and the hazard rate

$$\lambda_{x(\cdot)}(t) = -\frac{S'_{x(\cdot)}(t)}{S_{x(\cdot)}(t)},$$

which depends on a vector of stresses $x(\tau) = (x_1(\tau), ..., x_m(\tau))^T, \quad \tau \leq 0.$

The time-to-failure $T_{x(\cdot)}$ could be called the *resource* of the item. But the notion of the resource should not depend on $x(\cdot)$. Denote by

$$\Lambda_{x(\cdot)}(t) = \int_0^t \lambda_{x(\cdot)}(u)du$$

the accumulated hazard rate. Then the random variable $R = \Lambda_{x(\cdot)}(T_{x(\cdot)})$ has the standard exponential distribution with the survival function $S_R(t) = e^{-t}$, $t \geq 0$.

It takes values in the interval $[0, \infty)$ and doesn't depend on $x(\cdot)$. Note that $T_{x(\cdot)} = t$ if and only if $R = \Lambda_{x(\cdot)}(t)$. For all $x(\cdot)$ the moment t for items working under the stress $x(\cdot)$ is equivalent to the moment $\Lambda_{x(\cdot)}(t)$ for items working in conditions when the time-to-failure has the standard exponential distribution. So the number $\Lambda_{x(\cdot)}(t) \in [0, \infty)$ is called *the exponential resource used until the moment t under the stress $x(\cdot)$*. The concrete item which failed at the moment t under the stress $x(\cdot)$ used $\Lambda_{x(\cdot)}(t)$ of the resource until this moment.

Instead of the exponential resource one can define a resource with any probability distribution, so we can consider a whole class of resources. Really, suppose that G is some fixed survival function, strictly decreasing and continuous on $[0, \infty)$ and $H = G^{-1}$ is the inverse function of G.

The survival function of the random variable

$$R^G = f^G_{x(\cdot)}(T_{x(\cdot)}) = H(S_{x(\cdot)}(T_{x(\cdot)}))$$

is G and doesn't depend on $x(\cdot)$. The random variable R^G is called the *G-resource* and the number $f^G_{x(\cdot)}(t)$ is called the *G-resource used till the moment t*. In the particular case when $G = S_{x_0}$ is the survival function under the "normal" stress x_0 the moment t under the stress $x(\cdot)$ is equivalent to the moment $f^{S_{x_0}}_{x(\cdot)}(t)$ under the "normal" stress x_0 for all $x(\cdot)$.

The models of accelerated life will be formulated in dependence on the way of resource using. Note that the different resource can have different way of using. This is the cause of considering a *whole class of resources*. Denote by E some set of stresses.

4.2 Additive Accumulation of Damages Model and Its Submodels

Definition 4.2.1 The additive accumulation of damages (AAD) model [Bagdonavičius (1978)] holds on E if there exists a positive function r on E such that for all $x(\cdot) \in E$

$$\frac{\partial f^G_{x(\cdot)}(t)}{\partial t} = r[x(t)] \tag{4.1}$$

with the initial condition $f^G_{x(\cdot)}(0) = 0$.

This model means that the rate of G-resource using at the moment t depends only on the value of the stress at this moment.

The equation (4.1) implies

$$S_{x(\cdot)}(t) = G\left(\int_0^t r[x(\tau)]\, d\tau\right). \tag{4.2}$$

This model is also called the accelerated failure time (AFT) model [Cox and Oakes (1984)].

If $x(\tau) \equiv x = const$ then we have the model

$$S_x(t) = G\{r(x)t\}. \tag{4.3}$$

So the stress changes only the scale of distribution. Applicability of this model in accelerated life testing was first noted by Pieruschka (1961).

In the particular case of the step-stress of the form

$$x(\tau) = x_i,\ \tau \in [t_{i-1}, t_i),\ (i = 1, \cdots, m), \tag{4.4}$$

where $0 = t_0 < t_1 < \cdots < t_m = +\infty$, we have the following model:

$$S_{x(\cdot)}(t) = G\{\sum_{i=1}^{k} r(x_i)(t_i - t_{i-1}) + r(x_k)(t - t_{k-1})\}, \quad \text{if} \quad t \in [t_{k-1}, t_k).$$

In the following proposition some useful relations between the means and the quantiles of times-to-failure under constant and non-constant stresses are done.

Suppose that $x(\cdot)$ is a time-varying stress, $x(\tau)$ is a constant in time stress, equal to the value of the stress $x(\cdot)$ at the moment τ, and x_0 is some constant in time stress. Denote $t_p(x(\cdot))$ the p-quantile of the random variable $T_{x(\cdot)}$.

Proposition 4.2.1 [Bagdonavičius (1990)]. *Suppose that $x(\cdot)$, $x(\tau)$, $x_0 \in E$ for all $\tau \geq 0$ and the AAD model is true on E. Then*

$$\int_0^{t_p(x(\cdot))} \frac{d\tau}{t_p(x(\tau))} = 1. \tag{4.5}$$

If the means $\mathbf{E}\,(T(x(\cdot)))$, $\mathbf{E}\,(T(x(\tau)))$ exist then

$$\mathbf{E}\left(\int_0^{T(x(\cdot))} \frac{d\tau}{\mathbf{E}\,(T(x(\tau)))}\right) = 1. \tag{4.6}$$

The model (4.6) is the model of Miner (1949).

In the case of the stress (4.4) we have

$$\sum_{k=1}^{m} \frac{\mathbf{E}T_k(x(\cdot))}{\mathbf{E}T(x_k)} = 1, \tag{4.7}$$

where

$$T_k(x(\cdot)) = (T(x(\cdot)) - t_{k-1})I_{(t_{k-1} \leq T(x(\cdot)) < t_k)} + (t_k - t_{k-1})I_{(T(x(\cdot)) \geq t_k)}$$

is the life of an item, tested under the stress $x(\cdot)$, in the interval $[t_{k-1}, t_k)$.

In the case of the stress (4.4) the model (4.5) has the form

$$\sum_{i=1}^{k-1}(t_i - t_{i-1})/t_p(x_i) + (t_p(x(\cdot)) - t_{k-1})/t_p(x_k) = 1, \tag{4.8}$$

if $t_p(x(\cdot)) \in [t_{k-1}, t_k)$.

The model (4.8) is the *model of Peshes-Stepanova* [see, Kartashov (1979)]. So *all of the models (4.3), (4.5)–(4.8) are implied by the AAD model and illustrate properties of this model.*

Suppose that the rate of resource using at the moment t is proportional not only to some function of the stress at the moment t but also to some function of the resource used until this moment.

Definition 4.2.2 The modified AAD model is true on E if there exist a function r on E and a positive on $\mathbf{R}^+$ function q such that for all $x(\cdot) \in E$

$$\frac{\partial f^G_{x(\cdot)}(t)}{\partial t} = r[x(t)]\, q[f^G_{x(\cdot)}(t)]$$

with the initial condition $f_{x(\cdot)}(0) = 0$.

We show that the class of the modified AAD models coincides with the class of AAD models :

Proposition 4.2.2 *Suppose that the integral*

$$\int_0^x \frac{dv}{q(v)}$$

converges for all $x \geq 0$.

The modified AAD model is satisfied on E *for some resource* R^G *iff the AAD model is satisfied on* E *for some resource* R^{G_1}*. In this case*

$$H_1(p) = \int_0^{H(p)} \frac{dv}{q(v)},$$

where $H_1 = G_1^{-1}$, $H = G^{-1}$.

Specifying G (then the AAD model is parametric) in (4.2) we specify resource for which the rate of using at any moment t depends only on the value of the stress at this moment. If G is unknown the AAD model is semiparametric.

Following Viertl (1988) we shall do now some remarks on parametrization of the function $r(\cdot)$. If the equality (4.3) is true for all $x \in E_0$ then for all $x_1, x_2 \in E_0$ then

$$S_{x_2}(t) = S_{x_1}(\rho(x_1, x_2)t),$$

where the function $\rho(x_1, x_2) = r(x_2)/r(x_1)$ shows the degree of changing of the scale of distribution when the stress x_2 is used instead of the stress x_1. It is evident that $\rho(x, x) = 1$.

Suppose at first that x is unidimensional. The rate of scale changing can be defined by the *infinitesimal characteristic* [see Viertl (1988)]:

$$\delta(x) = \lim_{\Delta x \to 0} \frac{\rho(x, x+\Delta x) - \rho(x,x)}{\Delta x} = [\log r(x)]'.$$

So for all $x \in E_0$ the function $r(x)$ is given by the formula:

$$r(x) = r(x_0) \exp\left\{ \int_{x_0}^{x} \delta(v)\, dv \right\},$$

where $x_0 \in E_0$ is some fixed stress.

Suppose that $\delta(x)$ is *proportional to some known function* $u(x)$ *of the stress*:

$$\delta(x) = \alpha\, u(x), \quad \alpha > 0.$$

In this case

$$r(x) = e^{\beta_0 + \beta_1 z(x)}, \tag{4.9}$$

where $z(x)$ is some known function, β_0, β_1 are unknown parameters.

Example 4.2.1 $\delta(x) = \alpha$,
i.e., the rate of scale changing is constant. Then

$$r(x) = e^{\beta_0 + \beta_1 x},$$

where $\beta_1 > 0$. It is so called *log-linear model.*

Example 4.2.2 $\delta(x) = \alpha/x$.
Then

$$r(x) = e^{\beta_0 + \beta_1 \log x} = \alpha_1 x^{\beta_1},$$

where $\beta_1 > 0$. It is so called *power rule model.*

Example 4.2.3 $\delta(x) = \alpha/x^2$.
Then

$$r(x) = e^{\beta_0 + \beta_1 / x} = \alpha_1 e^{\beta_1 / x},$$

where $\beta_1 < 0$. It is so called *Arrhenius model.*

Example 4.2.4 $\delta(x) = \alpha/x(1-x)$.
Then

$$r(x) = e^{\beta_0 + \beta_1 \ln \frac{x}{1-x}} = \alpha_1 \left(\frac{x}{1-x} \right)^{\beta_1}, \quad 0 < x < 1,$$

where $\beta_1 > 0$. It is the *model of Meeker-Luvalle* (1995).

The Arrhenius model is widely used to model product life when the stress is the temperature, the power rule model - when the stress is voltage, mechanical loading, the log-linear model is applied in endurance and fatigue data analysis, testing various electronic components [see Nelson (1990)]. The model of Meeker-Luvalle is used when x is proportion of humidity.

If it is not very clear which of the first three models to choose, one can take more large class of models. For example, all these models are the particular cases of the class of models determined by

$$\delta(x) = \alpha\, x^{\gamma}$$

with unknown γ or, in terms of the function $r(x)$, by

$$r(x) = \begin{cases} e^{\beta_0 + \beta_1 (x^{\varepsilon} - 1)/\varepsilon}, & \text{if} \quad \varepsilon \neq 0; \\ e^{\beta_0 + \beta_1 \log x}, & \text{if} \quad \varepsilon = 0. \end{cases}$$

In this case the parameter ε must be estimated.

The model (4.9) can be generalized. One can suppose that $\delta(x)$ is a linear combination of some known functions of the stress:

$$\delta(x) = \sum_{i=1}^{k} \alpha_i u_i(x).$$

In such a case

$$r(x) = \exp\left\{\beta_0 + \sum_{i=1}^{k} \beta_i z_i(x)\right\},$$

where $z_i(x)$ are some known functions of the stress, $\beta_0, \ldots, \beta_k$ are unknown (perhaps, not all of them) parameters.

Example 4.2.5 $\delta(x) = 1/x + \alpha/x^2$.
Then

$$r(x) = e^{\beta_0 + \beta_1 \log x + \beta_2 / x} = \alpha_1 x e^{\beta_2 / x},$$

where $\beta_1 = 1$, $\beta_2 < 0$. It is so called *Eyring model*, applied when the stress x is the temperature.

Example 4.2.6 $\delta(x) = \sum_{i=1}^{k} \alpha_i / x^i$.

Then

$$r(x) = \exp\left\{\beta_0 + \beta_1 \log x + \sum_{i=1}^{k-1} \beta_i / x^i\right\}.$$

It is so called *generalized Eyring model.*

Suppose now that the stress $x = (x_1, \ldots, x_n)$ is multidimensional. Define [see Viertl (1988)] the infinitesimal characteristics $\delta_i(x)$ by the equalities

$$\delta_i(x) = \lim_{\Delta x_i \to 0} \frac{\rho(x, x + \Delta x_i e_i) - \rho(x, x)}{\Delta x_i} = \frac{\partial \log r(x)}{\partial x_i},$$

where $e_i = (0, \ldots, 1, \ldots, 0)$. The unity is the ith coordinate.

Generalizing the unidimensional case, $\delta_i(x)$ can be parametrized in the following manner

$$\delta_i(x) = \sum_{j=1}^{k_i} \alpha_{ij} u_{ij}(x),$$

where $u_{ij}(x)$ are known functions, α_{ij} are unknown constants. In this case

$$r(x) = \exp\left\{ \beta_0 + \sum_{i=1}^{m} \sum_{j=1}^{k_i} \beta_{ij} z_{ij}(x) \right\},$$

where $z_{ij}(x)$ are known functions, β_{ij} are unknown constants.

Example 4.2.7 $\delta_1(x) = 1/x_1 + (\alpha_{11} + \alpha_{12} x_2)/x_1^2$, and $\delta_2(x) = \alpha_{21} + \alpha_{22}/x_1$. Then

$$r(x) = \exp\left\{\beta_0 + \beta_1 \log x_1 + \beta_2 x_2 + \beta_3/x_1 + \beta_4 x_2/x_1\right\}.$$

It is so called *generalized Eyring model.* This model is used for certain semiconductor materials, when x_1 is the temperature and x_2 is the voltage.

Example 4.2.8 $\delta_i(x) = \alpha_i u_i(x_i)$, where u_i are known functions. Then

$$r(x) = \exp\left\{ \beta_0 + \sum_{j=1}^{k} \beta_j z_j(x_j) \right\},$$

where z_j are known functions. It is so called *generalized Arrhenius model.* It also called the log-linear model.

The AAD model can be considered as parametric, semiparametric or nonparametric. Parametric models, used in accelerated life testing, where thoroughly investigated [see books of Balakrishnan (1995), Mann, Schafer and Singpurwalla (1974), Meeker and Escobar (1998), Nelson (1990), Viertl (1988), reviews of Meeker and Escobar (1993), Nelson and Macarthur (1992), Nelson and Meeker (1991), Viertl and Spencer (1991)].

In this case the function G is taken from some class of distributions as Weibull, lognormal, loglogistic, etc. The function $r(x)$ is parametrized. Several groups of items are observed under different "higher" stress conditions $x_1, ..., x_m$ and the obtained data are used to estimate the unknown parameters β of the function $r(x)$ and parameters of the function G (if there are some different from β). The maximum likelihood method of estimation is used as a rule.

Semiparametric analysis of AAD model when $r(x)$ is parametrized was considered by Bagdonavičius and Nikulin (1998a), Basu and Ebrahimi (1987), Lin and Ying (1995), Robins and Tsiatis (1992), Schmoyer (1986, 1991), Sethuraman and Singpurwalla (1982), Shaked and Singpurwalla (1983), Tsiatis (1990), Ying (1993) and others.

Nonparametric estimation when r is not parametrized was considered by Bagdonavičius and Nikulin (1997e, 1998a,c).

4.3 Generalized Multiplicative Models

Definition 4.3.1 The generalized multiplicative (GM) model [Bagdonavičius and Nikulin (1994)] holds on E if there exist a positive function r on E and a survival function S_0 such that for all $x(\cdot) \in E$

$$\frac{\partial f^G_{x(\cdot)}(t)}{\partial t} = r[x(t)] \frac{\partial f^G_0(t)}{\partial t} \tag{4.10}$$

with the initial conditions

$$f^G_{x(\cdot)}(0) = f^G_0(0) = 0, \quad \textit{where} \quad f^G_0(t) = H(S_0(t)).$$

We'll call S_0 the *baseline survival function.*

This model means that the rate of resource using at the moment t is proportional to some "baseline rate". The proportionality constant is a function of the stress applied at the moment t.

Taking two stresses $x(\cdot)$, $y(\cdot) \in E$ we have

$$\frac{\partial f^G_{x(\cdot)}(t)}{\partial t} / \frac{\partial f^G_{y(\cdot)}(t)}{\partial t} = r(x(t))/r(y(t)).$$

The ratio of resource using at the moment t depends only on values of the stresses $x(\cdot)$ and $y(\cdot)$ at the moment t.

The equation (4.10) implies that

$$S_{x(\cdot)}(t) = G\left(\int_0^t r[x(\tau)]\, dH(S_0(\tau))\right). \tag{4.11}$$

If $x(\tau) \equiv x = const$ the model (4.11) implies

$$S_x(t) = G\left(r(x) H(S_0(t))\right). \tag{4.12}$$

Proposition 4.3.1 [Bagdonavičius and Nikulin (1995)]. *Suppose that $x(\cdot)$, $x(\tau)$, $x_0 \in E$ for all $\tau \geq 0$.*

If the GM model is true on E, then

$$\begin{aligned} S_{x(\cdot)}(t) &= G\left(\int_0^t \frac{H(S_{x(\tau)}(t))}{H(S_{x_0}(t))}\, dH(S_{x_0}(\tau))\right) \\ &= G\left(\int_0^t H(S_{x(\tau)}(\tau))\, d\, log\, H(S_{x_0}(\tau))\right). \end{aligned}$$

If $x(\cdot)$ is the step-stress of the form (4.4), $x_0 \in E$ is any fixed constant in time stress then

$$\begin{aligned} S_{x(\cdot)}(t) &= G\Big(\sum_{i=1}^{k} \rho(x_i)(H(S_{x_0}(t_i)) - H(S_{x_0}(t_{i-1}))) \\ &\qquad + \rho(x_k)(H(S_{x_0}(t)) - H(S_{x_0}(t_k)))\Big) \end{aligned}$$

if $t \in [t_{k-1}, r_k)$; here $\rho(x_i) = r(x_i)/r(x_0)$.

If

$$x(\tau) = \begin{cases} x_0, & 0 \leq \tau \leq t_1, \\ x_1, & \tau > t_1, \end{cases} \tag{4.13}$$

then

$$S_{x(\cdot)}(t) = \begin{cases} S_{x_0}(t), & 0 \leq \tau \leq t_1, \\ G\left(H(S_{x_0}(t_1)) + \rho(x_1)(H(S_{x_0}(t)) - H(S_{x_0}(t_1)))\right), & \tau > t_1. \end{cases} \tag{4.14}$$

Consider some submodels of considered models. Consider GM models with G specified

a) if the distribution of the resource R^G is exponential, i.e.

$$G(t) = \exp(-t)\mathbf{1}_{\{t \geq 0\}},$$

then

$$\frac{\partial f^G_{x(\cdot)}(t)}{\partial t} = \lambda_{x(\cdot)}(t),$$

i.e., the rate of resource using is the hazard rate. In this case the model GM can be formulated in the following way:

$$\lambda_{x(\cdot)}(t) = r[x(t)]\lambda_0(t),$$

where $\lambda_0(t)$ is the baseline hazard rate. We obtain the *proportional hazards* (Cox) model [Cox (1972)].

If $x(\cdot)$ is the step-stress of the form (4.13), we have

$$\lambda_{x(\cdot)}(\tau) = \begin{cases} \lambda_{x_0}(\tau), & 0 \le \tau \le t_1, \\ \rho\lambda_{x_0}(\tau), & \tau > t_1, \end{cases}$$

where $\rho = r(x_1)/r(x_0)$.

This particular case of the Cox model is called the *tampered failure rate* (TFR) model [Bhattacharyya and Soejoeti (1989)]. In terms of survival functions we have

$$S_{x(\cdot)}(t) = \begin{cases} S_{x_0}(t), & 0 \le t \le t_1, \\ S_{x_0}(t_1)\left(\frac{S_{x_0}(t)}{S_{x_0}(t_1)}\right)^{\rho}, & t > t_1. \end{cases}$$

b) If the distribution of the resource is loglogistic, i.e.,

$$G(t) = \frac{1}{1+t}\,\mathbf{1}_{\{t \ge 0\}},$$

then the GM model can be formulated in the following way:

$$\frac{\lambda_{x(\cdot)}(t)}{S_{x(\cdot)}(t)} = r[x(t)]\,\frac{\lambda_0(t)}{S_0(t)}.$$

If $x(\cdot)$ is the step-stress of the form (4.13) then we have

$$\frac{\lambda_{x(\cdot)}(t)}{S_{x(\cdot)}(t)} = \begin{cases} \frac{\lambda_{x_0}(t)}{S_{x_0}(t)}, & 0 \le t \le t_1, \\ \rho\frac{\lambda_{x_0}(t)}{S_{x_0}(t)}, & t > t_1 \end{cases}$$

or

$$S_{x(\cdot)}(t) = \begin{cases} S_{x_0}(t), & 0 \le t \le t_1, \\ S_{x_0}(t_1)\left[1 + \rho\left(\frac{S_{x_0}(t_1)}{S_{x_0}(t)} - 1\right)\right]^{-1}, & t > t_1. \end{cases}$$

If stresses are constant in time then we obtain the model

$$\frac{1}{S_x(t)} - 1 = r(x)\,(\frac{1}{S_0(t)} - 1).$$

It is the analogue of the logistic regression model which is used for analysis of dichotomous data when the probability of "success" in dependence of some factors is analyzed. The obtained model is near to the Cox model when t is small .

c) If the resource is lognormal, then

$$G(t) = \Phi(\log t),$$

where Φ is the distribution function of the standard normal law. If covariates are constant in time then in terms of survival functions the GM model can be written as follows:

$$\Phi^{-1}\left(S_x(t)\right) = \log\left(r(x)\right) + \Phi^{-1}\left(S_0(t)\right).$$

It is the generalized probit model [see Dabrowska and Doksum (1988)]. If $x(\cdot)$ is the step-stress of the form (4.13) we can use the formula (4.14) taking $H(p) = \exp\{\Phi^{-1}(p)\}$.

We can consider the resource R^G distributed on $\mathbb{R}$.

d) Taking $G(t) = \exp(-\exp(t))$ we obtain the model:

$$\frac{\lambda_{x(\cdot)}(t)}{\Lambda_{x(\cdot)}(t)} = r[x(t)]\,\frac{\lambda_0(t)}{\Lambda_0(t)},$$

where

$$\Lambda_{x(\cdot)}(t) = \int_0^t \lambda_{x(\cdot)}(\tau)d\tau,\ \Lambda_0(t) = \int_0^t \lambda_0(\tau)d\tau.$$

If $x(\cdot)$ is the step-stress of the form (4.13) then

$$S_{x(\cdot)}(t) = \begin{cases} S_{x_0}(t), & 0 \le t \le t_1 \\ g_{x_0,x_1}(t), & t > t_1, \end{cases}$$

where

$$g_{x_0,x_1}(t) = \exp\left\{-\Lambda_{x_0}(t_1)\left(\frac{\Lambda_{x_0}(t)}{\Lambda_{x_0}(t_1)}\right)^{\rho}\right\}.$$

e) Taking $G(t) = (1+e^t)^{-1}$,we obtain the model

$$\frac{\lambda_{x(\cdot)}(t)}{1 - S_{x(\cdot)}(t)} = r[x(t)]\,\frac{\lambda_0(t)}{1 - S_0(t)}.$$

This model is near to the Cox model when t is large. If $x(\cdot)$ is the step-stress of the form (4.13) then

$$S_{x(\cdot)}(t) = \begin{cases} S_{x_0}(t), & 0 \le t \le t_1 \\ h_{x_0,x_1}(t), & t > t_1, \end{cases}$$

where

$$h_{x_0,x_1}(t) = \left\{1 + v_{x_0}(t_1)\left(\frac{v_{x_0}(t)}{v_{x_0}(t_1)}\right)^{\rho}\right\}^{-1},$$

$$v_{x_0}(t) = \frac{1}{S_{x_0}(t)} - 1.$$

Other submodels can be considered [see Bagdonavičius and Nikulin (1995)].

At first glance it looks like there are too many GM models. It appears that it is not so.

Really, assume that a function G is continuous and strictly decreasing on $[0,\infty[$ and $G_1(u) = G((u/\theta)^p)$. Let $\{S_z,\ z \in [z_0, z_1] \subset \mathbf{R}^m\}$ be a class of continuous survival functions, $H = G^{-1} :]0,1] \to [0,\infty]$ and $H_1 = G_1^{-1}$ be the inverse functions of G and G_1 respectively. If

$$H(S_z(t)) = \lambda(z) H(S_{z_0}(t)),\ t \in \mathbf{R},\ z \in [z_0, z_1] \tag{4.15}$$

for some continuous strictly monotone function $\alpha : [z_0, z_1] \to \mathbf{R}^+$, then

$$H_1(S_z(t)) = \lambda^{1/p}(z) H_1(S_{z_0}(t)),\ t \in \mathbf{R},\ z \in [z_0, z_1]. \tag{4.16}$$

The inverse result also takes place:

Theorem 4.3.1 (Characterization of the GM model) [Bagdonavičius, Malov and Nikulin (1998)]. *Assume that a function G is continuous and strictly decreasing on $[0,\infty[$ and the equality (4.15) is satisfied. Then the equality (4.16) is satisfied iff $G_1(u) = G((u/\theta)^p)$, $u \in [0,\infty)$, for some positive constants θ and p.*

The theorem implies that, for example, the Cox model is a submodel of the GM model when G is not only standard exponential but when it is any exponential or two-parameter Weibull survival function. So submodels of the GM model form classes generated by classes of resource distributions which differ only by location and scale parameters.

The function $r(x)$ in the GM models can be parametrized in the same manner as the function $r(x)$ in the AAD model.

Parametric estimation for each specified GM model (11) can be done using the method of maximum likelihood. In the particular case of the Cox model semiparametric estimation is done by Cox (1972) using the partial likelihood approach. Semiparametric estimation for other GM models, when r is parametrized, S_0 is unknown and G is specified is done in Bagdonavičius & Nikulin (1997a). Asymptotic properties of estimators are investigated in Bagdonavičius and Nikulin (1997b).

The solution of the problem of semiparametric estimation when G is not specified motivated the following definition and proposition.

GM models suppose that the ratio of resource using rates at the moment t depends only on values of stresses applied at this moment. We can suppose that this ratio depends not only on values of stresses but also on resources used until the moment t.

Definition 4.3.2 The modified GM model is true on E if there exist a function r on E and a positive on $\mathbf{R}^+$ function q such that for all $x(\cdot) \in E$

$$\frac{\partial f^G_{x(\cdot)}(t)}{\partial t} = r[x(t)]\ q[f^G_{x(\cdot)}(t)]\ \frac{\partial f^G_0(t)}{\partial t} \tag{4.17}$$

with the initial condition $f_{x(\cdot)}(0) = 0,\ f_0(0) = 0$.

This model means that the rate of resource using at the moment t is proportional to some function of the stress at the moment t, to some function of the resource used until the moment t and to some baseline rate.

The model (4.17) implies that for all $x(\cdot), x_0(\cdot) \in E$

$$\frac{\partial f^G_{x(\cdot)}(t)}{\partial t} \Big/ \frac{\partial f^G_{x_0(\cdot)}(t)}{\partial t} = \frac{r\{x(t)\}}{r\{x_0(t)\}} \cdot \frac{q\{\Lambda_{x(\cdot)}(t)\}}{q\{\Lambda_{x_0(\cdot)}(t)\}}, \tag{4.18}$$

So the ratio of resource using rates at the moment t depends as on values of stresses as on resources used until this moment. Classes of GM and modified GM models coincide:

Proposition 4.3.2 *Suppose that the integral*

$$\int_0^x \frac{dv}{q(v)}$$

converges for all $x \geq 0$.

The modified GM model is satisfied on E for some resource R^G *iff the GM model is satisfied on E for some resource* R^{G_1}. *In this case*

$$f^{G_1}_{x(\cdot)}(t) = \int_0^{f^G_{x(\cdot)}(t)} \frac{dv}{q(v)}, \quad H_1(p) = \int_0^{H(p)} \frac{dv}{q(v)},$$

where $H_1 = G_1^{-1}$, $H = G^{-1}$.

So if the ratio of rates of resource R^G using at the moment t is depending on the values of stresses at this moment and on resources used until t then we can find resource R^{G_1} such that the ratio of the rates of G_1-resource using doesn't depend on the resources used until t and depends only on stresses applied at this moment.

In the case of the exponential resource the resource used until the moment t under the stress $x(\cdot)$ is the accumulated hazard rate $\Lambda_{x(\cdot)}(t)$ and (4.18) implies: for all $x(\cdot)$, $x_0(\cdot) \in E$

$$\frac{\lambda_{x(\cdot)}(t)}{\lambda_{x_0(\cdot)}(t)} = \frac{r\{x(t)\}}{r\{x_0(t)\}} \cdot \frac{q\{\Lambda_{x(\cdot)}(t)\}}{q\{\Lambda_{x_0(\cdot)}(t)\}}, \tag{4.19}$$

where r is some positive functional on E and q is a positive function in $[0, \infty[$.

Proposition 4.3.2 implies that the model (4.19) is satisfied on E iff there exist a resource R^G such that the GM model (4.11) is satisfied on E. In this case

$$f^G_{x(\cdot)}(t) = \int_0^{\Lambda_{x(\cdot)}(t)} \frac{dv}{q(v)}.$$

The model (4.19) with specified q is equivalent to the GM model with the survival function of the resource $G = H^{-1}$, where

$$H(u) = \int_0^{-\ln u} \frac{dv}{q(v)}.$$

So the class of models (4.19) with different q coincides with the class of GM models with different G.

If we wish to do semiparametric estimation in the GM model with unknown G we can do semiparametric estimation in the model (4.19) with unknown q.

Consider at first submodels of the model (4.19) with monotone q and with the next parametrizations

$$r\{x(t)\} = e^{\beta^T z(t)}, \quad q\{\Lambda_{x(\cdot)}(t)\} = e^{\gamma \Lambda_x(t)},$$

where $\beta = (\beta_1, ..., \beta_k)^T \in I\!R^k$ and $\gamma \in I\!R$ are unknown vector and scalar parameters, $z(\cdot) = z(x(\cdot))$ is some known function (possibly multidimensional) of the stress $x(\cdot)$. We shall use the same notation x for $z = z(x)$. Then we have:

THE GENERALIZED PROPORTIONAL HAZARDS (GPH) MODEL [Bagdonavičius and Nikulin (1998d)]:
for all $x(\cdot)$, $x_0(\cdot) \in E$

$$\frac{\lambda_{x(\cdot)}(t)}{\lambda_{x_0(\cdot)}(t)} = \exp\{\beta^T\{x(t) - x_0(t)\} + \gamma\{\Lambda_{x(\cdot)}(t) - \Lambda_{x_0(\cdot)}(t)\}\}. \tag{4.20}$$

So the ratio of the rates of resource using (i.e., the ratio of hazard rates) depends not only on the difference of stresses but also on the difference of resources used until the moment t. If we fix $x_0(\cdot) \in E$ and denote

$$\lambda_0(t) = \lambda_{x_0(\cdot)}(t) \exp\{-\beta^T x_0(t) - \gamma \Lambda_{x_0(\cdot)}(t)\},$$

then the model (4.20) can be written in the following form:

$$\lambda_{x(\cdot)}(t) = \exp\{\beta^T x(t) + \gamma \Lambda_{x(\cdot)}(t)\}\lambda_0(t). \tag{4.21}$$

We suppose that $\lambda_0(t)$ is unknown. If $\gamma = 0$, we have the proportional hazards (Cox) model. If $\gamma \neq 0$ then the resource used until the moment t is treated as an additional covariate.

The model (4.21) implies for all $x(\cdot) \in E$

$$S_{x(\cdot)}(t) = \begin{cases} \{1 - \gamma \int_0^t e^{\beta^T x(u)} d\Lambda_0(u)\}^{1/\gamma}, & \text{if } \gamma \neq 0, \\ \exp\{-\int_0^t e^{\beta^T x(u)} d\Lambda_0(u)\}, & \text{if } \gamma = 0, \end{cases}$$

with

$$\Lambda_0(t) = \int_0^t \lambda_0(u)du$$

and for all $x(\cdot)$, $x_0(\cdot) \in E$

$$S_{x(\cdot)}(t) = \begin{cases} \{1 + \int_0^t e^{\beta^T\{x(\tau)-x_0(\tau)\}} dS^{\gamma}_{x_0(\cdot)}(\tau)\}^{1/\gamma}, & \text{if } \ \gamma \neq 0, \\ \exp\{\int_0^t e^{\beta^T\{x(\tau)-x_0(\tau)\}} d\ln S_{x_0(\cdot)}(\tau)\}, & \text{if } \ \gamma = 0. \end{cases}$$

If $x(\tau) \equiv x = const, \quad x_0(t) \equiv x_0 = const$, then

$$S_x(t) = \begin{cases} \{1 + e^{\beta^T(x-x_0)}(S^{\gamma}_{x_0}(t) - 1)\}^{1/\gamma} & \text{if } \ \gamma \neq 0, \\ (S_{x_0}(t))^{e^{\beta^T(x-x_0)}}, & \text{if } \ \gamma = 0. \end{cases}$$

The function q can be non-monotonous in $\Lambda_{x(\cdot)}$. In this case we can consider some non-loglinear functions of $\Lambda_{x(\cdot)}$, for example, the parametrization

$$q\{\Lambda_{x(\cdot)}(t)\} = \exp\left\{\sum_{i=1}^{m} \gamma_i \Lambda^i_{x(\cdot)}(t)\right\}.$$

In the latter case we have the model:

$$\lambda_{x(\cdot)}(t) = \exp\left\{\beta^T x(t) + \sum_{i=1}^{m} \gamma_i \Lambda^i_{x(\cdot)}(t)\right\}\lambda_0(t). \tag{4.22}$$

Methods of semiparametric estimation and properties of estimators for the GPH model (4.21) and for the model (4.22) with unknown $\lambda_0(t)$ were considered in Bagdonavičius and Nikulin (1998d).

4.4 Generalized Additive and Additive-Multiplicative Models

Definition 4.4.1 The generalized additive (GA) model [Bagdonavičius and Nikulin (1995)] holds on E if there exist a function a on E and a survival function S_0 such that for all $x(\cdot) \in E$

$$\frac{\partial f^G_{x(\cdot)}(t)}{\partial t} = \frac{\partial f^G_0(t)}{\partial t} + a(x(t))$$

with the initial conditions $f^G_0(0) = f^G_{x(\cdot)}(0) = 0$.

So the stress influences additively the rate of resource using. The last equation implies that

$$S_{x(\cdot)}(t) = G\left(H(G_0(t)) + \int_0^t a(x(\tau))d\tau\right).$$

In the case of the exponential resource we obtain the *additive hazards model*:

$$\lambda_x(t) = \lambda_0(t) + a(x(t)).$$

The GM and GA models can be generalized.

Definition 4.4.2 The generalized additive-multiplicative (GAM) model [Bagdonavičius and Nikulin (1997f)] holds on E if there exist functions a and r (positive) on E and a survival function S_0 such that for all $x(\cdot) \in E$

$$\frac{\partial f^G_{x(\cdot)}(t)}{\partial t} = r[x(t)]\,\frac{\partial f^G_0(t)}{\partial t} + a(x(t))$$

with the initial conditions $f_0^G(0) = f^G_{x(\cdot)}(0) = 0$.

So the stress influences the rate of resource using as multiplicatively as additively. The last equation implies that

$$S_{x(\cdot)}(t) = \int_0^t r[x(\tau)]df_0^G(\tau) + \int_0^t a(x(\tau))d\tau.$$

In the particular case of the exponential resource we obtain the model

$$\lambda_{x(\cdot)}(t) = r[x(t)]\lambda_0(t) + a(x(t)).$$

It is the *additive-multiplicative hazard model* [Lin and Ying (1996)]. Submodels of GA and GAM models, methods of semiparametric estimation and properties of estimators for GA and GAM models are done in Bagdonavičius and Nikulin (1997d,f,g).

4.5 Models Describing the Influence of Stresses to the Shape and Scale of Distribution

Consider now the important model which doesn't lie in the class of the GAM models but includes the AAD model as the particular case.

Suppose that the constant in time stresses $x \in E$ change not only the scale but also the shape : for all $x \in E$

$$S_x(t) = S_{x_0}\left(\left(\frac{t}{\theta(x)}\right)^{\alpha(x)}\right). \tag{4.23}$$

Definition 4.5.1 The changing shape (CHSH) model [Bagdonavičius and Nikulin (1998a)] is true on E if there exist positive on E functions r and α such that for all $x(\cdot) \in E$

$$\frac{\partial f^G_{x(\cdot)}(t)}{\partial t} = r[x(t)]\ t^{\alpha(x(t))-1}. \tag{4.24}$$

The equality (4.24) implies

$$S_{x(\cdot)}(t) = S_{x0}\left(\int_0^t r[x(\tau)]\tau^{\alpha(x(\tau))-1}d\tau\right).$$

Nonparametric estimation in this model is done in [Bagdonavičius and Nikulin (1998a)]. Parametric estimation for some specified distributions can be found in Mann, Schafer and Singpurwalla (1974), Glaser (1984), Nelson (1984).

4.6 The Model of Sedyakin and Its Generalizations

Definition 4.6.1 The generalized model of Sedyakin [Bagdonavičius (1978)] is true on E if there exist a positive on $E \times \mathbf{R}^+$ function g such that for all $x(\cdot) \in E$

$$\frac{\partial f^G_{x(\cdot)}(t)}{\partial t} = g\left(x(t), f_{x(\cdot)}(t)\right) \tag{4.25}$$

with the initial condition $f_{x(\cdot)}(0) = 0$.

This model means that the rate of resource using at the moment t depends on the value of the stress at the moment t and the resource used until that moment.

Consider the stepwise stress of the form (4.4). In this case the solution $f_{x(\cdot)}(t)$ of the equation (4.25) satisfies the equalities

$$f_{x(\cdot)}(t) = f_{x_i}(t - t_{i-1} + t^*_{i-1}),\ \text{ if }\ t \in [t_{i-1}, t_i)\ (i = 1, 2, \ldots, m), \tag{4.26}$$

where t^*_i can be found solving the equations

$$f_{x_1}(t_1) = f_{x_2}(t^*_1), \ldots, f_{x_i}(t_i - t_{i-1} + t^*_{i-1}) = f_{x_{i+1}}(t^*_i)\ (t_0 = 0,\ i = 1, \ldots, m-1).$$

These equalities are equivalent to the equalities

$$S_{x(\cdot)}(t) = S_{x_i}(t - t_{i-1} + t^*_{i-1}), \quad \text{if} \quad t \in [t_{i-1}, t_i),$$

where

$$S_{x_1}(t_1) = S_{x_2}(t^*_1), \ldots, S_{x_i}(t_i - t_{i-1} + t^*_{i-1}) = S_{x_{i+1}}(t^*_i).$$

where t_i^* can be found solving the equations

$$f_{x_1}(t_1) = f_{x_2}(t_1^*), \ldots, f_{x_i}(t_i - t_{i-1} + t_{i-1}^*) = f_{x_{i+1}}(t_i^*) \; (t_0 = 0, \, i = 1, \ldots, m-1).$$

These equalities are equivalent to the equalities

$$S_{x(\cdot)}(t) = S_{x_i}(t - t_{i-1} + t_{i-1}^*), \quad \text{if} \quad t \in [t_{i-1}, t_i),$$

where

$$S_{x_1}(t_1) = S_{x_2}(t_1^*), \ldots, S_{x_i}(t_i - t_{i-1} + t_{i-1}^*) = S_{x_{i+1}}(t_i^*).$$

In the case $m = 2$ we obtain

$$S_{x(\cdot)}(t) = \begin{cases} S_{x_1}(t), & \text{if } t < t_1, \\ S_{x_2}(t - t_1 + t_1^*) & \text{otherwise,} \end{cases}$$

where $S_{x_1}(t_1) = S_{x_2}(t_1^*)$.

These equations imply that

$$\mathbf{P}\left\{T_{x(\cdot)} \geq t_1 + s \mid T_{x(\cdot)} \geq t_1\right\} = \mathbf{P}\left\{T_{x_2} \geq t_1^* + s \mid T_{x_2} \geq t_1^*\right\}.$$

We have come to the *model of Sedyakin* [see Sedyakin (1966)]. This model is also called the *cumulative exposure (CE) model* [Nelson (1980)].

There are no methods of estimation for this model.

What is the region of applications of this model? Suppose that the model is parametric and it is impossible to obtain the complete sample under the "normal" conditions of functioning of items. When the right censored data is used, the goodness-of-fit tests can test that the left tail of a survival distribution corresponds well the chosen model. But often the estimates of p-quantiles with p near the unity are needed and in the case of bad choice of the model big mistakes can be made. The utilization of the model of Sedyakin can help to solve this problem. If the stepwise stresses are used, it is possible to obtain failures of items at the end of life under the "normal" conditions and therefore to test if the right tail is from the class of specified distribution.

A test for the Sedyakin's model can be found in Bagdonavičius and Nikoulina (1997).

Similarly as GM model generalizes the AAD model the generalized Sedyakin's model can be generalized by the model :

$$\frac{\partial f_{x(\cdot)}^G(t)}{\partial t} = g\left(x(t), f_{x(\cdot)}(t)\right) \frac{\partial f_0^G(t)}{\partial t}$$

with the initial condition $f_{x(\cdot)}(0) = f_0(0) = 0$.

In practice can be and more general models. The rate of resource using can depend not only on a stress at the moment t and the resource used till this moment but also on the past. For example, some parameter of an item can achieve the same level under high stress or during gradual wear out and after these moments behave not identically even under identical conditions. In practice it would mean deterioration of items under the high stress.

Schabe and Viertl (1995) used some basic axiomatic assumptions to obtain the general model which in our notations can be written as follows:

$$\frac{\partial f^G_{x(\cdot)}(t)}{\partial t} = g(t, x(t)).$$

As Sedyakin's model it is too general to obtain estimation procedures.

4.7 The Heredity Hypothesis

A process of production is *unstable* if reliability of items produced in different time intervals are different. If items produced in some specified time interval are considered and the models G1, G2 or G3 are true on E_0, then for all $x_1, x_2 \in E_0$

$$\begin{aligned} S_{x_2}(t) &= G_{x_1}(\rho(x_1, x_2)t), \\ S_{x_2}(t) &= G(\rho(x_1, x_2)H(S_{x_1}(t))), \\ S_{x_2}(t) &= G(H(S_{x_1}(t) + b(x_1, x_2))), \end{aligned}$$

respectively.

Definition 4.7.1 If one of the models AAD, GM or GA is true, the process of production is unstable and the function $\rho(x_1, x_2)$ (the models AAD or GM) or $b(x_1, x_2)$ (the model GA) is invariant, going from one group of items to another one, then we'll say that the heredity hypothesis is satisfied.

If one of the models AAD, GM or GA and the heredity principle are true, then sufficiently large data can be accumulated during a long period of observations and good estimators of the functions $\rho(x_1, x_2)$ or $b(x_1, x_2)$ can be obtained. The reliability of newly produced items under the "normal" stress x_1 can be estimated from accelerated life data obtained under some accelerated stress $x_2 > x_1$, using the estimators $\hat{\rho}(x_1, x_2)$ or $\hat{b}(x_1, x_2)$.

Methods of estimation and hypothesis testing when a process of production is unstable were considered in Bagdonavičius and Nikulin (1997c).

We don't consider here degradation models. Interesting approach is done in Lu and Meeker (1993). See also Nelson (1990).

References

1. Bagdonavičius, V. (1978). Testing the hypothesis of the additive accumulation of damages, *Probability Theory and its Applications*, **23**, 403–408.

2. Bagdonavičius, V. (1990). Accelerated life models when the stress is not constant, *Kybernetika*, **26**, 289–295.

3. Bagdonavičius, V. and Nikulin, M. (1994). Stochastic models of accelerated life, In *Selected Topics on Stochastic Modelling* (Ed., R. Gutiérrez and M. J. Valderrama), World Scientists, Singapore, pp. 73–87.

4. Bagdonavičius, V. and Nikulin, M. (1995). Semiparametric models in accelerated life testing, *Queen's Papers in Pure and Applied Mathematics*, Queen's University, Kingston, Ontario, Canada, **98**, 70p.

5. Bagdonavičius, V. and Nikulin, M. (1997a). Transfer functionals and semiparametric regression models, *Biometrika*, **84**, 365–378.

6. Bagdonavičius, V. and Nikulin, M. (1997b). Asymptotic analysis of semiparametric models in survival analysis and accelerated life testing, *Statistics*, **29**, 261–283.

7. Bagdonavičius, V. and Nikulin, M. (1997c). Accelerated life testing when a process of production is unstable, *Statistics and Probability Letters*, **35**, 269–275.

8. Bagdonavičius, V. and Nikulin, M. (1997d). Statistical analysis of the generalized additive semiparametric model with random covariates. *Questiio*, **21**, 273–291.

9. Bagdonavičius, V. and Nikulin, M., (1997e). Sur l'application des stress en escalier dans les expériences accélérées, *Comptes Rendus, Academie des Sciences de Paris*, **325**, Série I, 523–526.

10. Bagdonavičius, V. and Nikulin, M. (1997f). Analysis of general semiparametric models with random covariates, *Romanian Journal of Pure and Applied Mathematics*, **42**, 351–369.

11. Bagdonavičius, V. and Nikulin, M. (1997g). Semiparametric estimation in the generalized additive-multiplicative model, In *Probability and Statistics*, 2 (Eds., I. A. Ibragimov and V. A. Sudakov), Proceedings of the Steklov Math. Institute, St. Petersburg, pp. 7–27.

12. Bagdonavičius, V. and Nikoulina, V. (1997). A goodness-of-fit test for Sedyakin's model, *Romanian Journal of Pure and Applied Mathematics*, **52**, 5–14.

13. Bagdonavičius, V., Malov, S. and Nikulin, M. (1998). On characterizations and semi-parametric regression estimation in Archimedean copula, *Journal of Applied Statistical Science* (to appear).

14. Bagdonavičius, V. and Nikulin, M. (1998a). On nonparametric estimation from accelerated experiments, In *Mathematical Methods in Reliability* (Eds., N. Limnios and D. Ionescu), Boston: Birkhaüser.

15. Bagdonavičius, V. and Nikulin, M. (1998b). Additive and multiplicative semiparametric models in accelerated life testing and survival analysis, *Queen's Papers on Pure and Applied Mathematics* (to appear).

16. Bagdonavičius, V. and Nikulin, M. (1998c). On application of step-stresses in accelerated life testing, *ESAIM* (to appear).

17. Bagdonavičius, V. and Nikulin, M. (1998d). Generalized proportional hazards model: Modified partial likelihood approach, *Lifetime Data Analysis* (submitted).

18. Balakrishnan, N. (1995). *Recent Advances in Life-Testing and Reliability*, Boca Raton, FL: CRC Press.

19. Basu, A. P. and Ebrahimi, N. (1982). Nonparametric accelerated life testing, *IEEE Transactions on Reliability*, **31**, 432–435.

20. Bhattacharyya, G. K. and Stoejoeti, Z. (1989). A tampered failure rate model for step-stress accelerated life model, *Communication in Statistics—Theory and Methods*, **18**, 1627–1643.

21. Cox, D. R. (1972). Regression models and life tables, *Journal of the Royal Statistical Society, Series B*, **34**, 187–220.

22. Cox, D. R. and Oakes, D. (1984). *Analysis of Survival Data*, London: Chapman and Hall.

23. Dabrowska, D. M. and Doksum, K. A. (1988). Partial likelihood in transformations models with censored data, *Scandinavian Journal of Statistics*, **15**, 1–23.

24. Glaser, R. E. (1984). Estimation for a Weibull accelerated life testing model, *Naval Research Logistics Quarterly*, **31**, 559–570.

25. Kartashov, G. D. (1979). *Methods of Forced (Augmented) Experiments* (in Russian), Moscow: Znaniye Press.

26. Lin, D. Y. and Ying, Z. (1994). Semiparametrical analysis of the additive risk model, *Biometrika*, **81**, 61–71.

27. Lin, D. Y. and Ying, Z. (1995). Semiparametric inference for the accelerated life model with time-dependent covariates, *Journal of Statistical Planning and Inference*, **44**, 47–63.

28. Lin, D. Y. and Ying, Z. (1996). Semiparametric analysis of the general additive-multiplicative hazard models for counting processes, *The Annals of Statistics*, **23**, 1712–1734.

29. Lu, C. J. and Meeker, W. Q. (1993). Using degradation measures to estimate a time-to-failure distribution, *Technometrics*, **35**, 161–174.

30. Mann, N. R., Schafer, R. E. and Singpurwalla, N. D. (1974). *Methods for Statistical Analysis of Reliability and Life Data*, New York: John Wiley & Sons.

31. Meeker, Jr., W. Q. and LuValle, M. J. (1995). An accelerated life test model based on reliability kinetics, *Technometrics*, **37**, 133–146.

32. Meeker, Jr., W. Q. and Escobar, L. A. (1993). A review of recent research and current issues in accelerating testing, *International Statistical Review*, **61**, 147–168.

33. Meeker, Jr., W. Q. and Escobar, L. A. (1998). *Statistical Methods for Reliability*, New York: John Wiley & Sons.

34. Miner, M. A. (1945). Cumulative damage in fatigue, *Journal of Applied Mechanics*, **12**, A159–A164.

35. Nelson, W. (1990). *Accelerated Testing*, New York: John Wiley & Sons.

36. Nelson, W. and Meeker, W. (1991). Accelerated testing: Statistical models, test plans, and data analysis, *Technometrics*, **33**, 236–238.

37. Nelson, W. and Macarthur, E. (1992). Accelerated testing: Statistical models, test plans, and data analysis, *Applied Statistics*, **41**, 224–225.

38. Pieruschka, E. (1961). Relation between lifetime distribution and the stress level causing failures, *LMSD-800440*, Lockhead Missiles and Space Division, Sunnyvale, California.

39. Robins, J. M. and Tsiatis, A. A. (1992). Semiparametric estimation of an accelerated failure time model with time dependent covariates, *Biometrika*, **79**, 311–319.

40. Sedyakin, N. M. (1966). On one physical principle in reliability theory, *Technical Cybernetics*, **3**, 80–87.

41. Sethuraman, J. and Singpurwalla, N. D. (1982). Testing of hypotheses for distributions in accelerated life tests, *JASA*, **77**, 204–208.

42. Schabe, H. and Viertl, R. (1995). An axiomatic approach to models of accelerating life testing, *Engineering Fracture Mechanics*, **30**, 203–217.

43. Schmoyer, R. (1986). An exact distribution-free analysis for accelerated life testing at several levels of a single stress, *Technometrics*, **28**, 165–175.

44. Schmoyer, R. (1991). Nonparametric analyses for two-level single-stress accelerated life tests, *Technometrics*, **33**, 175–186.

45. Shaked, M. and Singpurwalla, N. D. (1983). Inference for step-stress accelerated life tests, *Journal of Statistical Planning and Inference*, **7**, 295–306.

46. Tsiatis, A. A. (1990). Estimating regression parameters using linear rank tests for censored data, *Annals of Statistics*, **18**, 354–372.

47. Viertl, R. (1988). *Statistical Methods in Accelerated Life Testing*, Göttingen: Vandenhoeck & Ruprecht.

48. Viertl, R. and Spencer, F. (1991). Statistical methods in accelerated life testing, *Technometrics*, **33**, 360–362.

49. Ying, Z. (1993). A large sample study of rank estimation for censored regression data, *Annals of Statistics*, **21**, 76–99.

5

On Semiparametric Estimation of Reliability From Accelerated Life Data

V. Bagdonavičius and M. S. Nikulin

University of Vilnius, Lithuania,
Université de Bordeaux, Bordeaux, France &
Steklov Mathematical Institute, St. Petersburg, Russia

Abstract: Additive accumulation of damages model and its generalizations are considered. Semiparametric estimation procedure of the survival function under the normal stress from accelerated life testing data is proposed. Experiments with step-stresses with random switch up moments are considered. Asymptotic properties of estimation are investigated.

Keywords and phrases: Accelerated life testing, additive accumulation of damages, step-stresses, survival function

5.1 Introduction

Suppose that $S_{x(\cdot)}(t) = \mathbf{P}\{T_{x(\cdot)} > t\}$ is the reliability function of the time-to-failure $T_{x(\cdot)}$ under the stress $x(\tau), \tau \geq 0$. Denote x_0 the constant in time "normal" stress corresponding to the usual conditions of functioning of items, $f_x(t) = (S_{x_0}^{-1} \circ S_x)(t)$-the resource used until the moment t under the stress $x(\cdot)$ [see Bagdonavičius (1990)]. The moment t under the stress $x(\cdot)$ is equivalent to the moment $f_x(t)$ under the stress x_0.

Suppose that the constant in time stresses from some set of stresses E change only the scale and the shape of the time-to-failure distribution for all $x \in E$:

$$S_x(t) = S_{x_0}\left(\left(\frac{t}{\theta(x)}\right)^{\alpha(x)}\right). \tag{5.1}$$

In this case $f_x'(t) = r(x)\ t^{\alpha(x)-1}$, where $r(x) = \alpha(x)/\theta(x)^{\alpha(x)}$. So the model (1) means that the rate of resource using under the stress x is increasing, if $\alpha(x) > 1$, decreasing, if $0 < \alpha(x) < 1$ and constant, if $\alpha(x) = 1$. In the latter case the stress changes only the scale of distribution. The natural generalization of the model (5.1) to the case of time varying stresses is the model

$$S_{x(\cdot)}(t) = S_{x_0}\left(\int_0^t r[x(\tau)]\ \tau^{\alpha(x(\tau))-1} d\tau\right). \tag{5.2}$$

In the case $\alpha(x(\tau)) = 1$ for all $\tau \geq 0$ we have the additive accumulation of damages (AAD) model [see Bagdonavičius (1978)]:

$$S_{x(\cdot)}(t) = S_{x_0}\left(\int_0^t r[x(t)] d\tau\right), \tag{5.3}$$

where r is some positive function of a stress. It is also called the accelerated failure time model [Cox and Oakes (1984)].

If $x(t) \equiv x = const$, then

$$S_x(t) = S_{x_0}\left(r(x)t\right), \tag{5.4}$$

so the stress changes the scale of the time-to-failure distribution.

Consider the model (5.3) or the model (5.2) with unknown S_{x_0} and r or S_{x_0}, r and α, respectively. Suppose that time done for an experiment is t. If all mentioned functions are unknown, observations under the stress x_0 are needed in some time intervals for S_{x_0} estimation. If the variation coefficient of T_{x_0} is not large then failures of items may not be observed under the stress x_0 in the interval $[0, t]$. Then the following experiment can be done: at the beginning of an experiment failures of items can be observed under an accelerated stress x_1 and only when they begin to fail, for example, after the k-th failure, the stress x_1 can be changed by the "normal" stress x_0. After this failures can be observed and under the "normal" stress.

Semiparametric analysis of the AAD model was considered by Basu and Ebrahimi (1982), Lin and Ying (1995), Robins and Tsiatis (1992), Schmoyer (1986, 1991), Sethuraman and Singpurwalla (1982), Shaked and Singpurwalla (1983), Tsiatis (1990), Ying (1993). In these papers the form of the function r is specified. Asymptotic distribution of estimators of the reliability function under the "normal" stress was not considered. Lin, Robins, Tsiatis and Ying obtained the asymptotic distribution of parameters of the function $r(x)$ when it is parametrized. They considered censoring with uniformly bounded density function. So these results cannot be applied for the case of type I censoring which is natural in accelerated experiments. The asymptotic covariance matrices of their estimators depend on the derivative of the baseline hazard rate function, which cannot be well estimated.

Bagdonavičius and Nikulin (1997) considered semiparametric estimation in the AAD model when moments of switch up of step-stresses are fixed and non-random. We consider the case of random switch up moments for AAD and more general model (5.2). Use of random moments sometimes can be more useful because the time interval in which failures begin can be unknown.

5.2 Estimation in the AAD Model

Suppose that the model (5.3) is true. We assume that the functions G and $r(x)$ are unknown. Suppose that two groups of items are tested: the first group of n_1 items are tested under the constant in time accelerated stress x_1 and the complete sample $T_{11}, \cdots, T_{1n_1}$ or in terms of order statistics

$$T_{1,(1)} \leq ... \leq T_{1,(n_1)}$$

is obtained. The second group of n_2 items is tested under the stress x_1 until the k-th failure and after this moment under the "normal" stress x_0 until the moment t, i.e. under the stress

$$x_2(\tau) = x_1 I_{\{0 \leq \tau < T_{2,(k)}\}} + x_0 I_{\{\tau \geq T_{2,(k)}\}}$$

and the type I censored sample

$$T_{2,(1)} \leq \cdots T_{2,(k)} \leq \cdots \leq T_{2,(m_2)}$$

is obtained ($m_2 \leq n_2$).

The purpose is to estimate from these data the reliability function $S_0 = S_{x_0}$ and the mean value $\mathbf{E}T(x_0)$ under the normal stress x_0.

Denote by $S_1 = S_{x_1}$ and $S_2^{(n_2)} = S_{x_2(\cdot)}^{(n_2)}$ the reliability functions of the times-to-failure for the items of the first and the second groups, respectively, $A_2^{(n_2)} = -\ln S_2^{(n_2)}$-the accumulated failure rate function for the items of the second group. Then

$$S_2^{(n_2)}(\tau) = -\int_0^\infty \mathbf{P}\left(T_{2i} > \tau | T_{2,(k)} = u\right) \, dS_{T_{2,(k)}}(u),$$

where T_{2j} are the times to failure under the stress $x_2(\cdot)$ (possibly not all observed), $j = 1, \cdots, n_2$. We suppose that T_{2j} are absolutely continuous with positive densities on $[0, \infty)$. Then there exist a survival function S_2 such that

$$n_2^{1/2}(S_2^{(n)} - S_2) \xrightarrow{w} 0 \quad \text{as} \quad n_2 \to \infty \tag{5.5}$$

where $\xrightarrow{w}$ denotes weak convergence.

Denote by $N_1(\tau)$, $N_2(\tau)$ the numbers of observed failures in the interval $[0, \tau]$ and $Y_1(\tau)$, $Y_2(\tau)$-the numbers of items at risk (non-failured and non-censored) just before the moment τ for the items of the first and the second group, respectively, $r = r(x)$. Note that $r(x_0) = 1$. If the AAD model is true, the moment u under the stress x_0 is "equivalent" to the moments u/r and $(u/r)\wedge T_{2,(k)} + (u - rT_{2,(k)}) \vee 0$ under the stresses x_1 and $x_2(\cdot)$, respectively. So the pseudoestimator (depending on r) of the cumulated hazard rate function $A_0 = -\ln S_{x_0}$ is

$$\tilde{A}_0(s,r) = \int_0^s \frac{dN_1(\frac{u}{r}) + dN_2\left(\frac{u}{r}\wedge T_{2,(k)} + (u - rT_{2,(k)})\vee 0\right)}{Y_1(\frac{u}{r}) + Y_2\left(\frac{u}{r}\wedge T_{2,(k)} + (u - rT_{2,(k)})\vee 0\right)}. \tag{5.6}$$

The pseudoestimator of the reliability function S_0 is the product-integral

$$\tilde{S}_0(s,r) = \mathbf{\Pi}_{0\le u\le s}(1 - d\tilde{A}_0(u,r)).$$

Then the pseudoestimators for S_1 and S_2 are

$$\tilde{S}_1(s,r) = \tilde{S}_0(rs,r), \quad \tilde{S}_2(s,r) = \tilde{S}_0(r(s\wedge T_{2,(k)}) + (s - T_{2,(k)})\vee 0, r).$$

Under the AAD model the moment τ under the stress $x_2(\cdot)$ is "equivalent" to the moment $\tau\wedge T_{2,(k)} + \frac{\tau - T_{2,(k)}}{r}\vee 0$ under the stress x_1. So consider the logrank-type statistic

$$\begin{aligned}
\hat{U}(r) &= a_n^{-2}\int_0^\infty \left\{ \frac{Y_2(\tau)dN_1(\tau\wedge T_{2,(k)} + \frac{\tau - T_{2,(k)}}{r}\vee 0)}{Y_1(\tau\wedge T_{2,(k)} + \frac{\tau - T_{2,(k)}}{r}\vee 0) + Y_2(\tau)} \right.\\
&\quad - \left. \frac{Y_1(\tau\wedge T_{2,(k)} + \frac{\tau - T_{2,(k)}}{r}\vee 0)dN_2(\tau)}{Y_1(\tau\wedge T_{2,(k)} + \frac{\tau - T_{2,(k)}}{r}\vee 0) + Y_2(\tau)} \right\}\\
&= a_n^{-2}\left\{ \sum_{j=1}^{n_1} \frac{Y_2(T_{1,(j)}\wedge T_{2,(k)} + r(T_{1,(j)} - T_{2,(k)})\vee 0)}{Y_1(T_{1,(j)}) + Y_2(T_{1,(j)}\wedge T_{2,(k)} + r(T_{1,(j)} - T_{2,(k)})\vee 0)} \right.\\
&\quad - \left. \sum_{j=1}^{m_2} \frac{Y_1(T_{2,(j)}\wedge T_{2,(k)} + \frac{T_{2,(j)} - T_{2,(k)}}{r}\vee 0)}{Y_1(T_{2,(j)}\wedge T_{2,(k)} + \frac{T_{2,(j)} - T_{2,(k)}}{r}\vee 0) + Y_2(T_{2,(j)})} \right\};
\end{aligned}$$

here $a_n^2 = n_1 n_2/n$ is the normalizing constant.

The function $\hat{U}$ is a decreasing step function of r. If we assume that times to failure are absolutely continuous random variables with supports on $[0,\infty[$, then we have that $\hat{U}(0) > 0$, $\hat{U}(\infty) < 0$ with the probability 1.

The estimator of the parameter r is defined in the following manner:

$$\hat{r} = \hat{U}^{-1}(0) = \sup\{r : \hat{U}(r) \ge 0\}. \tag{5.7}$$

The cumulated failure-rate function A_0 and the survival function S_0 under the "normal" stress can be estimated by

$$\hat{A}_0(s) = \tilde{A}_0(s,\hat{r}), \quad \hat{S}_0(s) = \hat{S}_0(s,\hat{r}),$$

where the estimator $\hat{r}$ is obtained by the formula (5.7). The estimators $\hat{A}_0(s)$ and $\hat{S}_0(s)$ are

$$\hat{A}_0(s) = \int_0^s \frac{dN_1(\frac{u}{\hat{r}}) + dN_2\left[\frac{u}{\hat{r}} \wedge T_{2,(k)} + (u - \hat{r}T_{2,(k)}) \vee 0\right]}{Y_1(\frac{u}{\hat{r}}) + Y_2\left[\frac{u}{\hat{r}} \wedge T_{2,(k)} + (u - \hat{r}T_{2,(k)}) \vee 0\right]}$$

$$= \sum_{j:\, T_{1,(j)} \le \frac{s}{\hat{r}}} \frac{1}{Y_1(T_{1,(j)}) + Y_2[T_{1,(j)} \wedge T_{2,(k)} + \hat{r}(T_{1,(j)} - T_{2,(k)}) \vee 0]}$$

$$+ \sum_{j:\, T_{2,(j)} \le \frac{s}{\hat{r}} \wedge T_{2,(k)} + (s - \hat{r}T_{2,(k)}) \vee 0} \frac{1}{Y_1[T_{2,(j)} \wedge T_{2,(k)} + \frac{T_{2,(j)} - T_{2,(k)}}{\hat{r}} \vee 0] + Y_2(T_{2,(j)})}. \tag{5.8}$$

and

$$\hat{S}_0(s) = \prod_{(i,j) \in B(s)} \left(1 - \frac{1}{Y_1(T_{1,(i)}) + Y_2(T_{1,(i)} \wedge T_{2,(k)} + \hat{r}(T_{1,(i)} - T_{2,(k)}) \vee 0)}\right)$$
$$\times \left(1 - \frac{1}{Y_2(T_{2,(j)}) + Y_1(T_{2,(j)} \wedge T_{2,(k)} + \frac{T_{2,(j)} - T_{2,(k)}}{\hat{r}} \vee 0)}\right), \tag{5.9}$$

where

$$B(s) = \{(i,j) | \hat{r}T_{1,(i)} \le s;\ \hat{r}(T_{2,(j)} \wedge T_{2,(k)}) + (T_{2,(j)} - T_{2,(k)}) \vee 0 \le s\}.$$

The estimator of the mean time to failure under the normal stress x_0 :

$$\hat{\mathbf{E}}(T_{x_0}) = -\int_0^\infty s d\hat{S}_0(s) = -\sum_{i=1}^{n_1} \hat{r}T_{1,(i)} \Delta\hat{S}_0(\hat{r}T_{1,(i)}) - \sum_{i=1}^{r_2} \hat{r}T_{2,(i)} \Delta\hat{S}_0(\hat{r}T_{2,(i)})$$
$$- \sum_{i=r_2+1}^{m_2} (T_{2,(i)} - T_{2,(k)} + \hat{r}T_{2,(k)}) \Delta\hat{S}_0(T_{2,(i)} - T_{2,(k)} + \hat{r}T_{2,(k)}).$$

The observed likelihood function for r can be written as follows

$$L(r) = \prod_{i=1}^{n_1} [\tilde{S}_{x_1}(T_{1,(i)}-, r) - \tilde{S}_{x_1}(T_{1,(i)}, r)]$$
$$\times \prod_{j=1}^{m_2} [\tilde{S}_{x_2}(T_{2,(j)}-, r) - \tilde{S}_{x_2}(T_{2,(j)}, r)] \tilde{S}_{x_2}^{n_2 - m_2}(t, r)$$

or

$$L(r) = \prod_{i=1}^{n_1} [\tilde{S}_0(rT_{1,(i)}-, r) - \tilde{S}_0(rT_{1,(i)}, r)]$$
$$\times \prod_{j=1}^{m_2} [\tilde{S}_0(r(T_{2,(j)} \wedge T_{2,(k)}) + (T_{2,(j)} - T_{2,(k)}) \vee 0-, r)$$
$$- \tilde{S}_0(r(T_{2,(j)} \wedge T_{2,(k)}) + (T_{2,(j)} - T_{2,(k)}) \vee 0, r)]$$
$$\times \tilde{S}_0^{n_2 - m_2}(rT_{2,(k)} + t - T_{2,(k)}, r).$$

5.3 Properties of Estimators

Suppose that $k_i = n_i/n \to l_i \in (0,1)$, $k/n_2 \to p \in (0,1)$. Take

$$U(r) = \int_0^t \frac{-S_2(\tau)dS_1(\tau \wedge t_p + \frac{\tau - t_p}{r} \vee 0) + S_1(\tau \wedge t_p + \frac{\tau - t_p}{r} \vee 0)dS_2(\tau)}{l_1 S_1(\tau \wedge t_p + \frac{\tau - t_p}{r} \vee 0) + l_2 S_2(\tau)},$$

where $t_p = S^{-1}(1-p)$ is the p-quantile of the time-to-failure under the stress x_1. Denote by r_0 the true value of r under the AAD model. Then $U(r_0) = 0$.

Theorem 5.3.1 *Suppose that the densities $f_i = -S_i'$ are continuous and positive on $[0, \infty)$ $(i = 1, 2)$ and $U'(r_0) \neq 0$. Then under the AAD model*

$$a_n(\hat{r} - r_0) \xrightarrow{\mathcal{D}} N(0, \sigma_{r_0}^2),$$

where

$$\sigma_{r_0}^2 = \frac{1 - S_0(r_0 t_p + t - t_p) + l_1(1-p)b(r_0)(pb(r_0) - 2\ln(1-p))}{(U'(r_0))^2}$$

and

$$\begin{aligned} b(r_0) &= \frac{1 - 1/r_0}{\lambda_0(r_0 t_p)(1-p)}\{\alpha_0(r_0 t_p + t - t_p)S_0(r_0 t_p + t - t_p) \\ &- \int_0^t \alpha_0(r_0 t_p + \tau - t_p)dS_0(r_0 t_p + t - t_p)\}. \end{aligned} \tag{5.10}$$

PROOF. Note that

$$\begin{aligned} a_n(\hat{U}(r) - U(r)) &= a_n \Bigg\{ \int_0^t -\hat{\xi}_2(\tau, r)d\hat{S}_1(\tau \wedge T_{2,(k)} + \frac{\tau - T_{2,(k)}}{r} \vee 0) \\ &+ \int_0^t \xi_2(\tau, r)dS_1(\tau \wedge t_p + \frac{\tau - t_p}{r} \vee 0) \\ &+ \int_0^t \left[\hat{\xi}_1(\tau, r)d\hat{S}_2(\tau) - \xi_1(\tau, r)dS_2(\tau)\right] \Bigg\}, \end{aligned}$$

where

$$\hat{\xi}_1(\tau, r) = \frac{\hat{S}_1(\tau \wedge T_{2,(k)} + \frac{\tau - T_{2,(k)}}{r} \vee 0-)}{k_1 \hat{S}_1(\tau \wedge T_{2,(k)} + \frac{\tau - T_{2,(k)}}{r} \vee 0-) + k_2 \hat{S}_2(\tau-)},$$

$$\xi_1(\tau, r) = \frac{S_1(\tau \wedge t_p + \frac{\tau - t_p}{r} \vee 0)}{l_1 S_1(\tau \wedge t_p + \frac{\tau - t_p}{r} \vee 0) + l_2 S_2(\tau)},$$

$$\xi_2(\tau, r) = (1 - l_1 \xi_1(\tau, r))/l_2, \quad \hat{\xi}_2(\tau, r) = (1 - k_1 \hat{\xi}_1(\tau))/k_2$$

and $\hat{S}_i$ is the empirical survival function for the ith group of items :

$$\hat{S}_i(\tau) = 1 - N_i(\tau)/n_i, \quad \tau \geq 0, \quad i = 1, 2.$$

Suppose that $n_i/n \to l_i \in]0,1[$. Take $\tau \leq t$. Then using the equality

$$\begin{aligned} & n_2^{1/2}\left(\hat{S}_2(\tau) - S_2^{(n_2)}(\tau)\right) \\ = & -S_2^{(n_2)}(\tau) \int_0^\tau \frac{I(Y_2^{(n_2)}(u) > 0)}{\sqrt{n_2} S_2^{(n_2)}} \{dN_2^{(n_2)}(u) - Y_2^{(n_2)}(u) d\Lambda_2^{(n_2)}(u)\}, \end{aligned}$$

the Theorem 6.2.1 [Fleming and Harrington (1991, p. 231)] and the weak convergence (5.5), we have

$$a_n(\hat{S}_i - S_i) \xrightarrow{\mathcal{D}} \eta_i \quad \text{on} \quad D[0, s],$$

where $s \in [0, \infty[$ $(i = 1)$ or $s \in [0, t]$ $(i = 2)$, η_i is a mean zero Gaussian process such that for all $0 \leq s_1 \leq s_2 \leq s$

$$\mathbf{Cov}\,(\eta_i(s_1), \eta_i(s_2)) = l_{3-i}(1 - S_i(s_1)) S_i(s_2) \ (i = 1, 2).$$

For $s > t$ define $\eta_2(s) = 0$.

The asymptotic distribution of the empirical quantile $T_{2,(k)}$ can be written in terms of the process η_2:

$$a_n(T_{2,(k)} - t_p) \xrightarrow{\mathcal{D}} -\frac{\eta_2(t_p)}{S_2'(t_p-)} = -\frac{\eta_2(t_p)}{S_1'(t_p)}.$$

Using the functional delta method [Andersen et al. (1993)] we obtain that for all $u \leq t$

$$a_n\left(\hat{S}_1(\cdot \wedge T_{2,(k)} + \frac{\cdot - T_{2,(k)}}{r} \vee 0) - S_1(\cdot \wedge t_p + \frac{\cdot - t_p}{r} \vee 0)\right) \xrightarrow{\mathcal{D}}$$

$$-\frac{\eta_2(t_p)}{S_1'(t_p)} S_1'(t_p + \frac{\cdot - t_p}{r}) \frac{r-1}{r} I(\cdot > t_p) + \eta_1(\cdot \wedge t_p + \frac{\cdot - t_p}{r} \vee 0) := \eta_1^*(\cdot),$$

$$a_n(\hat{\xi}_1(\cdot) - \xi_1(\cdot)) \xrightarrow{\mathcal{D}} l_2\zeta_1(\cdot), \quad a_n(\hat{\xi}_2(\cdot) - \xi_2(\cdot)) \xrightarrow{\mathcal{D}} -l_1\zeta_1(\cdot) \quad \text{on} \quad D[0, u],$$

where

$$\zeta_1(\tau) = \frac{S_2(\tau)\eta_1^*(\tau) - S_1(\tau \wedge t_p + \frac{\tau - t_p}{r} \vee 0)\eta_2(\tau)}{(l_1 S_1(\tau \wedge t_p + \frac{\tau - t_p}{r} \vee 0) + l_2 S_2(\tau))^2}.$$

Using the functional delta method for stochastic integrals we obtain

$$a_n(\hat{U}(r) - U(r)) \xrightarrow{\mathcal{D}} \zeta(r),$$

where

$$\zeta(r) = \int_0^t \xi_1(\tau, r) d\eta_2(\tau) - \xi_2(\tau, r) d\eta_1^*(\tau) + \zeta_1(\tau) d(l_1 S_1(t_p + \frac{\tau - t_p}{r}) + l_2 S_2(\tau)).$$

Note that $\xi_1(\tau, r_0) = \xi_2(\tau, r_0) = 1$,

$$\zeta(r_0) = \eta_2(t) - \eta_1^*(t) + \int_0^t [\eta_1^*(\tau) - \eta_2(\tau)] d \ln S_2(\tau).$$

and $\mathbf{E}\zeta(r_0) = 0$. After some tedious calculation we obtain

$$\mathbf{Var}\zeta(r_0) = 1 - S_0(r_0 t_p + t - t_p) + l_1(1 - pb(r_0)(pb(r_0) - 2\ln(1-p)),$$

where $b(r_0)$ is done in the formula (5.10).

Under the assumptions of the theorem the function $U(r)$ is differentiable and decreasing on $[0, \infty[$ and $U(r_0) = 0$, the process $\zeta(r)$ is continuous at t_0. So there exist the inverse function U^{-1} in $]U(\infty), U(0)[$ and the Hadamar derivative of $U^{-1}(0)$ is

$$dU^{-1}(0)(h) = -\frac{h(U^{-1}(0))}{U'(U^{-1}(0))} = -\frac{h(r_0)}{U'(r_0)}$$

for all continuous on $]0, \infty[$ functions h [Andersen et al. (1993, Proposition II.8.4)].

Taking into account that $\hat{r} = \hat{U}^{-1}(0)$, $r_0 = U^{-1}(0)$,

$$a_n(\hat{U}(r_0) - U(r_0)) = a_n \hat{U}(r_0) \xrightarrow{\mathcal{D}} \zeta(r_0) \tag{5.11}$$

and using the functional delta method we obtain

$$a_n(\hat{r} - r_0) \xrightarrow{\mathcal{D}} -\frac{\zeta(r_0)}{U'(r_0)} := V(r_0)$$

and the equality (5.10).

Note that

$$\begin{aligned} r_0 U'(r_0) &= -(t - t_p)\lambda_0(r_0 t_p + t - t_p) S_0(r_0 t_p + t - t_p) \\ &\quad + \int_0^{t - t_p} v \lambda_0(r_0 t_p + v) dS_0(r_0 t_p + v), \end{aligned}$$

where $\lambda_0(u) = -S_0'(u)/S_0(u)$ is the failure rate under the "normal" stress. The proof is complete. ∎

Consider asymptotic properties of the estimator $\hat{S}_0$ done by (5.9).

Theorem 5.3.2 *Under the assumptions of the Theorem 5.3.1*

$$a_n(\hat{A}_0(s) - A_0(s)) \xrightarrow{\mathcal{D}} B(s, r_0)V(r_0) + \delta(s, r_0),$$

where

$$\delta(s, r_0) = -\frac{l_1\eta_2(s/r) + l_2\eta_2^*(s, r)}{S_0(s)}, \tag{5.12}$$

$$\eta_2^*(s, r) = \eta_2((s/r) \wedge t_p + (s - rt_p) \vee 0) - \eta_2(t_p)(1 - r)\frac{S_2'(t_p + s - rt_p)}{S_1'(t_p)}\mathbf{1}_{\{u > rt_p\}},$$

$$B(s, r_0) = -\lambda_0(s)\left[\frac{s}{r_0} + l_2(t_p - \frac{s}{r_0})\mathbf{1}_{\{r_0 t_p < s \le r_0 t_p + t - t_p\}}\right]. \tag{5.13}$$

PROOF. Consider at first the asymptotic distribution of the pseudoestimator (5.6). Denote for all $s \geq 0$

$$A_0(s, r) = -\int_0^s \frac{l_1 dS_1(\frac{u}{r}) + l_2 dS_2\left(((\frac{u}{r}) \wedge t_p + (u - rt_p) \vee 0) \wedge t\right)}{l_1 S_1(\frac{u}{r}) + l_2 S_2\left(((\frac{u}{r}) \wedge t_p + (u - rt_p) \vee 0) \wedge t\right\}}.$$

Then $A_0(s, r_0) = A_0(s) = -\ln S_{x_0}(s)$ is the cumulated hazard rate function under the normal stress x_0. Note that the pseudoestimator (5.6) can be written in the form

$$\tilde{A}_0(s, r) = -\int_0^s \frac{d\hat{Z}(u)}{\hat{Z}(u-)},$$

where

$$\hat{Z}(u) = k_1\hat{S}_1(\frac{u}{r}) \\ +k_2\hat{S}_2\left(((\frac{u}{r}) \wedge T_{2,(k)} + (u - rT_{2,(k)}) \vee 0) \wedge (rT_{2,(k)} + t - T_{2,(k)})\right).$$

Using the functional delta method for integrals we obtain

$$a_n(\tilde{A}_0(s, r) - A_0(s, r)) \to \delta(s, r_0), \tag{5.14}$$

where $\delta(s, r_0)$ is done by (5.12).

Using (5.11), (5.14) and the functional delta method we obtain for fixed $s \geq 0$

$$a_n(\tilde{A}_0(s, \hat{r}) - A_0(s, r_0)) \to B(s, r_0)V(r_0) + \delta(s, r_0),$$

where

$$B(s, r_0) = \frac{\partial}{\partial r}A_0(s, r_0)$$

is done by the formula (5.13). ∎

Corollary 5.3.1 *Under the assumptions of the theorem for all $s \geq 0$*

$$a_n(\hat{A}_0(s) - A_0(s)) \xrightarrow{\mathcal{D}} N(0, \sigma_s^2), \quad a_n(\hat{S}_0(s) - S_0(s)) \xrightarrow{\mathcal{D}} N(0, \sigma_s^2 S_0^2(s)),$$

where

$$\sigma_s^2 = B^2(s, r_0)\mathbf{Var}\, V(r_0) + 2B(s, r_0)\mathbf{Cov}\,(V(r_0), \delta(s, r_0)) + \mathbf{Var}\, \delta(s, r_0),$$

$$\mathbf{Var}\, V(r_0) = \sigma_{r_0}^2, \quad \mathbf{Cov}\,(V(r_0), \delta(s, r_0) = -\frac{\mathbf{Cov}\,(\zeta(r_0), \delta(s, r_0)}{U'(r_0)};$$

for $s \leq r_0 t_p$

$$\mathbf{Var}\, \delta(s, r_0) = \frac{l_1 l_2}{S_0(s)}(1 - S_0(s)),$$

$$\mathbf{Cov}\,(\delta(s, r_0), \zeta(r_0)) = -\frac{l_1 l_2}{S_0(s)}(1 - p)b(r_0);$$

for $r_0 t_p < s \leq r_0 t_p + t - t_p$

$$\mathbf{Var}\, \delta(s, r_0) = \frac{l_1 l_2}{S_0^2(s)}\{S_0(s)(1 - S_0(s)) + l_2 c^2 p(1 - p) + 2cl_2 S_0(s)p\},$$

$$\begin{aligned}&\mathbf{Cov}\,(\delta(s, r_0), \zeta(r_0)) \\ &= -\frac{l_1 l_2}{S_0(s)}\{S_0(s)[pb(r_0) + 1 - p] + c(1 - p)[pb(r_0) - \ln(1 - p)]\},\end{aligned}$$

where

$$c = \frac{(1 - \frac{1}{r_0})\lambda_0(s)S_0(s)}{\lambda_0(t_p)S_0(t_p)};$$

for $s > r_0 t_p + t - t_p$

$$\mathbf{Var}\, \delta(s, r_0) = \frac{l_1 l_2}{S_0^2(s)}\{l_1(1 - S_0(s))S_0(s) + l_2 c^2 p(1 - p)\},$$

$$\begin{aligned}\mathbf{Cov}\,(\delta(s, r_0), \zeta(r_0)) &= -\frac{l_1 l_2}{S_0(s)}\{S_0(s)\ln S_0(r_0 t_p + t - t_p) \\ &\quad + c(1 - p)[pb(r_0) - \ln(1 - p)]\}.\end{aligned}$$

Remark 5.3.1 Replacing the unknown functions S_0 and λ_0 by their consistent estimators $\hat{S}_0$ and $\hat{\lambda}_0$, we obtain consistent estimator $\hat{\sigma}_s^2$ of σ_s^2. The function λ_0 is estimated by

$$\hat{\lambda}_0(s) = b^{-1}\int_0^\infty K(\frac{s - u}{b})d\hat{A}_0(u),$$

where K is the kernel function with the window size b. For example, we can use Epanechnikov kernel function

$$K(x) = 0.75(1 - x^2)\mathbf{1}_{\{|x| \leq 1\}}.$$

Taking into consideration the definition of $\hat{A}_0(u)$ we have

$$\begin{aligned}
&\hat{\lambda}_0(s) \\
&= b^{-1}\int_0^\infty K\left(\frac{s-u}{b}\right)\frac{dN_1(u/\hat{r})+dN_2((u/\hat{r})\wedge T_{2,(k)}+(u-\hat{r}T_{2,(k)})\vee 0)}{Y_1(u/\hat{r})+Y_2[(u/\hat{r})\wedge T_{2,(k)}+(u-\hat{r}T_{2,(k)})\vee 0]} \\
&= b^{-1}\left\{\sum_{j=1}^{n_1}\frac{K\left(\frac{s-\hat{r}T_{1,(j)}}{b}\right)}{Y_1(T_{1,(j)})+Y_2[T_{1,(j)}\wedge T_{2,(k)}+\hat{r}(T_{1,(j)}-T_{2,(k)})\vee 0]}\right. \\
&\left.+\sum_{j=1}^{m_2}\frac{K(\frac{s-\hat{r}(T_{2,(j)}\wedge T_{2,(k)})-(T_{2,(j)}-T_{2,(k)})\vee 0}{b})}{Y_1(T_{2,(j)}\wedge T_{2,(k)}+\frac{T_{2,(j)}-T_{2,(k)}}{\hat{r}}\vee 0)+Y_2(T_{2,(j)})}\right\}.
\end{aligned}$$

So the approximate $(1-\alpha)$ confidence interval for $S_0(s)$ has the form

$$\left(1+\frac{1-\hat{S}_{x_0}(t)}{\hat{S}_{x_0}(t)}\exp\{\mp\hat{\sigma}_s w_{1-\alpha/2}\}\right)^{-1}.$$

5.4 Estimation, When Stresses Change the Shape of Distribution

Consider the model (5.2). Ideas of estimating are similar. Suppose, for example, that G and r are completely unknown and the data is the same. Then the pseudoestimator of the reliability function S_0 is

$$\begin{aligned}
&\tilde{S}_0(s;\theta,\alpha) \\
&= \prod_{(i,j)\in B(s)}\left(1-\frac{1}{Y_1(T_{1,(i)})+Y_2(T_{1,(i)}\wedge T_{2,(k)}+\theta(T_{1,(i)}^\alpha-T_{2,(k)}^\alpha)\vee 0)}\right) \\
&\times\left(1-\frac{1}{Y_2(T_{2,(j)})+Y_1[\left((T_{2,(j)}\wedge T_{2,(k)})^\alpha+\frac{T_{2,(j)}-T_{2,(k)}}{\theta}\vee 0\right)^{1/\alpha}]}\right),
\end{aligned}$$

where $\theta=r/\alpha$, $r=r(x_1)$, $\alpha=\alpha(x_1)$,

$$B(s)=\{(i,j)|\theta T_{1,(i)}^\alpha\le s;\ \theta(T_{2,(j)}\wedge T_{2,(k)})^\alpha+(T_{2,(j)}-T_{2,(k)})\vee 0\le s\}$$

and the observed likelihood function for θ and α can be written as follows

$$L(\theta,\alpha) = \prod_{i=1}^{n_1}[\tilde{S}_0(\theta T_{1,(i)}^\alpha-;\theta,\alpha)-\tilde{S}_0(\theta T_{1,(i)}^\alpha;\theta,\alpha)]$$

$$\times \prod_{j=1}^{m_2} [\tilde{S}_0(\theta(T_{2,(j)} \wedge T_{2,(k)})^\alpha + (T_{2,(j)} - T_{2,(k)}) \vee 0-; \theta, \alpha)$$
$$-\tilde{S}_0(\theta(T_{2,(j)} \wedge T_{2,(k)})^\alpha + (T_{2,(j)} - T_{2,(k)}) \vee 0; \theta, \alpha)]$$
$$\times \tilde{S}_0^{n_2-m_2}(\theta t_1^\alpha + t - T_{2,(k)}; \theta, \alpha).$$

The estimator of the reliability function S_0 is $\hat{S}_0(s) = \tilde{S}_0(s; \hat{\theta}, \hat{\alpha})$, where $\hat{\theta}$, $\hat{\alpha}$ are the maximum likelihood estimators.

5.5 Estimation in AFT Model, When G is Completely Unknown and r is Parametrized

Consider the "traditional" censored data, when several groups of items are observed under accelerated stresses without using the "normal one".

Suppose that k groups of $n_1, ..., n_k$ items are tested. The items of the ith group are tested under the accelerated stress x_i time t_i and the moments of failures $T_{i1} \leq T_{i2} \leq ... \leq T_{im_i}$ are observed, $m_i \leq n_i$ $(i = 1, ..., k)$.

The survival function under the stress x_i is $S_{x_i}(t) = G\{r(x_i)t\}$. If G and r are completely unspecified and G and $r(x_i)$ are estimated, the value $r(x_0)$ and consequently the survival function S_{x_0} can't be estimated. If we suppose, that G is completely unknown, the function r must be parametrized. In the case of power rule $(r(x) = \alpha x^\beta)$ or Arrhenius $(r(x) = \alpha e^{\beta/x})$ models the parameters α can be included in the function G as it is unknown. So on both cases $S_{x_i}(t) = G\{e^{\beta y_i} t\}$, where $y_i = \ln x_i$ or $y_i = \frac{1}{x_i}$ correspondingly. If G and β are estimated, the function S_{x_0} can be estimated, too.

Note that

$$S_{x_i}(t) = S_{x_0}\{e^{\beta z_i} t\},$$

where $z_i = y_i - y_0$.

Consider estimation of the survival function S_{x_0} in this model. Denote $N_i(\tau)$ the numbers of observed failures of the ith group in the interval $[0, \tau]$, $Y_i(\tau)$-the numbers of items at risk just before the moment τ $(i = 1, ..., k)$.

Similarly as in Section 5.2 a pseudoestimator (still depending on β) of the cumulated failure rate function $A_0 = -\ln S_0$ can be defined for $s \leq \max_i \{e^{\beta z_i} t_i\}$:

$$\tilde{A}_0(s, \beta) = \int_0^s \frac{d \sum_{i=1}^k N_i(e^{-\beta z_i} u)}{\sum_{i=1}^k Y_i(e^{-\beta z_i} u)}.$$

The pseudoestimator of the reliability function S_0 would be

$$\tilde{S}_0(s, \beta) = e^{-\tilde{A}_0(s, \beta)}$$

or, in the alternative way: for $s \le \max_i \{e^{\beta z_i} t_i\}$

$$\tilde{S}_0(s,\beta) = \prod_{(i,j)\in B} \left(1 - \frac{1}{\sum_{l=1}^{k} Y_l(e^{\beta(z_i - z_l)} T_{ij})}\right),$$

where

$$B(s) = \{(i,j) | T_{ij} \le e^{-\beta z_i} s\}.$$

The observed likelihood function has the form:

$$L(\beta) = \prod_{i=1}^{k}\prod_{j=1}^{m_i} [\tilde{S}_0(e^{\beta z_i} T_{i,j}-, \beta) - \tilde{S}_0(e^{\beta z_i} T_{i,j}, \beta)] \tilde{S}_0^{n_i - m_i}(e^{\beta z_i} t_i, \beta).$$

Analogously as in Section 5.2 we consider the score function:

$$\hat{U}(\beta) = \frac{1}{n} \sum_{i=1}^{k} z_i \int_0^\infty \{dN_i(\tau) - Y_i(\tau) d\tilde{A}_0(e^{\beta z_i}\tau, \beta)\}.$$

Define $\hat{\beta} = \sup \arg\min |\hat{U}(\beta)|$. Suppose that $n_j \to \infty$, $n_j/n \to l_j \in (0,1)$. Denote $I_j(\tau) = I\{\tau \le e^{\beta_0 z_j} t_j\}$,

$$\xi_1(t) = \sum_{j=1}^{k} z_j l_j I_j(t) V_j(e^{-\beta_0 z_j} t), \quad D_1(t) = \sum_{j=1}^{k} z_j l_j I_j(t),$$

$$\xi_2(t) = \sum_{j=1}^{k} l_j I_j(t) V_j(e^{-\beta_0 z_j} t), \quad D_2(t) = \sum_{j=1}^{k} l_j I_j(t),$$

$$B_1(t) = \sum_{j=1}^{k} z_j l_j I_j(t) S_j(e^{-\beta_0 z_j} t), \quad B_2(t) = \sum_{j=1}^{k} l_j I_j(t) S_j(e^{-\beta_0 z_j} t),$$

$$Q(t) = \frac{t\lambda_0(t) S_0(t) D_1(t) \zeta(\beta_0)}{D(\beta_0)} + \xi_2(t),$$

$$B(t) = \frac{B_1(t)}{B_2(t)}, z(t) = \frac{\xi_1(t)}{B_2(t)} - \frac{B_1(t)}{B_2^2(t)} \xi_2(t),$$

$$C(t) = D_2(t) - B_2(t), \quad \zeta(\beta_0) = \int_0^\infty \{B(u) d\xi_2(u) - d\xi_1(u) - z(u) dC(u)\},$$

$$D(\beta_0) = \sum_{i=1}^{k} l_i z_i \{\int_0^{t_i} \left[z_i - \frac{D_1(u)}{D_2(u)}\right] (u S_0(u) d\lambda_0(u) - dS_0(u))\},$$

$$\hat{A}(s) = \tilde{A}_0(s, \hat{\beta}).$$

Similarly as in Section 5.2 the following theorem can be proved.

Theorem 5.5.1 *Suppose that the densities $f_i = -S_i'$ are continuous and positive on $[0, +\infty[$, $(i = 1, 2, ..., k)$ and $D(\beta_0) > 0$. Then for all $s \leq \max_i \{e^{\beta z_i} t_i\}$*

$$\sqrt{n}(\hat{A}_0(s) - A_0(s)) \xrightarrow{\mathcal{D}} - \frac{Q(s)}{S_0(s) D_2(s)}$$
$$+ \int_0^s Q(u) \left\{ d\left(\frac{1}{S_0(u) D_2(u)}\right) - \frac{d(1 - S_0(u)) D_2(u)}{(S_0(u) D_2(u))^2} \right\}.$$

So under the assumptions of the Theorem 5.5.1 the convergence $\sqrt{n}(\hat{A}_0(s) - A_0(s)) \to N(0, \sigma_s^2)$ is true for all $s \leq \max_i \{e^{\beta z_i} t_i\}$ and the variance σ_s^2 can be easily calculated as in Section 5.3.

The failure rate λ_0 can be estimated similarly as in the Section 5.3 and the approximate confidence interval for $A_0(s)$ and $S_0(s)$ can be obtained.

References

1. Andersen, P. K., Borgan, O., Gill, R. D. and Keiding, N. (1993). *Statistical Models Based on Counting Processes*, New York: Springer-Verlag.

2. Bagdonavičius, V. (1978). Testing the hypothesis of the additive accumulation of damages, *Probability Theory and Its Applications*, **23**, 403–408.

3. Bagdonavičius, V. (1990). Accelerated life models when the stress is not constant, *Kybernetika*, **26**, 289–295.

4. Bagdonavičius V. and Nikulin, M. (1997). Sur l'application des stress en escalier dans les expériences accélérées, *Comptes Rendus, Academie des Sciences de Paris*, **325**, Série I, 523–526.

5. Bagdonavičius, V. and Nikulin. M. (1997). Analysis of general semiparametric models with random covariates, *Romanian Journal of Pure and Applied Mathematics*, **42**, 351–369.

6. Basu, A. P. and Ebrahimi, N. (1982). Nonparametric accelerated life testing, *IEEE Transactions on Reliability* , **31**, 432–435.

7. Fleming, T. R. and Harrington, D. P. (1991). *Counting Processes and Survival Analysis*, New York: John Wiley & Sons.

8. Gasser, T. and Müller, H. G. (1979). Kernel estimation of regression functions, In *Smoothing Techniques for Curve Estimation, Lecture Notes in Mathematics*, pp. 23–68, Berlin: Springer-Verlag.

9. Lin, D. Y. and Ying, Z. (1995). Semiparametric inference for accelerated life model with time dependent covariates, *Journal of Statistical Planning and Inference*, **44**, 47–63.

10. Robins, J. M. and Tsiatis, A. A. (1992). Semiparametric estimation of an accelerated failure time model with time dependent covariates, *Biometrika*, **79**, 311–319.

11. Schmoyer, R. (1986). An exact distribution-free analysis for accelerated life testing at several levels of a single stress, *Technometrics*, **28**, 165–175.

12. Schmoyer, R. (1991). Nonparametric analyses for two-level single-stress accelerated life tests *Technometrics*, **33**, 175–186.

13. Sethuraman, J. and Singpurwalla, N. D. (1982). Testing of hypotheses for distributions in accelerated life tests, *JASA*, **77**, 204–208.

14. Shaked, M. and Singpurwalla, N. D. (1983). Inference for step-stress accelerated life tests, *Journal of Statistical Planning and Inference*, **7**, 295–306.

15. Tsiatis, A. A. (1990). Estimating regression parameters using linear rank tests for censored data, *Annals of Statistics*, **18**, 354–372.

16. Ying, Z. (1993). A large sample study of rank estimation for censored regression data, *Annals of Statistics*, **21**, 76–99.

6

Analysis of Reliability Characteristics Estimators in Accelerated Life Testing

Leo Gerville-Reache and Valentina Nikoulina

Université de Bordeaux, Bordeaux, France

Abstract: The efficiency of nonparametric estimators with respect to parametric estimators for the accelerated failure time model is considered. This article gives the asymptotic properties of the parametric estimators and recall the properties of the nonparametric estimators proved by Bagdonavičius and Nikulin (1997).

Keywords and phrases: Accelerated failure time model, kernel function, nonparametric estimation, parametric estimation, stress, type I censored data, Bagdonavičius-Nikulin models

6.1 Introduction

Suppose that $S_{x(\cdot)}$ is the reliability function of the time-to-failure $T_{x(\cdot)}$ under the time varying stress $x : [0, \infty] \to B \subset I\!R^m$.

Consider the accelerated failure time (AFT) model [Bagdonavičius (1978)]:

$$S_{x(\cdot)}(t) = S_{x_0}\left\{\int_0^t r(x(\tau))d\tau\right\}, \tag{6.1}$$

where S_{x_0} is the time to failure under the usual constant in time stress x_0, r some positive function $r : B \to [0, \infty[$.

If $x(t) \equiv x = const$, then

$$S_x(t) = S_{x_0}\{r(x)t\}.$$

If the classical parametric and noparametric estimation procedures [Nelson (1990), and Nikulin (1995)] the function r is parametrized in the following way:

$$r(x(t)) = e^{\beta^T z(t)}$$

(1995)] the function r is parametrized in the following way:

$$r(x(t)) = e^{\beta^T z(t)}$$

where $r_0(x(t)) \equiv 1$, $\beta = (\beta_0, ..., \beta_m)^T$ is the vector of unknown parameters, $z(t) = (z_0(t), ..., z_m(t))^T$ is the vector of some known functions of stress.

We consider the case when the function r is completely unknown, and the experiment is as follows:

Suppose that two groups of items are tested:

the first group of n_1 items is tested under the constant in time accelerated stress x_1 and the complete sample $T_{11} \leq ... \leq T_{1n_1}$ are obtained;

the second group is tested under the stress

$$x_2(\tau) = \begin{cases} x_1, & \text{if} \quad 0 \leq \tau \leq t_1, \\ x_0, & \text{if} \quad t_1 < \tau \leq t \end{cases}$$

and the type I censored sample $T_{21} \leq ... \leq T_{2m_2}$ $(m_2 \leq n_2)$ is obtained.

Such experiment is useful when variation coefficient of time-to-failure under the usual stress x_0 is not large and the most of failures occur in some interval $[\tau_1, \tau_2]$, where τ_1 is larger than the time t given for experiment.

The items of the second group use much of their resource under the accelerated stress x_1 and after the moment t_1 even under the usual stress x_0 failures of items can be obtained (see, Figure 6.1).

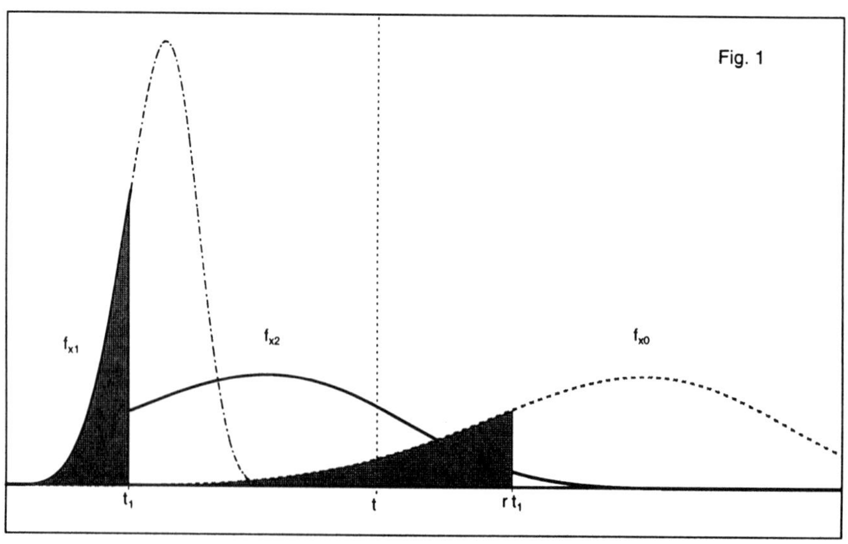

Figure 6.1: Test of tour groups of items. The first one under constant stress (x_1), and the second one under variable stress (x_2)

6.2 Parametric Estimation

Suppose that

$$S_{x_0}(t) = S_0\left((t/\theta)^{\alpha}\right),$$

where S_0 is some specified function, for example, if

$$S_0(t) = e^{-t}, \quad (1+t)^{-1}, \quad 1 - \Phi(\ln(t)),$$

then we obtain the classes of Weibull, loglogistic and lognormal distribution. So the model (6.1) can be written

$$S_{x(\cdot)}(t) = S_0\left(\left(\int_0^t r[x(\tau)]d\tau/\theta\right)^{\alpha}\right).$$

Under the model of the first section we can write

$$S_{x_1}(t) = S_0\left(\left(\frac{rt}{\theta}\right)^{\alpha}\right),$$

$$S_{x_2}(t) = S_0\left(((r(t_1 \wedge t) + (t - t_1) \vee 0)/\theta)^{\alpha}\right),$$

where $r = r(x_1)$.

Denote

$$\rho = \ln r, \quad \psi = \ln\theta, \quad S(t) = S_0(e^t), \quad f(t) = -S'(t), \quad \lambda(t) = f(t)/S(t).$$

Then

$$\begin{aligned}
S_{x_1}(t) &= S(\alpha(\ln t + \rho - \psi)); \\
S_{x_2}(t) &= \begin{cases} S(\alpha(\ln t + \rho - \psi)), & t \le t_1, \\ S(\alpha(\ln(e^{\rho}t_1 + t - t_1) - \psi)), & t > t_1; \end{cases} \\
f_{x_1}(t) &= f(\alpha(\ln t + \rho - \psi))\frac{\alpha}{t}; \\
f_{x_2}(t) &= \begin{cases} f(\alpha(\ln t + \rho - \psi))\frac{\alpha}{t}, & t \le t_1, \\ f(\alpha(\ln(e^{\rho}t_1 + t - t_1) - \psi))\frac{\alpha}{e^{\rho}t_1 + t - t_1}, & t > t_1. \end{cases}
\end{aligned}$$

Denote r_2 the random number of failures of items of the second group before the moment t_1. The likelihood function can be written as follows:

$$\begin{aligned}
L &= \prod_{j=1}^{n_1} f(\alpha(\ln T_{1j} + \rho - \psi))\frac{\alpha}{T_{1j}} \prod_{j=1}^{r_2} f(\alpha(\ln T_{2j} + \rho - \psi))\frac{\alpha}{T_{2j}} \\
&\quad \times \prod_{j=r_2+1}^{m_2} f(\alpha(\ln(e^{\rho}t_1 + T_{2j} - t_1) - \psi))\,\frac{\alpha}{e^{\rho}t_1 + T_{2j} - t_1} \\
&\quad \times S^{n_2 - m_2}(\alpha(\ln(e^{\rho}t_1 + t_2 - t_1) - \psi)),
\end{aligned}$$

Thus:

$$\begin{aligned} U_1(\alpha,\rho,\psi) &= \frac{\partial \ln L}{\partial \alpha} = \sum_{j=1}^{n_1} (\ln f)'(c(T_{1j}))\frac{c(T_{1j})}{\alpha} + \frac{n_1+m_2}{\alpha} \\ &+ \sum_{j=1}^{r_2} (\ln f)'(c(T_{2j}))\frac{c(T_{2j})}{\alpha} + \sum_{j=r_2+1}^{m_2} (\ln f)'(d(T_{2j}))\frac{d(T_{2j})}{\alpha} \\ &- (n_2-m_2)\lambda(d(t_2))\frac{d(t_2)}{\alpha}, \end{aligned}$$

$$\begin{aligned} U_2(\alpha,\rho,\psi) &= \frac{\partial \ln L}{\partial \rho} = \sum_{j=1}^{n_1} (\ln f)'(c(T_{1j}))\alpha + \sum_{j=1}^{r_2} (\ln f)'(c(T_{2j}))\alpha \\ &+ \sum_{j=r_2+1}^{m_2} \frac{(\ln f)'(d(T_{2j}))\alpha e^{\rho} t_1}{e^{\rho} t_1 + T_{2j} - t_1} - \sum_{j=r_2+1}^{m_2} \frac{e^{\rho} t_1}{e^{\rho} t_1 + T_{2j} - t_1} \\ &- \frac{(n_2-m_2)\lambda(d(t_2))\alpha e^{\rho} t_1}{e^{\rho} t_1 + t_2 - t_1}, \end{aligned}$$

$$\begin{aligned} U_3(\alpha,\rho,\psi) &= \frac{\partial \ln L}{\partial \psi} = -\alpha \left[\sum_{j=1}^{n_1} (\ln f)'(c(T_{1j})) + \sum_{j=1}^{r_2} (\ln f)'(c(T_{2j})) \right. \\ &+ \left. \sum_{j=r_2+1}^{m_2} (\ln f)'(d(T_{2j})) - (n_2-m_2)\lambda(d(t_2)) \right], \end{aligned}$$

where

$$c(u) = \alpha(\ln u + \rho - \psi), \quad d(u) = \alpha(\ln(e^{\rho} t_1 + u - t_1) - \psi).$$

In the case of Weibull, loglogistic and lognormal laws:

$$(\ln f)'(t) = 1 - e^t; \quad \frac{1-e^t}{1+e^t}; \quad -t,$$

respectively, and

$$\lambda(t) = e^t; \quad (1+e^{-t})^{-1}; \quad \frac{\varphi(t)}{1-\Phi(t)},$$

respectively. Here $\Phi(t)$ is the cumulative distribution function of the standard normal law and $\phi(t) = \Phi'(t)$.

If maximum likelihood estimators $\hat{\alpha}$, $\hat{\rho}$, $\hat{\psi}$ are obtained then the survival function estimator S_{x0} and the p-quantile estimator $t_p(x_0)$ are:

$$\hat{S}_{x0}(t) = S(\hat{\alpha}(\ln t - \hat{\psi})), \quad \hat{t}_p = \exp\left\{\hat{\psi} + \frac{1}{\hat{\alpha}} S^{-1}(1-p)\right\}.$$

In the case of the Weibull, loglogistic and lognormal laws:

$$S^{-1}(p) = \ln(-\ln(1-p)); \quad -\ln\left(\frac{1}{p} - 1\right); \quad \Phi^{-1}(1-p).$$

Denote $\mathbf{I}(\alpha,\rho,\psi) = (I_{ij}(\alpha,\rho,\psi))$ - (3×3) symmetric matrix of the following elements:

$$
\begin{aligned}
I_{11} &= -\frac{\partial^2 \ln L}{\partial \alpha^2} = -\frac{1}{\alpha^2}\Bigg\{\sum_{j=1}^{n_1}(\ln f)''(c(T_{1j}))[c(T_{1j})]^2 - n_1 - m_2 \\
&\quad + \sum_{j=1}^{r_2}(\ln f)''(c(T_{2j}))[c(T_{2j})]^2 + \sum_{j=r_2+1}^{m_2}(\ln f)''(d(T_{2j}))[d(T_{2j})]^2 \\
&\quad - (n_2-m_2)\lambda'(d(t_2))[d(t_2)]^2\Bigg\},
\end{aligned}
$$

$$
\begin{aligned}
I_{12} &= I_{21} = -\frac{\partial^2 \ln L}{\partial \alpha \partial \rho} = -\sum_{j=1}^{n_1}(\ln f)''(c(T_{1j}))c(T_{1j}) - \sum_{j=1}^{r_2}(\ln f)''(c(T_{2j})) \\
&\quad - \sum_{j=r_2+1}^{m_2}(\ln f)''(d(T_{2j}))d(T_{2j}) + (n_2-m_2)\lambda'(d(t_2))\ \frac{e^\rho t_1}{e^\rho t_1 + t_2 - t_1} \\
&\quad - \frac{1}{\alpha}\,U_2(\alpha,\rho,\psi) - \frac{1}{\alpha}\sum_{j=r_2+1}^{m_2}\frac{e^\rho t_1}{e^\rho t_1 + T_{2j} - t_1},
\end{aligned}
$$

$$
\begin{aligned}
I_{13} &= I_{31} = -\frac{\partial^2 \ln L}{\partial \alpha \partial \psi} = \sum_{j=1}^{n_1}(\ln f)''(c(T_{1j}))c(T_{1j}) + \sum_{j=1}^{r_2}(\ln f)''(c(T_{2j}))c(T_{2j}) \\
&\quad + \sum_{j=r_2+1}^{m_2}(\ln f)''(d(T_{2j}))d(T_{2j}) - (n_2-m_2)\lambda'(d(t_2))d(t_2) - \frac{1}{\alpha}U_3(\alpha,\rho,\psi),
\end{aligned}
$$

$$
\begin{aligned}
I_{22} &= -\frac{\partial^2 \ln L}{\partial \rho^2} = -\alpha^2\sum_{j=1}^{n_1}(\ln f)''(c(T_{1j})) - \alpha^2\sum_{j=1}^{r_2}(\ln f)''(c(T_{2j})) \\
&\quad - \alpha^2\sum_{j=r_2+1}^{m_2}(\ln f)''(d(T_{2j}))\left(\frac{e^\rho t_1}{e^\rho t_1 + T_{2j} - t_1}\right)^2 \\
&\quad - \sum_{j=r_2+1}^{m_2}[\alpha(\ln f)'(d(T_{2j})) - 1]\ \frac{e^\rho t_1(T_{2j}-t_1)}{(e^\rho t_1 + T_{2j} - t_1)^2} \\
&\quad + (n_2-m_2)\lambda'(d(t_2))\left(\frac{\alpha e^\rho t_1}{e^\rho t_1 + t_2 - t_1}\right)^2 \\
&\quad + (n_2-m_2)\lambda(d(t_2))\frac{\alpha e^\rho t_1(t_2-t_1)}{(e^\rho t_1 + t_2 - t_1)^2},
\end{aligned}
$$

$$
\begin{aligned}
I_{23} &= I_{32} = -\frac{\partial^2 \ln L}{\partial \rho \partial \psi} = \alpha^2\Bigg\{\sum_{j=1}^{n_1}(\ln f)''(c(T_{1j})) + \sum_{j=1}^{r_2}(\ln f)''(c(T_{2j})) \\
&\quad + \sum_{j=r_2+1}^{m_2}(\ln f)''(d(T_{2j}))\frac{e^\rho t_1}{e^\rho t_1 + T_{2j} - t_1}
\end{aligned}
$$

$$- (n_2 - m_2)\lambda'(d(t_2))\frac{e^\rho t_1}{(e^\rho t_1 + t_2 - t_1)}\Big\},$$

$$I_{33} = -\frac{\partial^2 \ln L}{\partial \psi^2} = -\alpha^2\left[\sum_{j=1}^{n_1}(\ln f)''(c(T_{1j})) + \sum_{j=1}^{r_2}(\ln f)''(c(T_{2j})) + \sum_{j=r_2+1}^{m_2}(\ln f)''(d(T_{2j})) - (n_2 - m_2)\lambda'(d(t_2))\right].$$

In the case of Weibull, loglogistic and lognormal laws:

$$(\ln f)''(t) = -e^t;\ \frac{-2e^t}{(1+e^t)^2};\ -1,$$

respectively, and

$$\lambda'(t) = e^t;\ \frac{e^t}{(1+e^t)^2};\ -t\frac{\varphi(t)}{1-\Phi(t)} + \left(\frac{\varphi(t)}{1-\Phi(t)}\right)^2,$$

respectively.

If n is large, the law of $(\hat\alpha, \hat\rho, \hat\psi)$ can be approximated by a normal law with mean (α, ρ, ψ) and the covariance matrix estimated by

$$\mathbf{I}^{-1}(\hat\alpha, \hat\rho, \hat\psi) = (I_{ls}(\hat\alpha, \hat\rho, \hat\psi))_{3\times 3}.$$

Denote

$$\hat Q_{x0}(t) = \ln\frac{\hat S_{x0}(t)}{1 - \hat S_{x0}(t)}.$$

The law of $(\hat Q_{x0}(t) - Q_{x0}(t))/\hat\sigma_{Q_0}$ is approximated by the normal law $N(0,1)$; here:

$$\hat\sigma_{Q_0} = \frac{S'(S^{-1}(\hat S_{x0}(t)))}{\hat S_{x0}(t)(1 - \hat S_{x0}(t))} \times\sqrt{(\ln t - \hat\psi)^2 I_{11}(\hat\alpha, \hat\rho, \hat\psi) - 2\hat\alpha(\ln t - \hat\psi)I_{13}(\hat\alpha, \hat\rho, \hat\psi) + \hat\alpha^2 I_{33}(\hat\alpha, \hat\rho, \hat\psi)}\ .$$

Then the $(1-\alpha)$-approximate confidence intervals for $Q_{x0}(t)$ and $S_{x0}(t)$ are $\hat Q_{x0} \pm \hat\sigma_{Q_0} w_{1-\alpha/2}$ and

$$\left(1 + \frac{1 - \hat S_{x0}(t)}{\hat S_{x0}(t)}\exp\{\mp\hat\sigma_{Q_0} w_{1-\alpha/2}\}\right)^{-1},$$

respectively.

Denote

$$\hat K_p(x_0) = \ln\hat t_p(x_0) = \hat\psi + \frac{1}{\hat\alpha}S^{-1}(1-p).$$

The law of $(\hat{K}_p(x_0) - K_p(x_0))/\hat{\sigma}_{K_p}$ is approximated by the normal law $N(0,1)$; here

$$\hat{\sigma}^2_{K_p} = \left(\frac{S^{-1}(1-p)}{\alpha^2}\right)^2 I_{11} - \frac{S^{-1}(1-p)}{\alpha^2} I_{13} + I_{33}.$$

Then the $(1-\alpha)$-approximate confidence intervals for $K_p(x_0) = \ln t_p(x_0)$ and $t_p(x_0)$ are

$$\hat{K}_p(x_0) \pm w_{1-\alpha/2}\hat{\sigma}_{K_p}$$

and

$$\hat{t}_p(x_0)\exp\{\pm\hat{\sigma}_{K_p} w_{1-\alpha/2}\}.$$

6.3 Nonparametric Estimation

Bagdonavičius and Nikulin (1997) considered nonparametric reliability estimation procedure for the plan of experimentation which was discussed in the first section.

Consider the model (6.1) and suppose that the above mentioned data are available, from results of Bagdonavičius and Nikulin we obtain that the estimator of the parameter r can be obtained in the following manner:

$$\hat{r} = \hat{U}^{-1}(0) = \sup\{r : \hat{U}(r) \geq 0\},$$

where:

$$\hat{U}(r) = a_n^{-2}\left\{\sum_{j:T_{1j}>t_1} \frac{Y_2(t_1 + r(T_{1j} - t_1))}{Y_1(T_{1j}) + Y_2(t_1 + r(T_{1j} - t_1))} - \sum_{j:T_{2j}>t_1} \frac{Y_1(t_1 + \frac{T_{2j}-t_1}{r})}{Y_1(t_1 + \frac{T_{2j}-t_1}{r}) + Y_2(T_{2j})}\right\},$$

$$n = n_1 + n_2, \quad a_n = \sqrt{\frac{n_1 n_2}{n}}$$

and

$$Y_1(\tau) = \sum_{j=1}^{n_1} I\{T_{1j} \geq \tau\}, \quad Y_2(\tau) = \left[\sum_{j=1}^{m_2} I(T_{2j} \geq \tau) + n_2 - m_2\right] I(\tau \leq t)$$

are the numbers of items at risk (non-failed and non-censored) just before the moment τ for the items of the first and the second group, respectively.

The function $\hat{U}$ is a decreasing step function of r. If we assume that T_i are absolutely continuous random variables with supports on $[0,\infty]$, we see that $\hat{U}(0) > 0$, $\hat{U}(\infty) < 0$, with the probability 1.

Denote A_{x_0} the failure rate under the stress x_0 and

$$A_{x_0}(t) = \int_0^t \lambda_{x_0}(u)du$$

the cumulated failure rate function. The estimator of S_{x_0} is

$$\hat{S}_0(s) = \hat{S}_{x_0}(s) = e^{-\hat{A}_0(s)},$$

where

$$\begin{aligned}\hat{A}_0(s) &= \sum_{j:\, T_{1j}\le \frac{s}{\hat{r}}} \frac{1}{Y_1(T_{1j}) + Y_2[T_{1j}\wedge t_1 + \hat{r}(T_{1j}-t_1)\vee 0]} \\ &+ \sum_{j:\, T_{2j}\le \frac{s}{\hat{r}}\wedge t_1 + (s-\hat{r}t_1)\vee 0} \frac{1}{Y_1[T_{2j}\wedge t_1 + \frac{T_{2j}-t_1}{\hat{r}}\vee 0] + Y_2(T_{2j})}.\end{aligned}$$

Under some assumptions they showed Bagdonavičius and Nikulin (1997) that

$$a_n(\hat{A}_0(s) - A_0(s)) \xrightarrow{D} N(0,\sigma_s^2), \quad a_n(\hat{S}_0(s) - S_0(s)) \xrightarrow{D} N(0,\sigma_s^2 S_0^2(s)),$$
$$\text{as} \quad n_i \to \infty, \quad n_i/n \to l_i \ (i=1,2),$$

where

$$\begin{aligned}\sigma_s^2 &= \left(\frac{l_1 s\lambda_0(s) + l_2(s\wedge(r_0t_1))\lambda_0(s\wedge(r_0t_1+t-t_1))}{r_0U'(r_0)}\right)^2 \\ &\times (S_0(t_1) - S_0(r_0t_1+t-t_1)) \\ &+ l_1l_2\left(\frac{l_1(1-S_0(s))}{S_0(s)} + \frac{l_2(1-S_0(s\wedge(r_0t_1+t-t_1)))}{S_0(s\wedge(r_0t_1+t-t_1))}\right),\end{aligned}$$

$\lambda_0 = \lambda_{x_0}$, $S_0 = S_{x_0}$, r_0 is the true value of r under the AFT model and

$$U'(r_0) = -(t-t_1)\lambda_0(r_0t_1+t-t_1)S_0(r_0t_1+t-t_1) + \int_0^{t-t_1} v\lambda_0(r_0t_1+v)dS_0(r_0t_1+v).$$

The $(1-\alpha)$-approximate confidence interval for $S_{x_0}(t)$ is

$$\left(1 + \frac{1-\hat{S}_{x_0}(t)}{\hat{S}_{x_0}(t)}\exp\{\mp\hat{\sigma}_s w_{1-\alpha/2}\}\right)^{-1},$$

where $\hat{\sigma}_s$ is obtained by changing S_0 and λ_0 by $\hat{S}_0(s)$ and

$$\hat{\lambda}_0(s) = b^{-1}\int_0^\infty K(\frac{s-u}{b})d\hat{A}_0(u),$$

where K is the kernel function with the window size b. For example, we can use Epanechnikov kernel function

$$K(x) = 0.75(1-x^2)\mathbf{1}_{\{|x|\le 1\}}.$$

We have

$$\hat{\lambda}_0(s) = b^{-1}\left\{\sum_{j=1}^{n_1}\frac{K\left(\frac{s-\hat{r}T_{1j}}{b}\right)}{Y_1(T_{1j}) + Y_2[T_{1j}\wedge t_1 + \hat{r}(T_{1j}-t_1)\vee 0]}\right.$$
$$\left.+\sum_{j=1}^{m_2}\frac{K\left(\frac{s-\hat{r}(T_{2j}\wedge t_1)+T_{2j}-t_1}{b}\right)}{Y_1(T_{2j}\wedge t_1 + \frac{T_{2j}-t_1}{\hat{r}}\vee 0) + Y_2(T_{2j})}\right\}.$$

6.4 Conclusion

It is clear that it's difficult to compare the asymptotic properties of those two estimators. Our goal now is to use numerical simulations to study the efficiency of those two estimators for finite size samples. The first results that we obtain for the Weibull distribution give the idea that even for small samples those estimators are good.

References

1. Bagdonavičius, V. and Nikulin, M. (1995). Semiparametric regression models in accelerated life testing, *Queen's Papers in Pure and Applied Mathematics*, **98**, Kingston, Canada.

2. Bagdonavičius, V. and Nikulin, M. (1997). Sur l'application des stress en escalier dans les expériences accélérées, *Comptes Rendus, Academie des Sciences de Paris*, **325**, Série I, 523–526.

3. Bagdonavičius, V. (1978). Testing the hypothesis of the additive accumulation of damages, *Probability Theory and Its Applications*, **23**, 403–408.

4. Nelson, W. (1990). *Accelerated Testing*, New York: John Wiley & Sons.

5. Singpurwalla, N. D. (1973a). Inference from accelerated life tests when observations are obtained from censored samples, *Technometrics*, **13**, 161–170.

6. Singpurwalla, N. D. (1973b). Inference from accelerated life tests using Arrhenius type re-parametrizations, *Technometrics*, **15**, 289–299.

7. Viertl, R. (1988). Statistical methods in accelerated life testing, Zurich: Vandenhoeck & Ruprecht.

7

Chi-Squared Goodness of Fit Test for Doubly Censored Data With Applications in Survival Analysis and Reliability

M. S. Nikulin and V. N. Solev

Université de Bordeaux, Bordeaux, France
Steklov Mathematical Institute, St. Petersburg, Russia

Abstract: We consider the problem of testing the goodness of fit of a parametric family $F(\cdot;\theta)$ of the survival distributions from some doubly censored data. We investigate Pearson-type chi-squared statistics which compare the Tsai and Growly (1985) estimator $\hat{F}_n(t)$ to the parametric MLE $F(t,\hat{\theta}_n)$.

Keywords and phrases: Chi-squared test, doubly censored data, subdistribution function, survival function

7.1 Introduction

Let X be a nonnegative random variable being interpreted as the lifetime under investigation and $J = [Z, Y]$ be a random interval which is independent from X. We suppose that Z, Y are the nonnegative random variables and $Z \leq Y$ a.s. The censoring mechanism is such that we observe only a random vector $(W, K) = \Gamma$, where

$$W = X \cdot \mathbf{1}_{[Z,Y]}(X) + Z \cdot \mathbf{1}_{[0,Z[}(X) + Y \cdot \mathbf{1}_{]Y,\infty[}(X),$$

$$K = 1 \cdot \mathbf{1}_{[Z,Y]}(X) + 2 \cdot \mathbf{1}_{[0,Z[}(X) + 3 \cdot \mathbf{1}_{]Y,\infty[}(X).$$

It is a usual situation in survival analysis where we observe an interval censored data. We note H, the distribution function of W, which is a nonnegative random variable and K is entire nonnegative random variable with conditional probability:

$$\pi_i(t) = \pi(t,i) = \mathbf{P}\{\mathbf{K} = \mathbf{i} \mid \mathbf{W} = \mathbf{t}\}, \qquad (\mathbf{i} = \mathbf{1}, \mathbf{2}, \mathbf{3}).$$

We denote by $\mathbf{1}_M(\cdot)$ the indicator function of a set M. Let $F_X(\cdot), F_Y(\cdot), F_Z(\cdot)$ be the distribution functions and $S_X(\cdot)$, $S_Y(\cdot)$, $S_Z(\cdot)$ be the survival functions of random variables X, Y, Z, respectively.

We assume (hypothesis H_0) that the distribution function of X belongs to a family of absolutely continuous functions $\{F_X(t;\theta),\ \theta \in \Theta\}$ where Θ is an open subset in the space R^d and denote by $p_X(t;\theta)$ the distribution density of X. We suppose that distributions of Z and Y are absolutely continuous with respect to Lebesgue measure and denote by

$$p_Y(t) = \frac{dF_Y(t)}{dt}, \quad p_Z(t) = \frac{dF_Z(t)}{dt}$$

the corresponding distribution densities.

In our case under H_0

$$H(t;\theta) = [F_Z(t) - F_Y(t)]F_X(t;\theta) + F_Y(t), \quad \theta = (\theta_1, ..., \theta_d) \in \Theta \subset R^d, \quad (7.1)$$

and the random vector Γ has the distribution $\mathbf{P}_\Gamma$ with density function $\pi(t,i) = \pi(t,i;\theta)$ with respect to the measure $dH \times d\mu$, where $d\mu$ is a counting measure on $\{1,2,3\}$

$$\pi(t,i;\theta) = \begin{cases} \frac{1}{h(t;\theta)} \cdot (F_Z(t) - F_Y(t)) \cdot p_X(t;\theta), & \text{if} \quad i = 1, \quad t \in \operatorname{supp} dH, \\ \frac{1}{h(t;\theta)} \cdot F_X(t;\theta) \cdot p_Z(t), & \text{if} \quad i = 2, \quad t \in \operatorname{supp} dH, \\ \frac{1}{h(t;\theta)} \cdot (1 - F_X(t;\theta)) \cdot p_Y(t), & \text{if} \quad i = 3, \quad t \in \operatorname{supp} dH, \end{cases} \quad (7.2)$$

where

$$h(t;\theta) = \frac{dH(t;\theta)}{dt}.$$

That is under H_0 the distribution $\mathbf{P}_\Gamma$ of random vector Γ has density function $p(t,i;\theta)$ with respect to the product of Lebesgue measure and counting measure μ :

$$p(t,i;\theta) = \begin{cases} [F_Z(t) - F_Y(t)] \cdot p_X(t;\theta), & \text{if} \quad i = 1, \\ F_X(t;\theta) \cdot p_Z(t), & \text{if} \quad i = 2, \\ (1 - F_X(t;\theta)) \cdot p_Y(t), & \text{if} \quad i = 3. \end{cases} \quad (7.3)$$

We suppose that $\Gamma_1 = (W_1, K_1), ..., \Gamma_n = (W_n, K_n)$ is an independent sample from the distribution $\mathbf{P}_\Gamma$ generated by $X_1, \ldots, X_n, J_1, \ldots, J_n$. Our goal is to estimate the distribution function $F_X(t) = F_X(t;\theta)$ of X using the data $\Gamma_1, \Gamma_2, ..., \Gamma_n$, $\Gamma_i \sim p(t,i;\theta)$.

Let $\hat{\theta}_n$ be MLE of the parameter θ based on a random sample $\Gamma_1, \Gamma_2, ..., \Gamma_n$ from the distribution with the density function $p(t,i;\theta)$. We construct the Tsai and Growly (1985) estimator $F_X^n(t)$ as a solution of the integral equation

$$F_X^n(t) = H^n(t) + F_X^n(t) \int_{u>t} \frac{dH_Z^n(u)}{F_X^n(u)} - (1 - F_X^n(t)) \int_{u \le t} \frac{dH_Y^n(u)}{1 - F_X^n(u)}, \quad (7.4)$$

where $H^n(t), H_Z^n(t), H_Y^n(t)$ are empirical versions of functions $H(t), H_Z(t), H_Y(t)$ and *subdistribution functions* $H_Z(t), H_Y(t)$ are defined by the relations

$$H_Z(t) = \int_{u \leq t} F_X(t;\theta)\, dF_Z(t), \quad H_Y(t) = \int_{u \leq t} (1 - F_X(t;\theta))\, dF_Y(t).$$

We put

$$U_n(t) = \sqrt{n}[F_X^n(t) - F_X(t,\theta)].$$

Gu and Zhang (1993) established under some regularity conditions the weak convergence of the process $U_n(t)$ to a Gaussian process. We shall analyze the asymptotic behaviour (when $n \to \infty$) of the process

$$U_n^\star(t) = \sqrt{n}[F_X^n(t) - F_X(t,\hat{\theta}_n)].$$

At first we shall prove that the process $U_n^\star(t)$ converges weakly (when $n \to \infty$) to a Gaussian process under the null hypothesis

$$H_0 : F_X(t) \in \{F_X(t;\theta), \theta \in \Theta\}.$$

Then we shall investigate the limiting behavior of the distribution of the Pearson-type chi-squared statistic which measure the closeness of process $F_X^n(t)$ and $F_X(t,\hat{\theta}_n)$ when n is large.

7.2 Weak Convergence of the Process $U_n(t)$

We use a notation:

$$\begin{aligned}
H_X(t) &= H_1(t) = \int_{u \leq t} (F_Z(u) - F_Y(u)) dF_X(u), \\
H_Z(t) &= H_2(t) = \int_{u \leq t} F_X(u) dF_Z(u), \\
H_Y(t) &= H_3(t) = \int_{u \leq t} (1 - F_X(u)) dF_Y(u), \\
p_j &= \mathbf{P}\{K = j\}, \quad q_j = 1 - p_j, \quad (j = 1, 2, 3).
\end{aligned}$$

It is obvious that the distribution functions $F_j(t)$ of random variables

$$W^j = W \cdot \mathbf{1}_{\{j\}}(K)$$

are defined by

$$F_j(t) = q_j + H_j(t), \quad (j = 1, 2, 3).$$

The distribution function $H(t) = \mathbf{P}\{W \leq t\}$ of random variable W is defined by

$$H(t) = H_1(t) + H_2(t) + H_3(t) = F_X(t)F_Z(t) + (1 - F_X(t))F_Y(t).$$

We define subsurvival functions $Q_j(t)$ by the relations

$$Q_j(t) = p_j - \int_{u \le t} dH_j(t), \quad (j = 1, 2, 3).$$

It is clear that the distribution functions F_X, F_Y, F_Z satisfy to the system of integral equations

$$\begin{cases} Q_1(t) = \int_{u>t}(F_Z(u) - F_Y(u))dF_X(u), \\ Q_2(t) = \int_{u>t} F_X(u)dF_Z(u), \\ Q_3(t) = \int_{u>t}(1 - F_X(u))dF_Y(u), \end{cases} \tag{7.5}$$

which is equivalent to the system

$$\begin{cases} H_1(t) = \int_{u \le t}(F_Z(u) - F_Y(u))dF_X(u), \\ H_2(t) = \int_{u \le t} F_X(u)dF_Z(u), \\ H_3(t) = \int_{u \le t}(1 - F_X(u))dF_Y(u). \end{cases} \tag{7.6}$$

The subsurvival functions $Q_j(t)$, $(j = 1, 2, 3)$, can be estimated by the empirical subsurvival functions $Q_j^{(n)}(t)$ defined by

$$Q_j^{(n)}(t) = \frac{1}{n}\sum_{i=1}^{n} \mathbf{1}_{[t,\infty]}(W_i) \cdot \mathbf{1}_{\{j\}}(K_i), \quad (j = 1, 2, 3).$$

It is evident that the empirical distribution function

$$H^{(n)}(t) = \frac{1}{n}\sum_{i=1}^{n} \mathbf{1}_{[0,t]}(W_i),$$

which estimates the distribution function $H(t)$, is connected with the empirical subdistribution functions

$$H_j^{(n)}(t) = p_j^{(n)} - Q_j^{(n)}(t)$$

by the relation

$$H^{(n)}(t) = H_1^{(n)}(t) + H_2^{(n)}(t) + H_3^{(n)}(t),$$

where

$$p_j^{(n)} = \frac{1}{n}\sum_{i=1}^{n} \mathbf{1}_{\{j\}}(K_i)$$

is the empirical estimator for p_j. We suppose that

$$F_Z(u) - F_Y(u) > 0, \quad (dF_X \ a.s.). \tag{7.7}$$

In this case

$$F_X(t) = \int_{u \le t} \frac{1}{F_Z(u) - F_Y(u)} dH_1(t). \tag{7.8}$$

We suppose also that

$$F_X(t) > 0, \quad (dF_Z\ a.s.), \quad 1 - F_X(t) > 0, \quad (dF_Y\ a.s.). \tag{7.9}$$

In this case from (7.6) it follows

$$F_Z(t) = \int_{u \le t} \frac{1}{F_X(u)} dH_2(u), F_Y(t) = \int_{u \le t} \frac{1}{1 - F_X(u)} dH_3(u). \tag{7.10}$$

It is easily to see that function $F_X(t)$ satisfies to the next equation

$$F_X(t) \int_{u>t} \frac{dH_2(u)}{F_X(u)} + (1 - F_X(t)) \int_{u \le t} \frac{dH_3(u)}{1 - F_X(u)} = H(t). \tag{7.11}$$

This equation can be rewritten in another form

$$F_X(t) = H(t) + F_X(t) \int_{u>t} \frac{dH_2(u)}{F_X(u)} - (1 - F_X(t)) \int_{u \le t} \frac{dH_3(u)}{1 - F_X(u)}. \tag{7.12}$$

It should be noted that a solution the equation (7.11) satisfies to the equation (7.12) only in the case when

$$\int_{u \ge 0} \frac{dH_2(u)}{F_X(u)} = 1.$$

The equation (7.12) can be rewritten in terms of the survival functions:

$$S_X(t) = Q(t) - S_X(t) \int_{u \le t} \frac{dQ_3(u)}{1 - S_X(u)} + (1 - S_X(t)) \int_{t<u} \frac{dQ_2(u)}{1 - S_X(u)},$$

where $Q(t) = 1 - H(t)$.

Tsai and Growly (1985) suggest to exploit the solution $S_X^{(n)}$ of this equation, when the functions Q, Q_2, Q_3 are replaced by their empirical versions $Q^{(n)}$, $Q_2^{(n)}$, $Q_3^{(n)}$:

$$S_X^{(n)}(t) = S^{(n)}(t) - S_X^{(n)}(t) \int_{u \le t} \frac{dQ_3^{(n)}(u)}{S_X^{(n)}(u)} + (1 - S_X^{(n)}(t)) \int_{t<u} \frac{dQ_2^{(n)}(u)}{1 - S_X^{(n)}(u)}.$$

Chang and Jang (1987), Chang (1990), Gu and Zhang (1993) established strong uniform consistency of the estimator $S_X^{(n)}$:

$$\|S_X^{(n)}(\cdot) - S_X(\cdot)\| \to 0 \quad a.s.$$

under the assumption (7.7). They proved also that $\sqrt{n}(S_X^{(n)} - S_X)$ converges weakly to a Gaussian process and computed the asymptotic variance of this process.

Denote

$$
\begin{aligned}
\Gamma_1(t,s) &= -\frac{\mathbf{1}_{[0,t]}(s)}{S_Y(s)-S_Z(s)}, \quad \Gamma_2(t,s) = \frac{1}{S_X(s)}\int_t^s \frac{dS_X}{S_Y-S_Z},\\
\Gamma_3(t,s) &= \frac{\mathbf{1}_{[0,t]}(s)}{1-S_X(s)}\int_0^{s\wedge t}\frac{dS_X}{S_Y-S_X},\\
g(t,s,ds) &= \frac{\mathbf{1}_{[0,t]}(s)dS_Y(s)}{S_X(s)}\int_t^s \frac{dS_X}{S_Y-S_Z}+\frac{dS_Z(s)}{1-S_X(s)}\int_0^{s\wedge t}\frac{dS_X}{S_Y-S_Z}.
\end{aligned}
$$

Let the functions $C_j(t,u)$ $(j=1,2,3)$ be the solutions of the integral equation

$$
C_j(t,u)=\Gamma_j(t,u)+\int_0^T g(t,s,ds)C_j(s,u), \quad (j=1,2,3).
$$

Chang (1990) proved that the asymptotic variance of the process $\sqrt{n}(S_X^{(n)}-S_X)$ has a form

$$
\begin{aligned}
&\int_0^t C_1^2(t,u)dF_1(u)+\int_0^t C_2^2(t,u)dF_2(u)+\int_0^t C_3^2(t,u)dF_3(u)\\
&\quad -\left\{\int_0^t [C_1(t,u)dF_1(u)+C_2(t,u)dF_2(u)+C_3(t,u)dF_3(u)]\right\}^2.
\end{aligned}
$$

Gu and Zhang (1993) established also the asymptotic efficiency of the estimator $S_X^{(n)}$ under mild conditions on the distributions of the censoring variables.

For the distribution function F_X we define [see details in Gu and Zhang (1993)] the linear operators A_F and B_F by the relations

$$
\begin{aligned}
\Big(A_F h\Big)(t) &= -\int_{u\le t}\frac{1-F(t)}{1-F(u)}h(u)\,dF_Y(u)+\int_{u>t}\frac{F(t)}{F(u)}h(u)\,dF_Z(u),\\
\Big(B_F g\Big)(t) &= g_1(t)+g_2(t)+g_3(t)-\int_{u\le t}\frac{1-F(t)}{1-F(u)}\,dg_3(u)+\int_{u>t}\frac{F(t)}{F(u)}\,dg_2(u),
\end{aligned}
$$

where $g(t)=(g_1(t),g_2(t),g_3(t))^T$.

Let us consider the linear operator R_F be defined by

$$
(R_F h)(t)=[F_Z(t)-F_Y(t)]\cdot h(t)+(A_F h)(t).
$$

Let $[a,b]$ be the smallest interval which contains supp dF_X. Define Banach space $D[a,b]$ (with the supremum norm $||\cdot||$) of all real-valued functions defined on $[a,b]$, which are right continuous and have left limits at $x\in]a,b]$ and right limit at a. We put

$$
\begin{aligned}
D_0[a,b] &= \{h:\ h\in D[a,b], h(b-)=0, h(a)=0\},\\
D_\star[a,b] &= \{h:\ (F_Z-F_Y)\cdot h\in D[a,b]\}, ||h||_\star = ||\,(F_Z-F_Y)\cdot h||,\\
D_0^{(3)} &= \{g:\ g(t)=(g_1(t),g_2(t),g_3(t))^T,\ g_j\in D[a,b],\\
&\qquad (j=1,2,3),\ B_{F_X}\,g\in D_0[a,b]\},\\
||g||_3 &= ||g_1||+||g_2||+||g_3||.
\end{aligned}
$$

Since the vector-function $g^n(t) = (H_X^n(t), H_Z^n(t), H_Y^n(t))^T$ is the empirical version of $g(t) = (H_X(t), H_Z(t), H_Y(t))^T$ it follows [see Billingsley (1977)] that the vector-valued process

$$G_n(t) = \sqrt{n}[g^n(t) - g(t)](t).$$

converges to some Gaussian process $G(t) = (G^1, G^2, G^3)^T(t)$ with

$$\mathbf{E}G^i(t) = 0, \text{ } and \text{ } \mathbf{E}G^i(t)G^j(s) = \delta_{ij}H_i(t \wedge s) - H_i(t)H_j(s), \ (i = 1, 2, 3). \tag{7.13}$$

Gu and Zhang proved under some regularity conditions that random process $U_n(t)$ weak converges in the space $D_\star[a, b]$ to the Gaussian process

$$U(t) = R_{F_X}^{-1} B_{F_X} G(t).$$

Solev (1997) generalized this result for stationary case.

7.3 The Weak Convergence of the Process $U_n^\star(t)$

Let $J = J(\theta; F_Z, F_Y)$ be the information matrix $J = ||J_{ij}||$,

$$\begin{aligned} J_{ij} = & -\int_0^\infty \frac{\partial^2 \ln p_X(t;\theta)}{\partial\theta_i \partial\theta_j} (F_Z(t) - F_Y(t))\, p_X(t;\hat{\theta})\, dt \\ & -\int_0^\infty \frac{\partial^2 \ln F_X(t;\theta)}{\partial\theta_i \partial\theta_j} F_X(t;\theta) p_Z(t)\, dt \\ & +\int_0^\infty \frac{\partial^2 \ln F_X(t;\theta)}{\partial\theta_i \partial\theta_j} (1 - F_X(t;\theta))\, p_Y(t)\, dt \ . \end{aligned}$$

We assume that MLE $\hat{\theta}_n$ exists and there is a decomposition

$$\sqrt{n}\left(\hat{\theta}_n - \theta\right) = J^{-1} V_n + o_{\mathbf{P}}(1), \tag{7.14}$$

where V_n is the normalized score vector

$$V_n = \frac{1}{\sqrt{n}} \sum_{i=1}^n \frac{\partial}{\partial\theta} \ln p(W_i, K_i; \theta). \tag{7.15}$$

We put

$$U_n^*(t) = \left(\frac{\partial}{\partial\theta} F_X(t;\theta)\right)^T \sqrt{n}(\hat{\theta}_n - \theta).$$

For random vector V_n there is a representation

$$\begin{aligned} V_n &= \sqrt{n}\biggl(\int_0^\infty \frac{\partial}{\partial\theta}\ln p_X(u;\theta)\,dH_X^n(u) + \int_0^\infty \frac{\partial}{\partial\theta}\ln F_X(u;\theta)\,dH_Z^n(u) \\ &\quad + \int_0^\infty \frac{\partial}{\partial\theta}\ln\,(1-F_X(u;\theta))\,dH_Y^n(u)\biggr). \end{aligned}$$

Since

$$\begin{aligned} \mathbf{0}_d &= \mathbf{E}_\theta V_n = \sqrt{n}\biggl(\int_0^\infty \frac{\partial}{\partial\theta}\ln p_X(u;\theta)\,dH_X(u) \\ &\quad + \int_0^\infty \frac{\partial}{\partial\theta}\ln F_X(u;\theta)\,dH_Z(u) + \int_0^\infty \frac{\partial}{\partial\theta}\ln\,(1-F_X(u;\theta))\,dH_Y(u)\biggr) \end{aligned}$$

we have

$$\begin{aligned} V_n &= \int_0^\infty \frac{\partial}{\partial\theta}\ln p_X(u;\theta)\,dG_n^1(u) \\ &\quad + \int_0^\infty \frac{\partial}{\partial\theta}\ln F_X(u;\theta)\,dG_n^2(u) + \int_0^\infty \frac{\partial}{\partial\theta}\ln\,(1-F_X(u;\theta))\,dG_n^3(u), \end{aligned} \tag{7.16}$$

where

$$\begin{aligned} G_n^1(t) &= \sqrt{n}[H_X^n(t)-H_X(t)], \quad G_n^2(t)=\sqrt{n}[H_Z^n(t)-H_Z(t)], \\ G_n^3(t) &= \sqrt{n}[(H_Y^n(t)-H_Y(t)]. \end{aligned}$$

So we can rewrite the coordinates of the random vector $V_n = (V_n^1,...,V_n^d)^T$ in the form

$$V_n^j = \int_0^\infty \frac{\partial}{\partial\theta_j}\left[\ln p(u;\theta)\right]^T\,dG_n(t), \qquad (j=1,...,d), \tag{7.17}$$

where

$$p(u;\theta) = (p(u,1;\theta), p(u,2;\theta), p(u,3;\theta))'.$$

From the result of Gu and Zhang (1993) it follows that

$$U_n(t) = R_{F_X^n}^{-1} B_{F_X^n}\, G_n(t).$$

Therefore it is easily to see from (7.17) that vector-valued process

$$\Phi_n(t) = (\Phi_n^1(t), \Phi_n^2)^T = (U_n(t), V_n)^T$$

weakly converges (when $n\to\infty$) to a Gaussian process

$$\Phi(t) = (\Phi^1(t), \Phi^2 t)^T = (U(t), V)^T,$$

where $U(t) \; = \; R_{F_X}^{-1} B_{F_X} \; G(t), \quad V = (V^1, ..., V^d)^T,$

$$V^j \; = \; \int_0^\infty \frac{\partial}{\partial \theta_j} [p(u;\theta)]^T \; dG(t), \quad (j = 1, ..., d). \tag{7.18}$$

There is the expansion

$$U_n^\star(t) \; = \; U_n(t) \; + \; U_n^*(t) \; + \; R_n(t), \tag{7.19}$$

where $R_n(t) \to 0$ when $n \to \infty$ in probability uniformly in t. So we conclude that the random process $U_n^\star(t)$ also weakly converges (when $n \to \infty$) to a Gaussian process $U^\star(t)$ and

$$U^\star(t) = R_{F_X}^{-1} B_{F_X} \; G(t) \; + \; \left[\frac{\partial}{\partial \theta} F_X(t;\theta)\right]^T \; J^{-1} \; \int_0^\infty \left[\frac{\partial}{\partial \theta} p(u;\theta)\right]^T \; dG(t), \tag{7.20}$$

where $G(t)$ is the Gaussian process which defined in (13). It would be noted that for simplicity we denote by $\frac{\partial}{\partial \theta} p(u;\theta)$ the matrix $||c_{ij}||$,

$$c_{ij} \; = \; \frac{\partial}{\partial \theta_j} p(u, i; \theta).$$

7.4 The Test Statistics

For a partition

$$0 \; < \; t_1 \; < \; t_2 \; < ... < t_r \; < \; T$$

we consider two random vectors

$$U_n \; = \; (U_n(t_1), ..., U_n(t_r))^T, \quad U_n^\star \; = \; (U_n^\star(t_1), ..., U_n^\star(t_r))^T.$$

We put

$$\mathbf{\Sigma}_n \; = \; \mathbf{Var} U_n, \qquad \mathbf{\Sigma}_n^\star \; = \; \mathbf{Var} U_n^\star.$$

It is clear that

$$\mathbf{\Sigma}_n^\star \; = \mathbf{\Sigma}_n \; - \; B' J^{-1} B, \qquad \text{where} \qquad B = ||b_{ij}||, \qquad b_{ij} = \frac{\partial}{\partial \theta_j} F_X(t_i;\theta).$$

Let $\hat{\mathbf{\Sigma}}_n, \hat{\mathbf{\Sigma}}_n^\star$ be the estimators for the covariance matrices $\mathbf{\Sigma}_n, \mathbf{\Sigma}_n^\star$ respectively which are obtained by replacing θ by MLE $\hat{\theta}_n$. We consider the Pearson-type statistics

$$X_n^2 \; = \; U_n^T \mathbf{\Sigma}_n^{-1} U_n, \qquad Y_n^2 \; = \; [U_n^\star]^T \, [\mathbf{\Sigma}_n^\star]^{-1} \, U_n.$$

The investigation of the asymptotic behavior of the distributions of this statistics may be carry on by the usual way. The details of such investigations were considered explicitly in Greenwood and Nikulin (1996), Nikulin (1973). Under the condition that $\mathbf{\Sigma}_n^\star$ is of rank r, the statistic Y_n^2 has a limiting χ_r^2 distribution, when $n \to \infty$.

Acknowledgement. This research was supported in part by INTAS-Kazakhstan grant 93-53 and by Russian Fond of Fundamental Research RAN.

References

1. Billingsley, P. (1977). *Convergence of Probability Measures*, New York: John Wiley & Sons.

2. Breslow, N. E. and Crowley, J. J. (1974). A large sample study of the life table and product limit estimates under random censorship, *Annals of Statistics*, **2**, 437–453.

3. Chang, M. N. and Yang, G. L. (1987). Strong consistency of a nonparametric estimator of the survival function with doubly censored data, *Annals of Statistics*, **15**, 1536–1547.

4. Chang, M. N. (1990). Weak convergence of self-consistent estimator of the survival function with doubly censored data, *Annals of Statistics*, **18**, 391–404.

5. Gu, M. G. and Zhang, C. H. (1993). Asymptotic properties of self-consistent estimators based on doubly censored data, *Annals of Statistics*, **21**, 611–624.

6. Greenwood, P. E. and Nikulin, M. S. (1996). *A Guide to Chi-Squared Testing*, New York: John Wiley & Sons.

7. Nikulin, M. S. (1973). On a chi-squared test for continuous distributions, *Theory of Probability and its Applications*, **18**, 638–639.

8. Nikulin, M. S. (1973). A chi-square test for continuous distributions with location and scale parameters, *Theory of Probability and its Applications*, **18**, 559–568.

9. Solev, V. N. (1997). Limiting behavior of some functionals of empirical processes based upon doubly censored data, *ESAIM P & S* (to appear).

10. Tsai, W. J. and Growlew, J. (1985). A large sample study of generalized maximum likelihood estimators from incomplete data via self-consistency, *Annals of Statistics*, **13**, 1317–1334.

8

Estimation of Kernel, Availability and Reliability of Semi-Markov Systems

B. Ouhbi and N. Limnios

Université de Technologie de Compiègne, Compiègne, France

Abstract: In this paper, we propose a new estimator for an absolutely continuous semi-Markov kernel based on its transition hazard rate estimator. We also give rigorous derivations of properties as strong consistency and weak convergence of this estimator. Estimators of Markov renewal matrix and semi-Markov transition matrix are also given as well as their asymptotic properties. From this estimator, we obtain estimators of reliability and availability and show that these ones are also strongly consistent.

Keywords and phrases: Semi-Markov process, semi-Markov kernel, Markov renewal matrix, semi-Markov transition matrix, availability, reliability

8.1 Introduction

Semi-Markov (SM) or Markov renewal processes (MRP) are a class of stochastic processes which generalize Markov jump processes as well as the renewal process, and can be applied in different situations (Weibull, Erlang, etc. distributions), and allow the computation of a large number of indicators. A fundamental question in semi-Markov problems is to know the semi-Markov kernel. There are many papers which have given estimators of the semi-Markov kernel. The aims of the present paper is to introduce a maximum likelihood estimator of SM kernels which have desirable asymptotic properties and do not present inconveniences when the censored time is little or the sample size is small. The central role here is played by the SM kernel hazard rate. In Ouhbi and Limnios (1996a), hazard rate associated to a SM kernel is estimated and its asymptotic properties are also given. This paper is devoted to estimate and to give asymptotic properties of SM kernel. An application of the previous results to the reliability analysis is done, we introduce estimators of reliability and availability of SM systems and show that these estimators are uniformly strongly consistent. Finally, we give a numerical application of our results to a three state semi-Markov system.

Let us consider a Markov renewal process, $(J, S) = (J_n, S_n)_{n \geq 0}$, defined on a probability complete space, where $(J_n)_{n\geq 0}$ is a Markov chain with values in E$= \{1, ..., s\}$, which is the state space of the process, and $(S_n)_{n\geq 0}$ are the jump times which will be in $\overline{\mathbb{R}}_+ = [0, \infty]$. J_0, J_1, ..., J_n,... are the consecutive states to be visited by the MRP and $X_1, X_2, ...$ defined by $X_n = S_n - S_{n-1}$, for $n \geq 1$, are the sojourn times in these states taking values in $\overline{\mathbb{R}}_+$.
A MRP can be completely determined if we know its initial law and its semi-Markov kernel (which we shall estimate) defined respectively by: $P(J_0 = k) = p(k)$ and

$$P(J_{n+1} = k, X_{n+1} \leq x | J_0, J_1, ..., J_n, X_1, X_2, ..., X_n) = Q_{J_n k}(x) \text{ a.s.},$$

for all $x \in \overline{\mathbb{R}}_+$ and $1 \leq k \leq s$. The probabilities

$$p_{ij} = Q_{ij}(\infty) \ \ (= \lim_{t \to +\infty} Q_{ij}(t))$$

are the transition probabilities of the Markov chain $(J_n)_{n\geq 0}$.
Suppose that $Q_{ij}(.)$ is absolute continuous with respect to the Lebesgue's measure and let $q_{ij}(.)$ to be its density and $f_{ij}(.)$ to be the density of $F_{ij}(.)$. We define, for all i and j in E, the instantaneous transition rate, $\lambda_{ij}(t)$, of a semi-Markov kernel by:

$$\lambda_{ij}(t) = \begin{cases} \frac{\frac{dQ_{ij}(t)}{dt}}{1 - \sum_{j\neq i} Q_{ij}(t)} & \text{if } p_{ij} > 0 \text{ and } \sum_{j\neq i} Q_{ij}(t) < 1 \\ 0 & \text{otherwise.} \end{cases}$$

Let us define also the cumulative hazard rate from state i to state j at time t by $\Lambda_{ij}(t) = \int_0^t \lambda_{ij}(u)du$ and the total cumulative hazard rate of state i at time t by $\Lambda_i(t) = \Sigma_{j=1}^s \Lambda_{ij}(t)$.

For i, $j \in E$ and $t \in [0, \infty)$ we have:

$$Q_{ij}(t) = \int_0^t \exp(-\Lambda_i(u))\lambda_{ij}(u)du.$$

Consider a regular subdivision $(v_k)_{0\leq k\leq M}$ of [0,T], with step $\Delta_T = \frac{T}{M}$ where M is a positive integer and approximate $\lambda_{ij}(t)$ by

$$\lambda^*_{ij}(t) = \sum_{k=0}^{M-1} \lambda_{ijk} 1_{[v_k, v_{k+1})}(t).$$

The maximum likelihood of λ_{ijk} is given by, cf. Ouhbi and Limnios (1996a),

$$\hat{\lambda}_{ijk} = \begin{cases} d_{ijk}/\nu_{ik} & \text{if } \nu_{ik} > 0 \\ 0 & \text{otherwise,} \end{cases}$$

where ν_{ik} is the sojourn time in state i on the interval time I_k, given for $N_T \geq 1$ by:

$$\nu_{ik} = \sum_{l=0}^{N_T-1} (X_{l+1} \wedge v_{k+1} - v_k) 1_{\{J_l=i, X_{l+1} \geq v_k\}} + (U_T \wedge v_{k+1} - v_k) 1_{\{J_{N_T}=i, U_T \geq v_k\}},$$

and

$$d_{ijk} = \sum_{l=0}^{N_T-1} 1_{\{J_l=i, J_{l+1}=j, X_{l+1} \in I_k\}}$$

is the number of transitions to state j from state i and for which the observed sojourn time in state i belongs to I_k.

8.2 Estimator of the Semi-Markov Kernel

In this section we will study the asymptotic properties of the semi-Markov kernel estimator given by:

$$\begin{aligned} \hat{Q}_{ij}(t, T) &= \int_0^t e^{-\hat{\Lambda}_i(s,T)} \hat{\lambda}_{ij}(s, T) ds \\ &= \Delta \sum_{\{k: 0 \leq v_k \leq t\}} e^{-\hat{\Lambda}_{ik}} \hat{\lambda}_{ijk} \end{aligned} \tag{8.1}$$

The method adopted here is to prove that the SM kernel estimator obtained from PEXE (piece wise exponential estimator) of the hazard rate functions is asymptotically uniformly equivalent to the empirical estimator $Q_{ij}^N(\cdot)$.

Theorem 8.2.1 *If $0 < \alpha < 1$ and all hazard rate functions $\lambda_{ij}(t)$, $i \in E$ and $j \in E$, are continuous, then the estimator of the semi-Markov kernel given by (8.1) is uniformly strongly consistent in the sense that, when T tends to infinity,*

$$\max_{i,j} \sup_{t \in [0,T]} |\hat{Q}_{ij}(t, T) - Q_{ij}(t)| \longrightarrow 0 \qquad a.s.$$

PROOF. Let i and j be two states. Since,

$$\begin{aligned} \hat{Q}_{ij}(t, T) - Q_{ij}^N(t) &= \sum_{\{k: 0 < v_k \leq t\}} \frac{e^{-\hat{\Lambda}_{ik}}}{\frac{\nu_{ik}}{N_i \Delta}} \frac{d_{ijk}}{N_i} - Q_{ij}^N(t) \\ &= \sum_{\{k: 0 < v_k \leq t\}} \left\{ \frac{e^{-\hat{\Lambda}_{ik}}}{\frac{\nu_{ik}}{N_i \Delta}} - 1 \right\} \frac{d_{ijk}}{N_i}. \end{aligned} \tag{8.2}$$

To prove the uniform strong consistency of the semi-Markov kernel, it is sufficient to prove that,

$$\sup_{t\in[0,T]}\left|\exp(-\hat{\Lambda}_i(t,T)) - \frac{\nu_i(t)}{N_i\Delta}\right| = O(N_i^{-1}).$$

To do this, note that,

$$\begin{aligned}\sup_{t\in[0,T]}\left|\exp(-\hat{\Lambda}_i(t)) - \frac{\nu_i(t)}{N_i\Delta}\right| &\le \sup_{t\in[0,T]}\left|\exp(-\hat{\Lambda}_i(t)) - \exp(-\Lambda_i(t))\right| \\ &+ \sup_{t\in[0,T]}\left|\exp(-\Lambda_i(t) - \frac{\nu_i(t)}{N_i\Delta}\right|\end{aligned}$$

From Ouhbi and Limnios (1996b), it is clear that the second term on the r.h.s. is a $O(N_i^{-1})$.

For the first term, we remark that,

$$\begin{aligned}\left|\hat{\Lambda}_i(t) - \Lambda_i(t)\right| &= \left|\int_0^t \frac{d\hat{H}_i(u,T)}{1-\hat{H}_i(u,T)} - \int_0^t \frac{dH_i(u)}{1-H_i(u)}\right| \\ &= \left|\int_0^t \frac{(\sum_{n=0}^{\infty}(\frac{\hat{H}_i(u,T)-H_i(u)}{1-H_i(u)})^n)d\hat{H}_i(u,T)}{1-H_i(u)} - \int_0^t \frac{dH_i(u)}{1-H_i(u)}\right| \\ &\le \left|\int_0^t \frac{d[\hat{H}_i(u,T)-H_i(u)]}{1-H_i(u)}\right| \\ &+ \left|\int_0^t \frac{(\sum_{n=1}^{\infty}(\frac{\hat{H}_i(u,T)-H_i(u)}{1-H_i(u)})^n)d\hat{H}_i(u,T)}{1-H_i(u)}\right| \\ &\le \left|\frac{\hat{H}_i(t,T)-H_i(t)}{1-H_i(t)}\right| + \left|\int_0^t [\hat{H}_i(u,T)-H_i(u)]\frac{dH_i(u)}{[1-H_i(u)]^2}\right| \\ &+ \left|\int_0^t \frac{(\sum_{n=2}^{\infty}(\frac{\hat{H}_i(u,T)-H_i(u)}{1-H_i(u)})^n)d\hat{H}_i(u,T)}{1-H_i(u)}\right| \\ &= O(N_i^{-1}),\end{aligned}$$

Then from the previous lemma, we conclude that the estimator of the semi-Markov kernel is asymptotically uniformly a.s. equivalent to the empirical estimator for semi-Markov kernel for which the uniform strong consistence was proved in Ouhbi and Limnios (1996b). ■

Theorem 8.2.2 *For all $i,\ j\in E$ and $t\in[0,T]$ fixed, $N_i^{1/2}(\hat{Q}_{ij}(t,T) - Q_{ij}(t))$ converges in law, as T tends to infinity, to a zero mean normal random variable with variance $Q_{ij}(t)(1-Q_{ij}(t))$.*

PROOF. To prove the weak convergence of the MLE of the semi-Markov kernel, we see from (8.2) that,

$$\sqrt{N_i}\hat{Q}_{ij}(t,T) = \sqrt{N_i}Q_{ij}^N(t) + O(N_i^{-1/2}),$$

So, $\sqrt{N_i}[\hat{Q}_{ij}(t,T) - Q_{ij}(t)]$ has the same limit in law, as T tends to infinity, as

$$\frac{\sum_{l=1}^N [1_{\{J_l=j, X_{ijl}\le t\}} - Q_{ij}(t)]1_{\{J_{l-1}=i\}}}{\sqrt{N_i}}.$$

The result is then a consequence of the central limit theorem [cf. Pyke and Schaufelle (1964)], applied to the function $f(r,s,x) = [1_{\{s=j,x\le t\}} - Q_{ij}(t)]1_{\{r=i\}}$.

■

8.3 Estimation of the Markov Renewal Matrix and Its Asymptotic Properties

The Markov renewal matrix is of considerable importance when studying the behaviour of MRP. The purpose of this section is to give an estimator of this matrix and its asymptotic properties.

Let $\psi(t)$ be the Markov renewal matrix defined by

$$\psi(t) = \sum_{l=0}^{\infty} Q^{(l)}(t),$$

where

$$Q_{ij}^{(0)}(t) = \begin{cases} 1 & \text{if } i=j \text{ and } t>0 \\ 0 & \text{otherwise,} \end{cases}$$

$$Q_{ij}^{(1)}(t) = Q_{ij}(t)$$

and for all $n > 1$,

$$Q_{ij}^{(n)}(t) = \begin{cases} \sum_{k=1}^{s} \int_0^t Q_{kj}^{(n-1)}(t-u) Q_{ik}(du) & \text{if } t>0 \\ 0 & \text{otherwise.} \end{cases}$$

Let us recall that since the state space, E, is finite, $\psi(t)$ is element wise finite for every $t \ge 0$, i.e. the MRP is normal [cf. Limnios and Oprisan (1997)].

We consider the estimator $\hat{\psi}(t)$ of the Markov renewal matrix $\psi(t)$ defined by

$$\hat{\psi}(t) = \sum_{l=0}^{\infty} \hat{Q}^{(l)}(t,T).$$

All results of this section are given without proofs which can be found in Ouhbi and Limnios (1997).

Theorem 8.3.1 *Under the assumptions of Theorem 8.2.1, the estimator of the Markov renewal process is uniformly strongly consistent on compact $[0, L]$, for $L \in \mathbb{R}_+$, in the sense that,*

$$\max_{i,j} \sup_{t\in[0,L]} |\hat{\psi}_{ij}(t,T) - \psi_{ij}(t)| \longrightarrow 0 \qquad p.s. \text{ as } T \longrightarrow \infty.$$

Theorem 8.3.2 *For all i, $j \in E$ and $t \in [0,T]$, $\sqrt{T}\Delta\psi_{ij}(t)$ converges in law to a normal random variable with mean zero and variance*

$$\sigma_{ij}^2(t) = \sum_{l=1}^{s}\sum_{k=1}^{s} \mu_{ll}\{[\psi_{il} * \psi_{kj}]^2 * Q_{lk} - [\psi_{il} * \psi_{kj} * Q_{lk}]^2\}(t),$$

where μ_{ll} is the mean recurrence time of state l.

8.4 Estimation of the Semi-Markov Transition Matrix and Its Properties

In applications of SM process, it is often of interest to study the asymptotic behaviour properties of the SM transition matrix. According to our knowledge, there is no work on this estimation problem. The purpose of this section is to give an estimator of the SM transition matrix and to study its asymptotic behaviour as T tends to infinity.

Let P(.) be the semi-Markov transition matrix function defined by:

$$P_{ij}(t) = P[J_{N_t} = j | J_0 = i] \qquad i,\ j \in E,$$

and let us also denote

$$B_{inkj} = \sum_{l=1}^{\infty}\sum_{r=1}^{l} Q_{in}^{(r-1)} * Q_{kj}^{(l-r)} \text{ and } B_{ij} = \sum_{n=1}^{s}\sum_{k=1}^{s} p(i)B_{nijk} * (I - diag(Q\mathbf{1})_{kk}).$$

$diag(\cdot)$ is a diagonal matrix of ith entry $\sum_{j=1}^{s} Q_{ij}(t)$ and $\mathbf{1} = (1,1,...,1)^t$. It is known, cf. Pyke (1961), that

$$P_{ij}(t) = 1_{\{i=j\}}\left(1 - \sum_{k=1}^{s} Q_{ik}(t)\right) + \sum_{k\in E}\int_0^t P_{kj}(t-s)Q_{ik}(ds).$$

By solving the above Markov renewal equation, it can be seen that, in matrix notation,

$$P(t) = (I - Q(t))^{(-1)} * (I - diag(Q(t)\mathbf{1})). \tag{8.3}$$

We propose the following estimator for P(.):

$$\hat{P}(t,T) \ = \ (I - \hat{Q}(t,T))^{(-1)} * (I - diag(\hat{Q}(t,T)\mathbf{1})). \tag{8.4}$$

Theorem 8.4.1 *Under the assumption of Theorem 8.2.1, the estimator of the semi-Markov matrix is uniformly strongly consistent in [0,L], for all $L \in \mathbb{R}_+$, in the sense that*

$$\max_{i,j} \sup_{t\in[0,L]} |\hat{P}_{ij}(t,T) - P_{ij}(t)| \longrightarrow 0 \qquad a.s., \text{ when } T \longrightarrow \infty.$$

PROOF. Let us consider the matrices $B(t) = I - diag(Q(t)\cdot \mathbf{1})$ and $\hat{B}(t,T) = I - diag(\hat{Q}(t,T)\cdot \mathbf{1})$, then $P(t) = [\psi * B](t)$ where $*$ is the Stieltjes convolution product and $\hat{P}(t,T) = [\hat{\psi} * \hat{B}](t,T)$ its estimator.

Let $i, j \in E$ be fixed then,

$$\begin{aligned}
\sup_{t\in[0,L]} |\hat{P}_{ij}(t,T) - P_{ij}(t)| &= \sup_{t\in[0,L]} |(\hat{\psi} * \hat{B})_{ij}(t,T) - (\psi * B)_{ij}(t)| \\
&\leq \sup_{t\in[0,L]} |(\hat{\psi}_{ij}(t,T) - \psi_{ij})(t)| \\
&+ \sup_{t\in[0,L]} |(\hat{\psi}_{ij}(t,T) - \psi_{ij})(t)| \cdot diag(\hat{Q}(t,T)\cdot \mathbf{1}) \\
&+ \sup_{t\in[0,L]} |diag(\hat{Q} - Q)_{jj}(t,T)| \psi_{ij}(L).
\end{aligned}$$

Since s, the number of states, is finite, the process is normal and therefore $\psi_{ij}(t)$ is finite and from Theorems 8.2.1 and 8.3.1 the uniform strong consistency of the estimators of the semi-Markov kernel and of the Markov renewal function in $[0, L]$, we see that $diag(\hat{Q} - Q)_{jj}(t,T)$ and $\Delta\psi_{ij}(t,T)$ on $[0, L]$ converges a.s. to zero as T tends to infinity. ■

Theorem 8.4.2 *For all i, $j \in E$ and $t \in [0,T]$, $\sqrt{T}(\hat{P}_{ij}(t,T) - P_{ij}(t))$ converges in law to a zero mean normal random variable with variance*

$$\begin{aligned}
\sigma^2 &= \sum_{n=1}^{s}\sum_{k=1}^{s} \mu_{nn} \left[\left(1 - \sum_{l=1}^{s} Q_{il}\right) * B_{inkj} - \psi_{ij} 1_{\{n=j\}}\right]^2 * Q_{nk}(t) \\
&- \left\{\left[\left(1 - \sum_{l=1}^{s} Q_{il}\right) * B_{inkj} - \psi_{ij} 1_{\{n=j\}}\right] * Q_{nk}(t)\right\}^2.
\end{aligned}$$

PROOF. From (8.4), we see that,

$$\begin{aligned}
\sqrt{T}\Delta P_{ij} &= \sqrt{T}[\hat{\psi}_{ij} * (I - diag(\hat{Q}\mathbf{1}))_{jj} - \psi_{ij} * (I - diag(Q\mathbf{1}))_{jj} \\
&= \sqrt{T}[(\hat{\psi}_{ij} - \psi_{ij}) * (I - diag(Q\mathbf{1}))_{jj} - \psi_{ij} * diag([\hat{Q} - Q]\mathbf{1})_{jj}] \\
&- (\hat{\psi}_{ij} - \psi_{ij}) * diag([\hat{Q} - Q]\mathbf{1})_{jj}. \qquad (8.5)
\end{aligned}$$

As in Ouhbi and Limnios(1997), it can be seen that $\sqrt{T}(\hat{\psi}_{ij} - \psi_{ij}) * diag([\hat{Q} - Q]\mathbf{1})_{jj}$ converges, in probability, when T tends to infinity, to zero. So, $\sqrt{T}\Delta P_{ij}(t,T)$ has the same limit as

$$\sqrt{T}[(\hat{\psi}_{ij} - \psi_{ij}) * (I - diag(Q\mathbf{1}))_{jj} - \psi_{ij} * diag([\hat{Q} - Q]\mathbf{1})_{jj}].$$

From Theorem 8.3.1, it has the same limit as,

$$\sqrt{T}\left[\left(1-\sum_{l=1}^{s} Q_{il}\right) * \left(\sum_{n=1}^{s}\sum_{k=1}^{s} B_{inkj} * \Delta Q_{nk}\right) - \psi_{ij} * \left(\sum_{l=1}^{s} \Delta Q_{jl}\right)\right]$$
$$= \sum_{n=1}^{s}\sum_{k=1}^{s} \sqrt{T}\left[\left(1-\sum_{l=1}^{s} Q_{il}\right) * B_{inkj} * \Delta Q_{nk} - \sum_{k=1}^{s} \sqrt{T}\psi_{ij} * \Delta Q_{jk}.\right]$$

Let f be a real function defined on $E \times E \times \mathbb{R}_+$ by

$$f(r,m,x) = \left[\left(1-\sum_{l=1}^{s} Q_{il}\right) * B_{inkj} - \psi_{ij} 1_{\{n=j\}}\right] 1_{\{r=n\}}(1_{\{m=k,x\leq t\}} - Q_{nk}).$$

So, $\sqrt{T}\Delta P_{ij}(t,T) = \sqrt{T} W_f(t)$, where

$$W_f(t) = \sum_{n=1}^{s}\sum_{k=1}^{s} \frac{\sum_{l=1}^{N_n} f(J_{l-1}, J_l, X_l)}{N_n},$$

and by Pyke and Schaufele's central limit theorem, we get the desired results. ∎

8.5 Reliability and Availability Estimation

In reliability, the state space, E, is naturally partitioned into two sets, U and D, where U is the set of working (or 'up') states, and D is the set of repair (or 'down') states, i.e. $E = U \cup D$ and $U \cap D = \emptyset$. Finite semi-Markov reliability models whose state space is partitioned in the above manner will be considered here. The transition from one state to another means, physically speaking, the failure or the repair of one of the components of the system. In the set of up states, U, the system is operational. No service is delivered if the system is in the set of down states, D. However, a repair will return the system from D to U. To model this situation, it will be assumed that the MRP is irreducible. The quantities examined here are the system's availability and reliability defined respectively by $A(t) = P[J_{N_t} \in U]$ and $R(t) = P[J_{N_s} \in U, \text{ for all } s \leq t]$.

8.5.1 Availability

In the sequel, for a matrix X, we will denote X^U its restriction on U.

Availability and reliability of a semi-Markov system can be expressed in closed forms as follows cf. Ouhbi and Limnios (1996b)

$$A(t) = \sum_{i \in E} p(i) A_i(t),$$

where $A_i(t)$ is the system's availability given that it starts in state i at the initial time. So, in matrix form

$$A(t) = p.P(t).\mathbf{1}_{\mathbf{s},\mathbf{s_1}},$$

where p is the initial law and $\mathbf{1}_{\mathbf{s},\mathbf{s_1}} = (1, 1, ..., 1, 0, 0, ..., 0)^t$ an s-dimensional column vector with the first s_1 elements are 1's and the others 0's. s_1 is the number of up states. Let $\hat{P}$ be the estimator of the semi-Markov transition matrix P given by:

$$\hat{P}(t, T) = (I - \hat{Q}(t, T))^{(-1)} * (I - diag(\hat{Q}(t, T)\mathbf{1})).$$

Then we propose the following estimator for the availability of the system,

$$\hat{A}(t, T) = \hat{p}.\hat{P}(t, T).1_{s,s_1}.$$

8.5.2 Reliability

The reliability R(t) of the system is given by:

$$R(t) = \sum_{i \in U} p(i) R_i(t).$$

Let $R_i(t)$ to be the conditional probability that the first failure does not occur up to time t, given that the process started from state $i \in U$. So

$$R(t) = p_U.P^U(t).\mathbf{1},$$

where

$$P^U(t) = (I - Q^U(t))^{(-1)} * (I - diag(Q(t)\mathbf{1})^U),$$

and $\mathbf{1} = (1, 1, ..., 1)^t$.

Let $\hat{Q}$ be the MLE of the semi-Markov kernel Q. Then we propose the following estimator for $P^U(.)$:

$$\hat{P}^U(t, T) = (I - \hat{Q}^U(t, T))^{(-1)} * (I - diag(\hat{Q}(t, T)\mathbf{1})^U)$$

The reliability estimator of the system is:

$$\hat{R}(t, T) = \hat{p}_U.\hat{P}^U(t, T).1.$$

8.5.3 Asymptotic properties of the estimators

In this paragraph, we will prove uniform strong consistency of the estimators of reliability and availability.

Theorem 8.5.1 *Under assumption of Theorem 8.2.1, the estimator of the availability of the semi-Markov system is uniformly strongly consistent in the sense that, for* $L \in \mathbb{R}_+$

$$\sup_{t\in[0,L]} |\hat{A}(t,T) - A(t)| \longrightarrow 0 \qquad a.s. \text{ as } T \longrightarrow \infty.$$

PROOF. This theorem is a direct consequence of the uniform strong consistence of the estimator of the semi-Markov transition matrix. ■

Theorem 8.5.2 *Under assumption of Theorem 8.2.1, the estimator of the reliability of the semi-Markov system is uniformly strongly consistent in the sense that, for* $L \in \mathbb{R}_+$

$$\sup_{t\in[0,L]} |\hat{R}(t,T) - R(t)| \longrightarrow 0 \qquad a.s., \text{ when } T \longrightarrow \infty.$$

8.6 Application

In this paragraph, we will apply our results to a three states semi-Markov process. The transitions between states are given in Figure 8.1, below. The sojourn time in state 1 before going to state 2 is exponential with parameter $\lambda_1 = 0.1$. When state 2 is entered, the next state will be 1 or 3, if the next state to be visited is 3 then the sojourn time is Weibull, $W(\alpha_2, \beta_2)$, with $\alpha_2 = 0.1$ and $\beta_2 = 2$, else the next state to be entered is 1 and the sojourn time in this state is Weibull, $W(\alpha_1, \beta_1)$ with $\alpha_1 = 0.3$ and $\beta_1 = 2$. When state 3 is entered, the next state to be visited is state 1 and the sojourn time in state 3 is exponential with parameter $\lambda_2 = 0.2$. We also suppose that states 1 and 2 are up states and state 3 is a down one, so $U = \{1, 2\}$ and $D = \{3\}$.

Results concerning reliability and availability of the previous semi-Markov system are given in Figure 8.2.

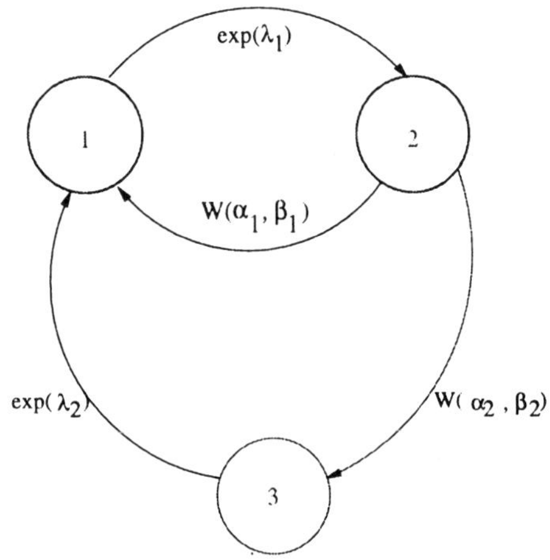

Figure 8.1: A three state semi-Markov system

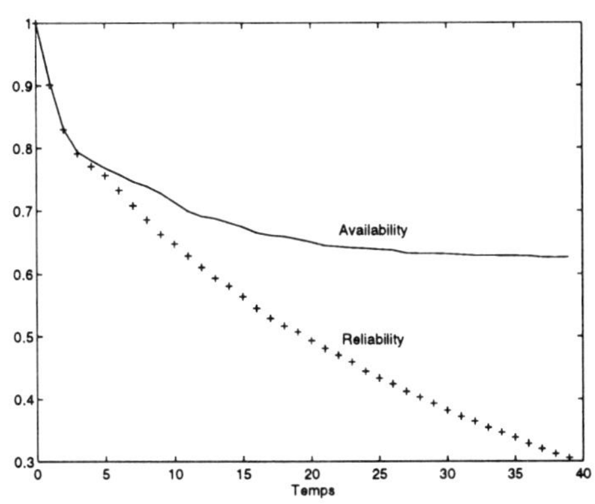

Figure 8.2: Reliability and availability of the semi-Markov system

---

References

1. Billingsley, P. (1961). Statistical inference for Markov process, The University of Chicago Press.

2. Pyke, R. and Schaufele, R. (1964). Limit theorems for Markov renewal process, *Annals of Mathematical Statistics*, **35**, 1746–1764.

3. Ouhbi, B. and Limnios, N. (1996a). Estimation du taux de hasard associé au noyau du processus semi-markovien et application en fiabilité, XXVIII Journées de statistique, 27-30 Mai, Québec, Canada, 573–577.

4. Ouhbi, B. and Limnios, N. (1996b). Non-parametric estimation for semi-Markov systems with application to reliability analysis, *Applied Stochastic Models and Data Analysis*, **12**, 209–220.

5. Ouhbi, B. and Limnios, N. (1997). Comportement asymptotique de la matrice de renouvellement markovien, *C. R. Acad. Sci. Paris*, **325**, Série I, 921–924.

6. Limnios, N. and Oprisan, Gh. (1997). On an additive functional of a semi-Markov process with arbitrary state space and applications, *Research Report*, UTC/DMA, 97.4, June.

7. Limnios N. (1997), Dependability analysis of semi-Markov systems, *Reliability Engineering and System Safety*, **55**, 203–207.

PART II

PROBABILISTIC METHODS

9

Stochastic Models of Systems in Reliability Problems

Vladimir S. Korolyuk

Ukrainian Academy of Science, Kiev, Ukraine

Abstract: Problems of singular perturbation of a reducible invertible operator are applied to reliability analysis of stochastic systems. Heuristic principles of phase merging and diffusion approximation are described.

Keywords and phrases: Stochastic system, singular perturbed operator, reliability analysis, phase merging, diffusion approximation

9.1 Introduction

This approach presents a survey of asymptotic methods in reliability analysis which were developed in the Ukrainian probabilistic school founded by the renowned outstanding scientist B. V. Gnedenko.

In the same time it seems to be an unrealistic task. There exist various analytical approaches which provide and ensure the effective tools in reliability analysis. The excellent sources are the *Handbook of Reliability Engineering* (Editor Igor A. Ushakov) and the monograph *Probabilistic Reliability Engineering* (1995) written by B. V. Gnedenko and I. A. Ushakov.

Therefore, I would like to focus specifically on the asymptotic analysis of stochastic systems based on a solution of the problem of singular perturbation for reducible invertible operators. The asymptotic analysis of stochastic analysis in series scheme, limit theorems and algorithms of phase merging and averaging as well as diffusion approximation and their application in reliability analysis and queueing theory are considered in our books (see References). These books contain extended references of works dedicated to asymptotic analysis of stochastic systems which can be represented by random evolutions.

9.2 Reliability Problem for a Redundant System

The reliability problem for a redundant system, considered by Gnedenko (1964a,b), has stimulated the emergence of the new asymptotical methods of phase merging for Markov and semi-Markov processes and, subsequently, the development of the analysis of stochastic systems in random medium. The main tool in this analysis is the solution of singular perturbation problems for reducibly invertible operators. This is the topic of my lecture dedicated to my unforgettable teacher B. V. Gnedenko.

9.2.1 Repairable duplicated system

The evolution of a *repairable duplicated system* with two identical working devices and one repairing facility is determined by the working time α of every device which has an arbitrary distribution function $F(t)$ as well as the service time β of the repairing facility which has distribution function $G(t)$.

The repairman problem seeks to establish the distribution function of the time to failure τ. The solution of this problem was given by Gnedenko (1964 a,b) in the Laplace transform mode

$$Ee^{-qs\tau} = \psi(qs)/[1 - g(qs)], \tag{9.1}$$

where

$$g(s) := \int_0^\infty e^{-st} G(t) dF(t).$$

Note that $q = 1 - g(0) = \mathcal{P}\{\beta > \alpha\}$ is the probability of a failure on every working interval.

The explicit form of the solution provides the limiting result obtained by Gnedenko (1975) and Soloviev (1964).

$$\mathcal{P}\{q\tau > t\} \to e^{-t/a} \quad \text{as} \quad q \to 0, \quad a := E\alpha.$$

The main tool in the proof of the limit theorem is the asymptotical representation of the denominator in the explicit relation for the Laplace transform of the distribution function of time to failure τ. Note that under condition $q \to 0$, the denominator tends to zero as well.

$$1 - g(qs) = q[1 + as] + o(q).$$

This is the *singularity property* of formula (9.1).

Gnedenko (1964a) considered also a *repairable system with two different devices*. The working times α_k and the service times β_k have different distribution functions

$$F_k(t) = \mathcal{P}\{\alpha_k < t\} \quad \text{and} \quad G_k(t) = \mathcal{P}\{\beta_k < t\}, \quad k = 1, 2.$$

Now common working time τ_k, $k = 1, 2$, is dependent on the starting number of the working device.

The explicit representation for the generating functions

$$\varphi_k(s) = Ee^{-s\tau_k}, \quad k = 1, 2,$$

can be obtained as a solution of the following linear equation represented in a matrix form

$$Q(s)\varphi = \psi, \quad \varphi := (\varphi_1(s), \varphi_2(s)) \tag{9.2}$$

$$Q(s) := \begin{bmatrix} 1 & -g_1(s) \\ -g_2(s) & 1 \end{bmatrix}$$

$$g_1(s) := \int_0^\infty e^{-st} G_2(t) dF_1(t), \quad g_2(s) := \int_0^\infty e^{-st} G_1(t) dF_2(t).$$

The problem of *singular perturbation* is arisen under consideration of the system in the series scheme with a small series parameter $\varepsilon > 0$.

By this the probability of the failure are supposed to be the following:

$$q_k^\varepsilon = 1 - g_k^\varepsilon(0) = \varepsilon q_k + o(\varepsilon), \quad k = 1, 2.$$

The problem of singular perturbation for an asymptotical solution of the system is clarified by an asymptotical representation of the matrix Q:

$$Q(\varepsilon s) = Q_0 + \varepsilon Q_1(s) + o(\varepsilon),$$

$$Q_0 := \begin{bmatrix} 1 & -1 \\ -1 & 1 \end{bmatrix}, \quad Q_1(s) := \begin{bmatrix} 0 & q_1 + sa_1 \\ q_2 + sa_2 & 0 \end{bmatrix}.$$

The improper matrix Q_0 has the determinant which is equal to zero.

The repairman problem for a redundant system can be considered as the *problem for sojourn time* of system in the working states $\{1, 2\}$ before failure in the absorbing state 0.

The *generalized repairman problem for renewal systems* was formulated in 1965 as the sojourn time in a subset of states for a semi-Markov random process with absorption [Korolyuk (1965)].

9.2.2 Sojourn time in a subset of states

Let $\kappa^\varepsilon(t)$, $t \geq 0$, be a semi-Markov process on a measurable phase space $X = X_0 \cup \{0\}$ given by the semi-Markov kernel [Korolyuk and Turbin (1993)]

$$Q^\varepsilon(x, B, t) = P_\varepsilon(x, B) G_x(t).$$

The transition probabilities $P_\varepsilon(x, B)$ of the imbedded Markov chain are represented as follows:

$$P_\varepsilon(x, B) = P(x, B) - \varepsilon P_1(x, B), \tag{9.3}$$

where the stochastic kernel $P(x, B)$ is the transition probabilities of uniformly ergodic Markov chain on subset X_0.

The perturbing kernel $P_1(x, B)$ provides absorption of an initial Markov chain with probabilities

$$P_\varepsilon(x, \{0\}) = \varepsilon P_1(x, X_0).$$

Introduce the sojourn time in subset X_0

$$\tau_x^\varepsilon := \inf\{t : \kappa^\varepsilon(t) = 0/\kappa^\varepsilon(0) = x\}, \quad x \in X_0.$$

The generating function $\varphi_\varepsilon(x, s) := Ee^{-\varepsilon s \tau_x^\varepsilon}$ is determined by a solution of the Markov renewal equation [Korolyuk (1993)]

$$[I - Q^\varepsilon]\varphi_\varepsilon = \varepsilon m_\varepsilon. \tag{9.4}$$

In view of (9.3) the equation (9.4) has the following asymptotical representation:

$$[I - P + \varepsilon Q_1 + \varepsilon\theta^\varepsilon]\varphi_\varepsilon = \varepsilon m_\varepsilon,$$

where $Q_1 := P_1 + smP$ and the operator θ^ε satisfies negligible condition

$$\|\theta^\varepsilon\| \to 0 \quad \text{as} \quad \varepsilon \to 0.$$

Operator $Q := I - P$ possesses the *reducibly invertible* property [Korolyuk and Turbin (1993), Korolyuk and Swishchuk (1995), Korolyuk and Korolyuk (1997)].

The Banach space B can be represented as the direct sum

$$B = N_Q \oplus \mathrm{R}_Q \tag{9.5}$$

of the null-space $N_Q := \{\varphi : Q\varphi = 0\}$ and of the space of values $\mathrm{R}_Q := \{\varphi : Q\psi = \varphi\}$. Decomposition (9.5) means that there exists the projector Π onto null-space N_Q and the potential operator R_0 defined by the following relation

$$\mathrm{R}_0 := [Q + \Pi]^{-1} - \Pi, \tag{9.6}$$

satisfying the following properties:

$$Q\mathrm{R}_0 = \mathrm{R}_0 Q = I - \Pi, \quad \Pi\mathrm{R}_0 = \mathrm{R}_0\Pi = \emptyset.$$

That is, the potential R_0 is a reducible inverse operator to the operator Q. The general solution of equation

$$Q\varphi = \psi$$

can be represented as follows:

$$\varphi = \mathrm{R}_0\psi + \varphi_0, \quad \varphi_0 \in N_Q.$$

For simplicity in what follows it is supposed that the operator Q and its potential R_0 are bounded. In the alternative cases some additional refinement is needed in asymptotic analysis of singular perturbation problems.

9.3 Problems of Singular Perturbation

The diverse scheme of asymptotical analysis of stochastic systems can be reduced to the problem of singular perturbation of a reducible invertible operator, which can be formulated in the following way. For a given vector $\psi \in B$ the asymptotic solution

$$\varphi^\varepsilon = \varphi + \varepsilon\varphi_1$$

of the equation

$$[\varepsilon^{-1}Q + Q_1]\varphi^\varepsilon = \psi + \theta^\varepsilon$$

is constructed with the asymptotically negligible term θ^ε:

$$\|\theta^\varepsilon\| \to 0 \quad \text{as} \quad \varepsilon \to 0.$$

Such a problem arises due to an asymptotic inversion of singular perturbed operator:

$$[\varepsilon^{-1}Q + Q_1]^{-1} = Q^0 + \varepsilon Q^1 + \dots$$

There exist many situations which cannot be classified [see, for example, Korolyuk (1993)]. Meanwhile, it is possible to extract some logically complete variants [Korolyuk and Swishchuk (1995), Korolyuk and Korolyuk (1997)].

The classification of problems of singular perturbation is based on properties of a *contracted* operator $\hat{Q}_1$ determined by the following relation

$$\hat{Q}_1\Pi = \Pi Q_1 \Pi. \tag{9.7}$$

The contracted operator $\hat{Q}_1$ acts on the contracted null-space $\hat{N}_Q$.

Example 9.3.1 Let Q be a generator of the Markov ergodic process with a finite number of ergodic classes: $X = \cup_{k=1}^N X_k$ and let $\pi_k(dx)$, $1 \le k \le N$, be stationary distributions on X_k, $1 \le k \le N$. The projector Π onto null-space N_Q acts as follows:

$$\Pi\varphi(x) = \sum_{k=1}^N \hat{\varphi}_k I_k(x), \quad \hat{\varphi}_k := \int_{X_k} \varphi(x)\pi_k(dx),$$

here

$$I_k(x) := \begin{cases} 1, & x \in X_k, \\ 0, & x \notin X_k. \end{cases}$$

The contracted null-space $\hat{N}_Q$ is an N-dimensional Euclidean space of vectors $\hat{\varphi} = (\hat{\varphi}_k, 1 \le k \le N)$.

Let a perturbing operator Q_1 get as follows:

$$Q_1\varphi(x) = \int_X Q_1(x, dy)\varphi(y), \quad x \in X.$$

Then the contracted operator $\hat{Q}_1$ on $\hat{N}_Q$ is defined according to relation (9.7) by the matrix

$$\hat{Q}_1 = [q_{kr} : \ 1 \le k, r \le N],$$

where

$$q_{kr} := \int_{X_k} \pi_k(dx) Q_1(x, X_r)$$

and

$$\hat{Q}_1\hat{\varphi} := \left(\sum_{r=1}^{N} q_{kr}\hat{\varphi}_r, 1 \le k \le N \right).$$

There are three logically complete variants:

(i) $\hat{Q}_1$ is invertible: there exists $\hat{Q}_1^{-1}$;

(ii) $\hat{Q}_1$ is zero-operator: $\hat{Q}_1\hat{\varphi} = 0$ for all $\hat{\varphi} \in \hat{N}_Q$;

(iii) $\hat{Q}_1$ is reducibly invertible: there exists null-space $\hat{N}_{\hat{Q}_1} \subset \hat{N}_Q$ such that

$$\hat{N}_Q = \hat{N}_{\hat{Q}_1} \oplus \hat{\mathrm{R}}_{\hat{Q}_1}.$$

There exists also the potential operator $\hat{\mathrm{R}}_0 = [\hat{Q}_1 + \hat{\Pi}]^{-1} - \hat{\Pi}$, where $\hat{\Pi}$ is the projector onto $\hat{N}_{\hat{Q}_1}$ which is defined by the following relation

$$\hat{\Pi}\hat{\varphi} = \hat{\varphi}\hat{1}, \quad \hat{\varphi} \in \hat{N}_{\hat{Q}_1}.$$

Here $\hat{1}$ is a unit vector in $\hat{N}_{\hat{Q}_1}$.

The solutions of singular perturbation problems in these three variants are given [Korolyuk and Turbin (1993), Korolyuk and Swishchuk (1995), Korolyuk and Korolyuk (1997)]. Here we introduce only two of them.

Proposition 9.3.1 *Let the contracted operator $\hat{Q}_1$ be invertible: $\exists\, \hat{Q}_1^{-1}$. Then the asymptotic representation*

$$[\varepsilon^{-1}Q + Q_1](\varphi + \varepsilon\varphi_1) = \psi + \theta^\varepsilon$$

can be realized by the following relations:

$$\begin{aligned} \hat{Q}_1\hat{\varphi} &= \hat{\psi}, \\ \varphi_1 &= \mathrm{R}_0(\psi - Q_1\varphi), \\ \theta^\varepsilon &= \varepsilon Q_1 \mathrm{R}_0(\psi - Q_1\varphi). \end{aligned}$$

Proposition 9.3.2 *Let the contracted operator* $\hat{Q}_1$ *be a zero-operator:*

$$\hat{Q}_1\hat{\varphi} = 0, \ \forall\hat{\varphi} \in \hat{N}_Q.$$

Let in addition the operator $Q_0 = Q_2 - Q_1 \mathrm{R}_0 Q_1$ *after contraction on the space* $\hat{N}_Q$ *have the inverse operator* $\hat{Q}_0^{-1}$.

Then the asymptotic representation

$$[\varepsilon^{-2}Q + \varepsilon^{-1}Q_1 + Q_2](\varphi + \varepsilon\varphi_1 + \varepsilon^2\varphi_2) = \psi + \theta^\varepsilon$$

can be realized by the following relations:

$$\begin{aligned} \hat{Q}_0\hat{\varphi} &= \hat{\psi}, \\ \varphi_1 &= -\mathrm{R}_0 Q_1\varphi, \\ \varphi_2 &= \mathrm{R}_0(\psi - Q_0\varphi), \\ \theta^\varepsilon &= \varepsilon[Q_2\varphi_1 + [Q_1 + \varepsilon Q_2]\varphi_2]. \end{aligned}$$

The remarkable fact is that every solution of a singular perturbation problem has an appropriate interpretation in analysis of stochastic systems.

9.4 Analysis of Stochastic Systems

Let us start with a rather simple example.

9.4.1 Phase merging scheme

Let $x_\varepsilon(t)$, $t \geq 0$, be a Markov jump process on a measurable split space $(X, \mathcal{X})$:

$$X = \cup_{v \in V} X_v, \quad X_v \cap X_{v'} = \emptyset, \quad v \neq v',$$

given by the generator

$$Q^\varepsilon\varphi(x) = q(x)\int_X P_\varepsilon(x, dy)[\varphi(y) - \varphi(x)].$$

The stochastic kernel $P_\varepsilon(x, dy)$ is represented as follows:

$$P_\varepsilon(x, B) = P(x, B) + \varepsilon P_1(x, B),$$

where the stochastic kernel $P(x, B)$ is coordinated with the splitting of phase space (9.8) as follows:

$$P(x, X_v) = \begin{cases} 1, & x \in X_v, \\ 0, & x \notin X_v. \end{cases}$$

In addition, the support Markov chain x_n, $n \geq 0$, with the transition probabilities $P(x, B)$ is supposed to be uniformly ergodic in every class X_v, $v \in V$, with stationary distribution $\pi_v(dx)$, $v \in V$. The Markov process $x_\varepsilon(t)$, $t \geq 0$, spends a long time under a small enough $\varepsilon > 0$ in every class X_v, $v \in V$, and under condition

$$p_v = \int_{X_v} \pi_v(dx) P_1(x, X_v) > 0, \quad v \in V,$$

sooner or later leaves each class X_v of ergodicity of support Markov process $x(t)$, $t \geq 0$, defined by the generator

$$Q\varphi(x) = q(x) \int_X P(x, dy)[\varphi(y) - \varphi(x)].$$

The asymptotic behavior of the Markov process $x_\varepsilon(t)$, $t \geq 0$ as $\varepsilon \to 0$ can be investigated by using the martingale characterization of Markov process $x^\varepsilon(t) = x_\varepsilon(t/\varepsilon)$ in the following form:

$$\mu_t^\varepsilon = \varphi(x^\varepsilon(t)) - \int_0^t L^\varepsilon \varphi(X^\varepsilon(s)) ds,$$

$$L^\varepsilon \varphi(x) = [\varepsilon^{-1} Q + Q_1] \varphi(x),$$

where

$$Q_1 \varphi(x) := q(x) \int_X P_1(x, dy) \varphi(y).$$

The phase merging effect is realized by Proposition 9.3.1 for test functions

$$\varphi^\varepsilon(x) = \varphi(v(x)) + \varepsilon \varphi_1(x),$$

where $v(x) = v$ for $x \in X_v$, is a merging function on X corresponding to the splitting (9.8).

The martingale characterization for the limit Markov process

$$\hat{x}(t) = P\text{-} \lim_{\varepsilon \to 0} v(x^\varepsilon(t))$$

on the merged phase space $(V, \mathcal{V})$ is as follows:

$$\mu_t = \hat{\varphi}(\hat{x}(t)) - \int_0^t \hat{Q}_1 \hat{\varphi}(\hat{x}(s)) ds$$

with the generator

$$\hat{Q}_1 \hat{\varphi}(v) := \hat{q}(v) \int_V \hat{P}(v, dv')[\hat{\varphi}(v') - \hat{\varphi}(v)], \tag{9.8}$$

where

$$\begin{aligned}
\hat{P}(v, r) &:= \int_{X_v} \pi_v(dx) P_1(x, X_r), \; X_r = \cup_{v \in r} X_v, \\
\hat{q}(v) &:= q_v \cdot p_v, \\
q_v^{-1} &:= \int_{X_v} \pi_v(dx)/q(x).
\end{aligned}$$

9.4.2 Heuristic principles of phase merging

Application of phase merging algorithms to real stochastic system require some complicated and tedious calculations. However, there exist some *heuristic phase merging principles* which give the same result as the phase merging algorithms without tiresome technical investigations. Such principles can be considered as some "evolutional laws" for stochastic systems. Heuristic principles can be applied under some reasonable assumptions [Korolyuk and Korolyuk (1997, Section 3.7)].

Finiteness. A system consists of a finite number of elements (devices, subsystems *etc.*).

Independence. Elements of system are functioning independently from each other.

Semi-Markov property. Functioning of every element is described by a Markov renewal process.

Ergodicity. Every element is considered in a steady-stable regime.

Persistence. The probabilities of failure are rather small comparing with renewal probabilities.

Heuristic phase merging principles are formulated in the following way:

1. **Lack of memory.** The common working time of the system to stoppage τ is determined by the exponential distribution

$$\mathcal{P}\{\tau > t\} = e^{-\Lambda t}, \quad t \geq 0.$$

The stoppage intensity is independent of the initial state of the system.

2. **Superposition of stoppages.** The intensity of system stoppage is determined by sum of intensities of the system stoppages in the renewal state of every element

$$\Lambda = \sum_{k=1}^{N} \Lambda_k.$$

3. **Independence of stoppages.** The system stoppages for every element is determined by the stoppage rule of a renewal process with thin out:

$$\Lambda_k = q_k \lambda_k,$$

where q_k is the probability of the system stoppage for k-th element and λ_k is the stationary intensity of working time for k-th element.

The action of heuristic principles is illustrated by the analysis of two real stochastic systems.

Protective system

There is one working device and one protective facility. The functioning of the working device is described by a simple renewal process with the distribution function of working time $G_k(t) = \mathcal{P}\{\beta_k \le t\}$, $k = 1, 0$, where β_1 is the action time and β_0 is the renewal time of the protective facility. The stoppage moment of system is occurred when the renewal time interval of the protective facility contains a renewal moment of the working device. According to the lack of memory principle the stoppage time of the system has the exponential distribution

$$\mathcal{P}\{\tau > t\} = e^{-\Lambda t}, \quad t \ge 0.$$

The intensity of stoppage λ is determined by the independence stoppage principle

$$\Lambda = q\lambda_1, \quad \lambda_1 = 1/E\beta_1.$$

The stoppage probability q in the renewal period of the protective facility is determined, according to the ergodicity assumption, as follows

$$q = \mathcal{P}\{\alpha^* < \beta_0\} = \lambda \int_0^\infty \bar{F}_0(t)\bar{G}(t)dt,$$

where α^* is the stationary remaining time with the distribution function

$$G^*(t) = \mathcal{P}\{\alpha^* \le t\} = \lambda \int_0^t \bar{G}(s)ds, \quad \lambda = 1/E\alpha.$$

Under natural persistence assumption $E\beta_0 << E\alpha$, the intensity of stoppage for the protective system can be estimated as follows:

$$\Lambda \sim E\beta_0/E\alpha E\beta_1.$$

This formula can be useful in the optimization problem.

Double renewal system

There are two working devices whose functions are described by the independent alternative renewal processes with the given distribution functions of working and repairing times

$$G_i^{(k)}(t) = \mathcal{P}\{\theta_i^{(k)} \le t\}, \quad i = 0, 1, \quad k = 1, 2,$$

where $\theta_1^{(k)}$ is the working time of the k-th device and $\theta_0^{(k)}$ is the repairing time of the k-th device. Such a system is called a *"two lifts system"*.

According to the heuristic phase merging principles the stoppage intensity of the system is determined by

$$\Lambda = q_1\lambda_1 + q_2\lambda_2, \quad \lambda_k = 1/E\theta_1^{(k)}, \quad k = 1, 2.$$

The stoppage probability q_k, $k = 1, 2$, are determined under condition that the working time of k-th device is finished during the repairing time of the other device. Therefore, we can obtain

$$q_1 = \mathcal{P}\{\theta_1^{(2)*} < \theta_0^{(1)}\}, \quad q_2 = \mathcal{P}\{\theta_1^{(1)*} < \theta_0^{(2)}\}.$$

The stationary remaining times $\theta_1^{(k)*}$ have the following distributions

$$\mathcal{P}\{\theta_1^{(k)*} \le t\} = \lambda_k \int_0^t \bar{G}_1^{(k)}(s)ds, \quad k = 1, 2.$$

Under the natural persistence assumptions

$$E\theta_0^{(k)} << E\theta_1^{(k)}, \quad k = 1, 2,$$

the intensity of stoppages for the double renewal system can be estimated as follows

$$\Lambda \sim [E\theta_0^{(1)} + E\theta_0^{(2)}]/E\theta_1^{(1)}E\theta_1^{(2)}.$$

The phase merging algorithms are the basis of the heuristic principles of phase merging. The question is arisen, how the phase merging principles can be justified in another way.

9.5 Diffusion Approximation Scheme

Algorithms of diffusion approximation occupy a special place among variety of methods of analysis based on ideas of the simplified description of mathematical models. The simplifying effect of such algorithms is based on the fact that all original parameters characterizing the evolution of a stochastic system transforms into three functions which determine a diffusion process: centering shift, diffusion drift and variance. At the same time in certain conditions of sufficiently high intensities of state transitions of the original system, the diffusion approximation presents a simplifying variant of system evolution with a practically sufficient degree of precision. The efficiency of ADA is based on the fact that the basic characteristics of diffusion approximation: centering shift, diffusion drift and variance have rather simple expression through the basic parameters of the system. ADA is based on limit theorems but the efficiency of ADA easily checks the conditions which do not correspond to the forming of the ADA limit theorems.

The algorithm of diffusion approximation is considered for Markov process of birth-and-death type which can be useful in reliability analysis of redundant systems (the more general form ADA see [Korolyuk and Korolyuk (1997, Chapter 5)], another approach see [Anisimov and Lebedev (1992)]).

The *basic assumptions for ADA* are the following:

1. Stochastic system is described by a Markov process $\nu(t)$, $t \geq 0$, on the real line, which is given by the intensity functions $\lambda_n(\nu) = n\lambda(\nu)$ and $\mu_n(\nu) = n\mu(\nu)$. The transition probabilities of the normalized process $\kappa_n(t) := \nu_n(nt)/n$ are determined as $\Delta \to 0$:

$$\mathcal{P}\{\kappa_n(t+0) = v \pm 1/n | \kappa_n(t) = v\} = \begin{cases} \Delta\lambda(v) \\ \Delta\mu(v) \end{cases} + o(\Delta). \tag{9.9}$$

2. The velocity of average drift

$$C(v) := \lambda(v) - \mu(v) \tag{9.10}$$

has an equilibrium point p defined by a solution of the equation

$$C(p) = 0. \tag{9.11}$$

Algorithm of diffusion approximation is given by the following relations

$$\kappa_n(t) \sim p + \zeta(t)\sqrt{n}, \tag{9.12}$$

where the Ornstein–Uhlenbeck diffusion process is defined by the generator

$$\varphi(u) = ua\varphi^{'}(u) + \frac{1}{2} B^2\varphi^{''}(u),$$

or, by the stochastic differential equation

$$d\zeta(t) = a\zeta(t)dt + Bdw_t.$$

The drift coefficient

$$a = C^{'}(p), \tag{9.13}$$

and the variance

$$B^2 = \lambda(p) + \mu(p). \tag{9.14}$$

Approximate relation (9.12) means the weak convergence

$$[\kappa_n(t) - p]\sqrt{n} \Rightarrow \zeta(t) \quad \text{as} \quad n \to \infty.$$

Supply energy system [Feller (1958, Vol. I, Chapter XVII)]

There are n identical devices functioning simultaneously and independently. Evolution of every device is described by an alternative renewal process with exponentially distributed working and repairing times. The intensities of working and repairing times are λ and μ respectively.

The evolution of system can be described by the Markov process of birth-and-death type $\nu_n(t)$, $t \geq 0$, which is the number of repairing devices in the instant time t. Such system can be interpreted in various way. In Feller (1958) every device is a supply energy welder. The intensity functions of the process $\nu_n(t)$ are the following

$$\lambda_n(v) = n(1-v)\lambda, \quad \mu_n(v) = nv\mu, \quad 0 \leq v \leq 1.$$

Therefore, the intensity functions for the normalized process $\kappa_n(t) = \nu_n(nt)/n$ are

$$\lambda(v) = (1-v)\lambda, \quad \mu(v) = v\mu, \quad 0 \leq v \leq 1.$$

Hencc, the velocity function is

$$C(v) = (1-v)\lambda - v\mu = \lambda - v(\lambda + \mu).$$

The equilibrium point according to (9.11) is

$$p = \lambda/(\lambda + \mu),$$

the drift coefficients according to (9.13) is

$$a = C^{'}(p) = -(\lambda + \mu).$$

The variance according to (9.14) is

$$B^2 = \lambda(p) + \mu(p) = 2\lambda\mu/(\lambda + \mu).$$

The algorithm of diffusion approximation gives the following asymptotical relation

$$\nu_n(t) \sim np + \zeta(t)\sqrt{n},$$

where the Ornstein–Uhlenbeck diffusion process is determined by the drift coefficient a and the variance B^2. Note that the stationary distribution of the process $\zeta(t)$ is normal. Hence the normalized supply energy process $\kappa_n(t)$ has asymptotically normal distribution with mean value p and variance

$$\sigma_n^2 = B^2/na = \frac{2\rho}{(1+\rho)^2/n}, \quad \rho = \mu/\lambda.$$

Repairman system with limited service [Derzko and Korolyuk (1997)]

As in the previous case there are n identical devices which is functioning simultaneously and independently. The intensities of working and repairing times are λ and μ respectively. But now there are only $r < n$ repairing facilities. The Markov process of birth-and-death type $\nu_n(t)$, $t \geq 0$, is a number of repairing or waiting for repairing devices. Considering the normalized process

$\kappa_n(t) = \nu_n(nt)/n$ under diffusion approximation scheme let us assume that $r = nr_0$, $r_0 < 1$, is fixed as $n \to \infty$. Now the intensity functions of jumps for the normalized process $\kappa_n(t)$ are given by the following relations:

$$\lambda(v) = (1 - v)\lambda, \quad 0 \leq v \leq 1,$$

$$\mu(v) = \begin{cases} v\mu, & 0 \leq v \leq r_0, \\ r_0\mu, & r_0 \leq v \leq 1. \end{cases}$$

Therefore, the velocity function

$$C(v) = \lambda(v) - \mu(v) = \begin{cases} \lambda - v(\lambda + \mu), & v \leq r_0, \\ \lambda - r_0\mu - v\lambda, & r_0 \leq v, \end{cases}$$

is peace-linear on the interval $[0, 1]$. The equilibrium point of the velocity function is dependent on a value of

$$C(r_0) = \lambda - r_0(\lambda + \mu).$$

For $r_0 > \lambda/(\lambda + \mu)$, that is $C(r_0) > 0$, the equilibrium point is

$$p = \lambda/(\lambda + \mu).$$

That is the same value as for the supply energy system. The diffusion approximation of the normalized process $\kappa_n(t)$ is given by the relations (9.12)–(9.14). For $r_0 < \lambda/(\lambda + \mu)$, that is $C(r_0) < 0$, the equilibrium point is

$$p = 1 - r_0\mu/\lambda.$$

The algorithm of diffusion approximation gives the following relations:

$$\nu_n(t) \sim np + \zeta(t)\sqrt{n}.$$

The Ornstein–Uhlenbeck diffusion process $\zeta(t)$ is determined by the drift coefficient

$$a = C'(p) = -\lambda,$$

and the variance

$$B^2 = \lambda(p) + \mu(p) = 2r_0\mu.$$

The more detail analysis of the repairman system with limited service see in Derzko and Korolyuk (1997). Algorithms of diffusion approximation can be applied to the analysis of the general repairman problem [Korolyuk, Derzko and Korolyuk (1998)].

References

1. Gnedenko, B. V. (1964a). On spare duplication, *Technical Cybernetics*, **4**, 3–12.

2. Gnedenko, B. V. (1964b). On duplication with renewal, *Technical Cybernetics*, **5**, 111–118.

3. Soloviev, A. D. (1964). Asymptotical distribution of the lifetime of a duplicated element, *Technical Cybernetics*, **5**, 119–121.

4. Gnedenko, D. B. and Soloviev A. D. (1975). A general model for standby with renewal, *Technical Cybernetics*, **6**.

5. Korolyuk, V. S. (1965). The sojourn time of semi-Markov process in a subset of states, *Ukrainian Mathematics Journal.*

6. Korolyuk, V. S. and Turbin, A. F. (1993). *Mathematical Foundation of the State Lumping of Large Systems*, Dordrecht, The Netherlands: Kluwer Academic Publishers.

7. Korolyuk, V. S. and Swishchuk, A. V. (1995). *Evolution of Systems in Random Media*, Boca Raton, FL: CRC Press.

8. Korolyuk, V. S. and Korolyuk, V. V. (1997). *Stochastic Models of Systems*, Dordrecht, The Netherlands: Kluwer Academic Publishers.

9. Anisimov, V. V. and Lebedev, E. A. (1992). *Stochastic queueing networks*, Kyiv. Lybid. (in Ukrainian).

10. Feller, W. (1958). *An Introduction to Probability Theory and its Applications*, Vol. I, New York: John Wiley & Sons.

11. Derzko, N. A. and Korolyuk, V. V. (1997). Repairman problem with limited service, *Preprint*, University of Toronto, Canada.

12. Korolyuk, V. S., Derzko, N. A. and Korolyuk, V. V. (1998). *Markovian repairman problems. Classification and Approximation.*

10

Markovian Repairman Problems. Classification and Approximation

Vladimir S. Korolyuk, Nicolas A. Derzko and Vladimir V. Korolyuk

Institute of Mathematics, Kiev, Ukraine

Abstract: Classification of Markov repairable systems are given by the equilibrium point of the velocity drift. The diffusion approximation of the normalized queueing process by the Ornstein–Uhlenbeck process is established.

Keywords and phrases: Repairable system, equilibrium point, diffusion approximation

10.1 Introduction

The repairman problems can be considered as basic models of stochastic systems in the reliability theory. In the same time the repairman problem is the elementary queueing system occurring according to the birth-and-death Markov process as a mathematical model of stochastic system. A Markovian model of a repairman problem is given by the following parameters. There are n identical working devices which operate independently and simultaneously. The working time of every device has the exponential distribution with intensity λ. There are also $r \leq n$ repairing facilities which serve to repair working devices when they are broken. In the case of $r = n$ a repairman system is a supply energy system [Feller (1958)]. The service time of every facility has the exponential distribution with intensity μ. Besides, there are m spare devices which are used for the replacement of broken devices. In the case of $m = 0$ a repairman system is a system with limited service [Derzko and Korolyuk (1997)]. Therefore, there are at most n devices operating at any time. The queueing process $\nu_n(t)$ in a repairman problem is described by the number of devices undergoing or waiting to be repaired at the moment of time t.

Under above formulated condition the queueing process $\nu_n(t)$ is a Markov jump process with the intensities of jumps given by the following relations. The intensities of jump value +1 are

$$\lambda_n(k) = \begin{cases} n\lambda, & k \leq m \\ (n+m-k)\lambda, & k \geq m \end{cases} \tag{10.1}$$

The intensities of jump value -1 are

$$\mu_n(k) = \begin{cases} k\mu, & k \leq r \\ r\mu, & k \geq r \end{cases} \tag{10.2}$$

The formulae (10.1) and (10.2) completely determine the behaviour of the queueing process in the repairmen problem. This formulae can be used in simulation of trajectories of the queueing process $\nu_n(t)$ applying the computer program based on the stochastic representation of the Markov jump process [Korolyuk and Korolyuk (1997, Section 2.1)].

Note, that the queueing process $\nu_n(t)$ takes integer values in the interval

$$0 \leq \nu_n(t) \leq n+m.$$

Therefore, the phase state space of values of the queueing process is the following finite set of integer numbers

$$E_n = \{0, 1, \dots, n+m\}.$$

The boundary states 0 and $n+m$ are reflecting barriers that is meant, that the queueing process $\nu_n(t)$ transpose from state 0 to state 1 and from state $n+m$ to state $n+m-1$ with the probability 1.

Hence, the queueing process $\nu_n(t)$ has the steady-stable distribution. But, it is evident, that the calculation of this stationary distribution can be realized only in some very particular cases. It is evident also, that the calculation of the global characteristics of behaviour of the queueing process on the large enough time interval can be realized by using simulation technique only for a fixed real valued initial parameters n, m, r, λ and μ which determine the queueing process.

The more effective approach in analysis of the repairman problem is an asymptotical investigation of the queueing process $\nu_n(t)$ as $n \to \infty$. As usually in the mathematical application the limit behaviour of mathematical model can be used effectively for real stochastic system which is determined finite values of the series parameters.

As it is well-known, the limit behaviour of the queueing process as usually is described by application of the average and diffusion approximation schemes. In the repairman problems there are such results but, as it is seems to us, there are another models which can be investigated in the asymptotical analysis.

The various situations can be considered under different relations between parameters which determine a repairman system.

10.2 Classification of Repairman Models

In the averaging scheme the queueing process is considered in the following normalized form:

$$\kappa_n(t) = \nu_n(tn)/n, \quad t \geq 0. \tag{10.3}$$

Further, let us assume that the parameters m and r are represented as follows:

$$m = nm_0, \quad r = nr_0, \tag{10.4}$$

m_0 and r_0 are fixed as $n \to \infty$.

The phase space of states of the normalized queueing process $\kappa_n(t)$ is a finite numbers of points on the interval $[0, 1 + m_0]$:

$$E_n^\kappa = \{\nu_k = k/n; \quad 0 \leq k \leq n + m\}. \tag{10.5}$$

The intensities of jump values, according to (10.1)–(10.3), are represented by the following functions:

$$\lambda(v) = \begin{cases} \lambda, & v \leq m_0, \\ (1 + m_0 - v)\lambda, & v \geq m_0, \end{cases} \tag{10.6}$$

and

$$\mu(v) = \begin{cases} v\mu, & v \leq r_0, \\ r_0\mu, & v \geq r_0. \end{cases} \tag{10.7}$$

The velocity of jumps is determined by the relation

$$C(v) = \lambda(v) - \mu(v). \tag{10.8}$$

The velocity function $C(v)$ is a piece-linear function of variable $v \in [0, 1 + m_0]$. The classification of repairman models in asymptotical analysis is realized according to the property of the velocity function $C(v)$.

The main distinction is given by the inequalities

$$\text{I.} \quad r_0 < m_0; \qquad \text{II.} \quad r_0 > m_0. \tag{10.9}$$

The velocity function has the following representation in the case I:

$$C_I(v) = \begin{cases} \lambda - v\mu, & 0 \leq v \leq r_0, \\ \lambda - r_0\mu, & r_0 \leq v \geq m_0, \\ (1 + m_0)\lambda - r_0\mu - v\lambda, & m_0 \leq v \leq 1 + m_0, \end{cases} \tag{10.10}$$

and in the case II:

$$C_{II}(v) = \begin{cases} \lambda - v\mu, & 0 \le v \le m_0, \\ (1+m_0)\lambda - v(\lambda+\mu), & m_0 \le v \ge r_0, \\ (1+m_0)\lambda - r_0\mu - v\lambda, & r_0 \le v \le 1+m_0. \end{cases} \tag{10.11}$$

The main property of the velocity function which is used in the classification of repairman problems is the position of an equilibrium point p defined by a solution of the equation

$$C(p) = 0. \tag{10.12}$$

It is evident that the solution of equation (10.12) is dependent only on the ratio $\rho = \lambda/\mu$. In both cases (10.9) there are three distinct positions of equilibrium point p (see Tables 10.1 and 10.2).

Table 10.1: $r_0 < m_0$

No	1	2	3
ρ	$\rho < r_0$	$\rho = r_0$	$\rho > r_0$
p	ρ	$[r_0, m_0]$	$1 + m_0 - r_0/\rho$

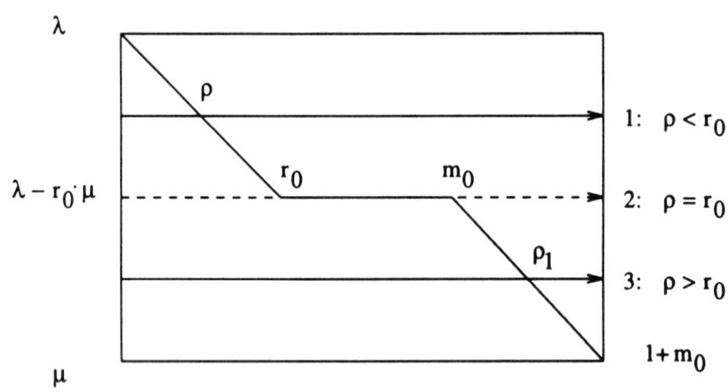

Figure 10.1: Velocity function

Table 10.2: $r_0 \ge m_0$, $r_0^* = r_0/(1 + m_0 - r_0)$

No	1	2	3
ρ	$\rho \le m_0$	$m_0 < \rho \le r_0^*$	$\rho > r_0^*$
p	ρ	$(1+m_0)\rho/(1+\rho)$	$1 + m_0 - r_0/\rho$

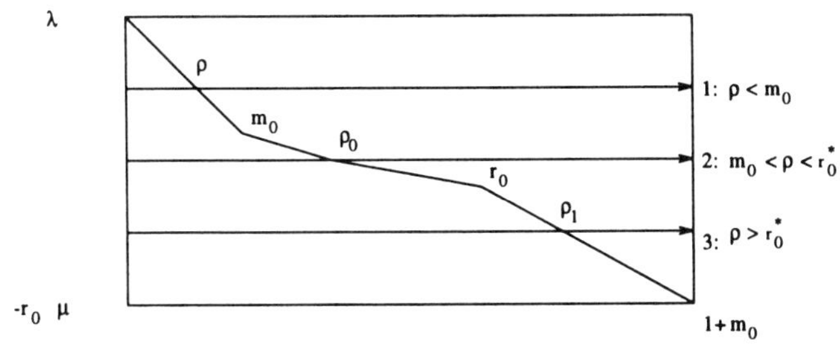

Figure 10.2: Velocity function

10.3 Asymptotical Analysis of Queueing Process

The asymptotical behaviour of the normalized queueing process $\kappa_n(t)$ is determined by the variants of Tables 10.1 and 10.2 [see Iglehart (1965), Korolyuk and Vavrikovich (1988)].

Proposition 10.3.1 *The normalized queueing process $\kappa_n(t) = \nu_n(tn)/n$ weakly converges to the equilibrium point p determined by a solution of the equation (10.12):*

$$\nu_n(tn)/n \Rightarrow p \quad as \quad n \to \infty. \tag{10.13}$$

The proof of Proposition 10.3.1 is based on the averaging theorem for Markov jump processes [Korolyuk and Korolyuk (1997, Chapter 5)].

Certainly, the case I_2 is not included in the convergence scheme (10.13).

Remark 10.3.1 The averaging theorem in [Korolyuk and Korolyuk (1997, Chapter 5)], can be used in asymptotical analysis of the normalized queueing process in a transient regime.

Diffusion approximation scheme

Asymptotical approach in analysis of queueing processes of repairman systems is based on the diffusion approximation for centered normalized queueing process in the following form:

$$\zeta_n(t) = [\nu_n(tn) - np]/\sqrt{n}. \tag{10.14}$$

Proposition 10.3.2 *The centered normalized queueing process $\zeta_n(t)$ weakly converges to the diffusion process of the Ornstein–Uhlenbeck type $\zeta(t)$ which is determined by its generator*

$$L\varphi(u) = a(u)\varphi^{'}(u) + \sigma^2\varphi^{''}(u), \tag{10.15}$$

where the drift coefficient $a(u)$ is determined by

$$a(u) = C^{'}(p)u, \tag{10.16}$$

and the variance σ^2 is determined by

$$\sigma^2 = B(p) = \lambda(p) + \mu(p). \tag{10.17}$$

The statement of Proposition 10.3.2 is a consequence of the diffusion approximation scheme for Markov jump processes (described in [Korolyuk and Korolyuk (1997, Chapter 5)]).

Remark 10.3.2 The diffusion approximation of the centered normalized queueing process is effective on the interval indicated in Tables 10.1 and 10.2 which contains the equilibrium point.

Remark 10.3.3 The diffusion approximation scheme can be used in analysis of transient regime in repairman problem [Korolyuk and Korolyuk (1997, Chapter 5)].

Remark 10.3.4 The queueing process for a repairman model has various diffusion approximations under different assumptions on asymptotic relations between parameters n, m and r. In a view of this circumstance, the optimization of diffusion approximation scheme is arisen.

Collective limit theorem

The proofs of Propositions 10.3.1 and 10.3.2 are based on the collective limit theorem [Korolyuk and Korolyuk (1997, Chapter 5.5)] for the normalized centered queueing process

$$\zeta^{\varepsilon}(t) = \varepsilon\nu^{\varepsilon}(t/\varepsilon^2) - \varepsilon^{-1}p(t), \tag{10.18}$$

where the queueing process $\nu^{\varepsilon}(t)$ is determined by the intensity functions of jumps ± 1 under condition $\nu^{\varepsilon}(t) = k$

$$\begin{aligned} \lambda_{\varepsilon}(k) &= \lambda(\varepsilon^2 k) + \varepsilon\lambda_1(\varepsilon^2 k), \\ \mu_{\varepsilon}(k) &= \mu(\varepsilon^2 k) + \varepsilon\mu_1(\varepsilon^2 k) \end{aligned} \tag{10.19}$$

with some given functions $\lambda(v)$, $\mu(v)$, $\lambda_1(v)$, $\mu_1(v)$.

Collective limit theorem [Korolyuk and Korolyuk (1997) Theorem 5.4)].

Let the intensities of jumps of queueing process $\nu^\varepsilon(t)$ be set by relation (10.19) with continuously differentiable functions $\lambda(v)$ and $\mu(v)$ having the bounded first derivatives and continuous functions $\lambda_1(v)$ and $\mu_1(v)$.

Let there exists a positive solution of the evolutional equation

$$\begin{aligned} dp(t)/dt &= C(p(t)), \\ C(v) &= \lambda(v) - \mu(v) \end{aligned}$$

with the initial condition $p(0) = p_0 \geq 0$. Let the initial values of queueing process converges in probability in such a way:

$$\zeta^\varepsilon(0) = \varepsilon\nu^\varepsilon(0) - \varepsilon^{-1}p_0 \Rightarrow \zeta^0 \quad \text{as} \quad \varepsilon \to 0.$$

Then the normalized centered queueing process (10.18) converges weakly as $\varepsilon \to 0$ to the diffusion process $\zeta^0(t)$:

$$\varepsilon\nu^\varepsilon(t/\varepsilon^2) - \varepsilon^{-1}p(t) \Rightarrow \zeta^0(t) \quad \text{as} \quad \varepsilon \to 0.$$

The generator of the limit diffusion process $\zeta^0(t)$ is determined by the relation

$$L_t^0\varphi(u) = a(t,u)\varphi^{'}(u) + \frac{1}{2}\, B(p(t))\varphi^{''}(u),$$

where

$$\begin{aligned} a(t,u) &= uC^{'}(p(t)) + b^{'}(p(t)), \\ b^1(v) &= \lambda_1(v) - \mu_1(v), \\ B(v) &= \lambda(v) + \mu(v). \end{aligned}$$

The initial condition is $\zeta^0(0) = \zeta^0$.

Corollary 10.3.1 *Let the velocity function $C(v)$ have an equilibrium point p which is determined by a solution of the equation*

$$C(p) = 0.$$

Then, under condition of the collective limit theorem for initial values:

$$\varepsilon^2\nu^\varepsilon(0) \Rightarrow p \quad as \quad \varepsilon \to 0,$$

the normalized centered queueing process

$$\zeta_p^\varepsilon(t) = \varepsilon\nu^\varepsilon(t/\varepsilon^2) - \varepsilon^{-1}p \Rightarrow \zeta_p^0(t) \quad as \quad \varepsilon \to 0.$$

The limit diffusion process $\zeta_p^0(t)$ is determined by the generator

$$L_p^0\varphi(u) = a(u)\varphi^{'}(u) + \frac{1}{2}\, B\varphi^{''}(u),$$

where

$$\begin{aligned} a(u) &= uC'(p) + b^1(p), \\ b^1(v) &= \lambda_1(v) - \mu_1(v), \\ B &= \lambda(p) + \mu(p). \end{aligned}$$

The initial condition is

$$\varepsilon\nu^{\varepsilon}(0) - \varepsilon^{-1}p \Rightarrow \zeta_p^0(0) \quad as \quad \varepsilon \to 0.$$

It is evident, that the statement of Proposition 10.3.1 is the direct consequence of the Proposition 10.3.2. To prove Proposition 10.3.2 we have to represent the intensities of jumps of the queueing process $\nu_n(t)$ for the repairman system in the form (10.19). The main assumption of Proposition 10.3.2 is that $m = nm_0$ and $r = nr_0$ with fixed values of m_0 and r_0. Hence, under the condition of Proposition 10.3.2 the intensity functions (10.1) and (10.2) are represented in the following form:

$$\lambda_n(k) = \lambda(\varepsilon^2 k), \quad \mu_n(k) = \mu(\varepsilon^2 k),$$

where $\varepsilon^2 = 1/n$ and the functions $\lambda(v)$ and $\mu(v)$ are given by the relations (10.6) and (10.7) respectively. Hence, in the relation (10.19)

$$\lambda_1(v) \equiv \mu_1(v) \equiv 0.$$

Now the normalized centered queueing process (10.14) is represented as follows:

$$\zeta_n(t) = \varepsilon\nu_n(t/\varepsilon^2) - \varepsilon^{-1}p,$$

with equilibrium point of the velocity function p.

The statement of Proposition 10.3.2 provides the corollary of the collective limit theorem.

References

1. Iglehart, D. L. (1965). Limiting diffusion approximation for the many server queue and repairman problem, *Journal of Applied Probability*, **2**, 429–441.

2. Korolyuk, V. S. and Vavrikovich, L. V. (1988). Diffusion approximation of the renewal reserved Markov system, *Cybernetic*, **5**, 97–100 (in Russian).

3. Korolyuk, V. S. and Korolyuk, V. V. (1997). *Stochastic Models of Systems*, Dordrecht, The Netherlands: Kluwer Academic Publishers.

4. Feller, W. (1958). *An Introduction to the Probability Theory and its Applications*, Vol. 1, New York: John Wiley & Sons.

5. Derzko, N. A. and Korolyuk, V. V. (1997). Repairman system with limited service, *Preprint* University of Toronto.

11

On Limit Reliability Functions of Large Systems. Part I

Krzysztof Kołowrocki

Gdynia Maritime Academy, Gdynia, Poland

Abstract: In the reliability investigation of large systems the problem of the complexity of their reliability functions appears. This problem may be approximately solved by the assuming that the number of the system components tends to infinity and finding the limit reliability function of the system. It is closely related to the limit theorems in the extreme value theory discussed in many publications. The solution for simple series and parallel systems is well known and given for instance in Barlow and Proschan (1975). The limit reliability functions of more complex series-parallel and parallel-series systems with identical components are discussed in Kołowrocki (1993). The results on limit reliability functions of series-parallel and parallel- series systems with different components are given in Kołowrocki (1994). In the paper the classes of limit reliability functions of any homogeneous and nonhomogeneous large series, parallel, series-parallel and parallel-series system are presented. Moreover, practically useful lemmas and their applications are given. Systems with multi-state components are more general and play an important role in the reliability practice. Therefore in the paper limit reliability functions of large homogeneous multi-state series, parallel, series-parallel and parallel-series systems are given. As a summary of this part of the paper a hypothesis on the class of possible limit reliability functions of considered multi-state systems is stated.

Keywords and phrases: Large scale systems, asymptotic reliability, multi-state systems

11.1 Introduction

Suppose that E_i, $i = 1, 2, \ldots, n$, are components of a system and X_i are independent random variables representing the lifetimes of E_i. We investigate limit distributions of a standardized random variable $(X - b_n)/a_n$, where X is the lifetime of the system and $a_n > 0$, $b_n \in (-\infty, \infty)$ are some suitably chosen numbers. And, since $P((X-b_n)/a_n > x) = P(X > a_n x + b_n) = R_n(a_n x + b_n)$, where $R_n(x)$ is the reliability function of the system, then we assume the following definition.

Definition 11.1.1 A reliability function $\Re(x)$ is called the limit reliability function of the system if there exist norming constants $a_n > 0$, $b_n \in (-\infty, \infty)$ such that

$$\lim_{n \to \infty} R_n(a_n x + b_n) = \Re(x) \qquad \text{for } x \in C_{\Re},$$

where $C_{\Re}$ is the set of continuity points of $\Re(x)$.

Hence, for sufficiently large n, we get the following approximate formula $R_n(x) \simeq \Re((x - b_n)/a_n)$.

Remark 11.1.1 We assume that the lifetime distributions do not necessarily have to be concentrated on the interval $[0, \infty)$. Then, a reliability function does not have to satisfy the usually demanded condition $R(x) = 1$ for $x \in (-\infty, 0)$. It is a generalization of the usually used concept of a reliability function. This generalization is convenient in the theoretical considerations. At the same time, from the achieved results about the generalized reliability functions, the same properties of the usually used reliability function appear.

Definition 11.1.2 We call the system homogeneous if its lifetimes X_i have an identical distribution function $F(x) = P(X_i \leq x)$, $x \in (-\infty, \infty)$, i.e., if its components E_i have the same reliability function $R(x) = 1 - F(x)$, $x \in (-\infty, \infty)$.

The simplest reliability structures are series and parallel systems.

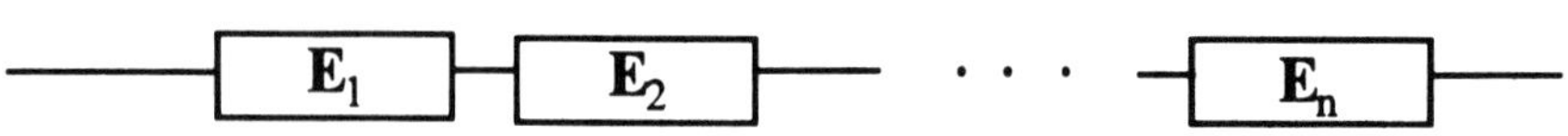

Figure 11.1: The scheme of a homogeneous series system

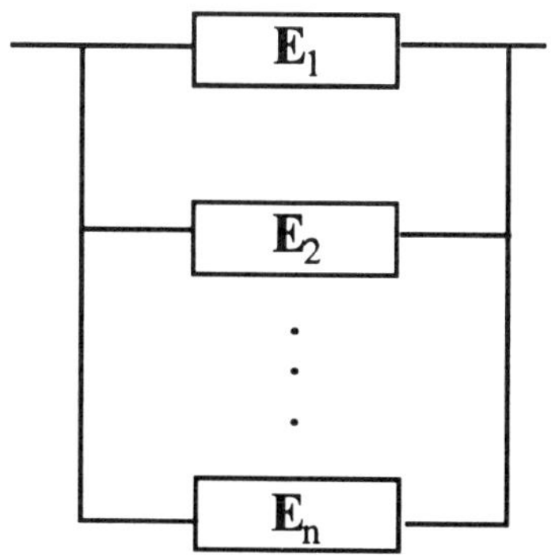

Figure 11.2: The scheme of a homogeneous parallel system

It is well known that the reliability functions of homogeneous series and parallel systems are respectively given by

$$\bar{R}_n(x) = [R(x)]^n \qquad \text{and} \qquad R_n(x) = 1 - [F(x)]^n, \tag{11.1}$$

where n is the number of system components.

The next simple systems are series-parallel and parallel-series homogeneous systems.

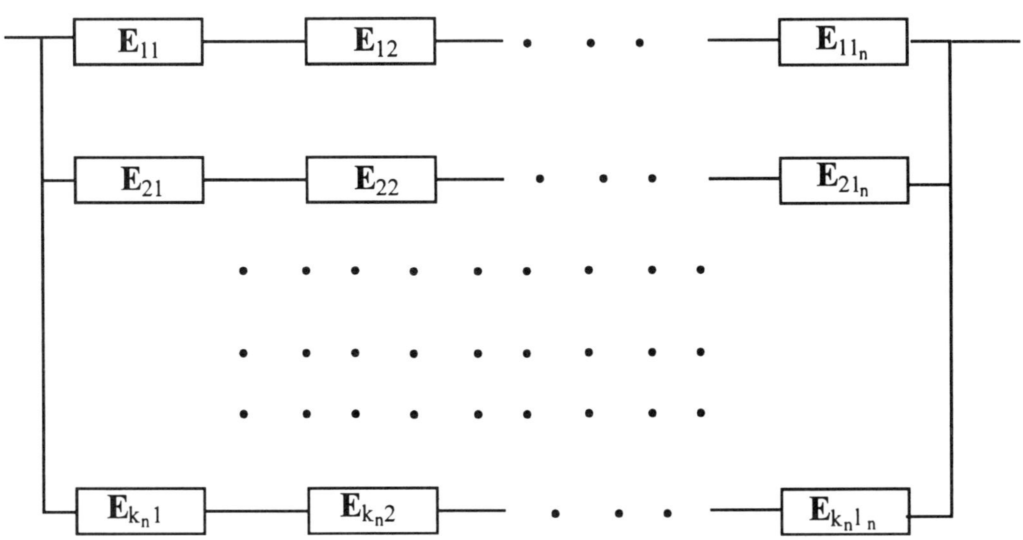

Figure 11.3: The scheme of a homogeneous series-parallel system

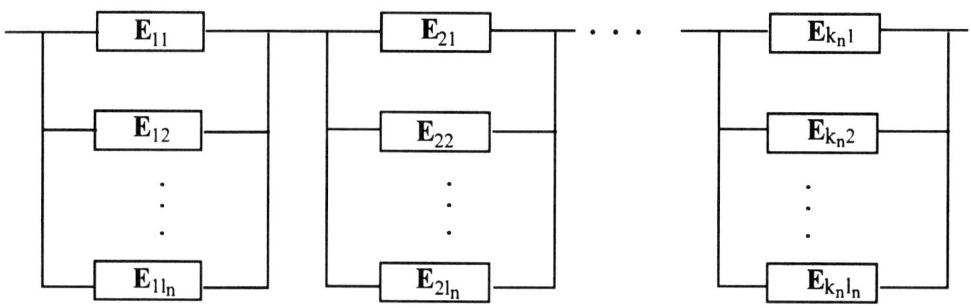

Figure 11.4: The scheme of a homogeneous parallel-series system

The reliability functions of these systems are respectively given by

$$R_n(x) = 1 - [1 - [R(x)]^{l_n}]^{k_n} \quad \text{and} \quad \bar{R}_n(x) = [1 - [F(x)]^{l_n}]^{k_n}, \tag{11.2}$$

where l_n is the number of series components and k_n is the number of parallel components (subsystems) in the first case and vice versa in the second case.

Definition 11.1.3 A series (parallel) system is called nonhomogeneous if it is composed of a, $1 \leq a \leq n$, different kinds of components and the fraction of the i-th kind component in the system is equal to q_i, where $q_i > 0$, $\sum_{i=1}^{a} q_i = 1$. Moreover, $R^{(i)}(x) = 1 - F^{(i)}(x)$, $i = 1, 2, \ldots, a$, is the reliability function of the i-th kind component.

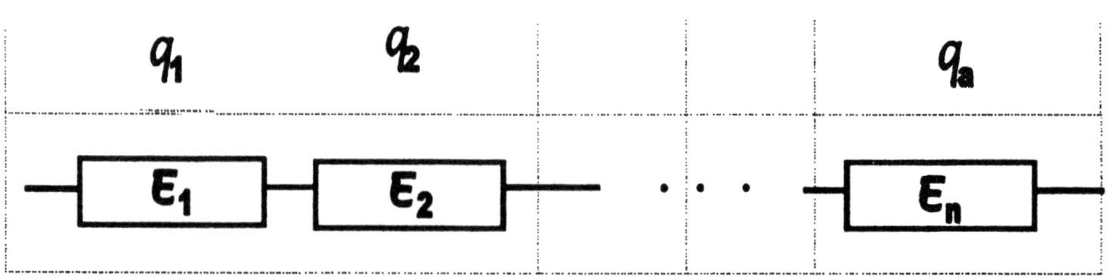

Figure 11.5: The scheme of a nonhomogeneous series system

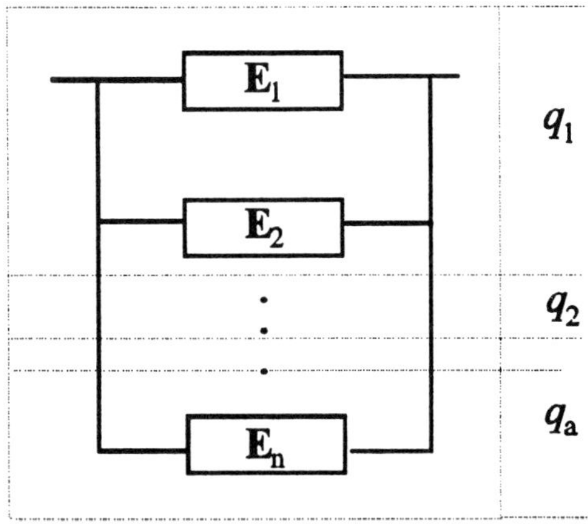

Figure 11.6: The scheme of a nonhomogeneous parallel system

It is easy to motivate that the sequences of reliability functions of nonhomogeneous series and parallel systems are respectively given by

$$\bar{R}'_n(x) = \prod_{i=1}^{a} (R^{(i)}(x))^{q_i n} \quad \text{and} \quad R'_n(x) = 1 - \prod_{i=1}^{a} (F^{(i)}(x))^{q_i n}. \tag{11.3}$$

Definition 11.1.4 A series-parallel (parallel-series) system is called nonhomogeneous if it is composed of a, $1 \leq a \leq k_n$, $k_n \in N$, different kinds of series (parallel) subsystems and the fraction of the i-th kind subsystem in the system is equal to q_i, where $q_i > 0$, $\sum_{i=1}^{a} q_i = 1$. Moreover, the i-th kind series (parallel) subsystem consists of e_i, $1 \leq e_i \leq l_n$, $l_n \in N$, kinds of components with reliability functions $R^{(i,j)}(x) = 1 - F^{(i,j)}(x)$, $j = 1, 2, \ldots, e_i$ and the fraction of the j-th kind component in this subsystem is equal to p_{ij}, where $p_{ij} > 0$ and $\sum_{j=1}^{e_i} p_{ij} = 1$.

The reliability functions of the series-parallel and parallel-series nonhomogeneous systems are respectively given by

$$R'_n(x) = 1 - \prod_{i=1}^{a} [1 - (R^{(i)}(x))^{l_n}]^{q_i k_n}, \tag{11.4}$$

where $R^{(i)}(x) = \prod_{j=1}^{e_i} (R^{(i,j)}(x))^{p_{ij}}$, $i = 1, 2, \ldots, a$,

$$\bar{R}'_n(x) = \prod_{i=1}^{a} [1 - (F^{(i)}(x))^{l_n}]^{q_i k_n}, \tag{11.5}$$

where $F^{(i)}(x) = \prod_{j=1}^{e_i} (F^{(i,j)}(x))^{p_{ij}}$, $i = 1, 2, \ldots, a$.

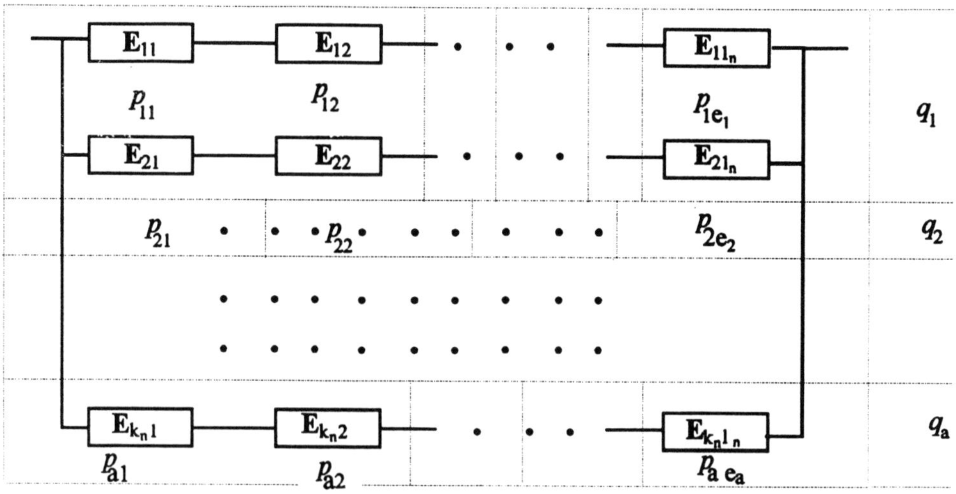

Figure 11.7: The scheme of a nonhomogeneous series-parallel system

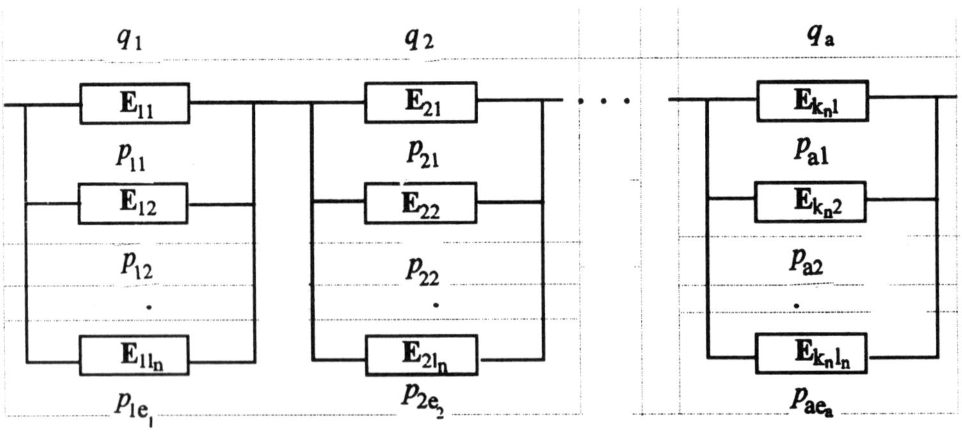

Figure 11.8: The scheme of a nonhomogeneous parallel-series system

11.2 Limit Reliability Functions of Homogeneous Systems

In Barlow and Proschan (1975), it is proved that the only possible limit reliability functions are:

$$\bar{\Re}_1(x) = \begin{cases} \exp[-(-x)^\alpha] & \text{for } x < 0, \\ 0 & \text{for } x \geq 0, \ \alpha > 0, \end{cases}$$

$$\bar{\Re}_2(x) = \begin{cases} 1 & \text{for } x < 0, \\ \exp[-x^\alpha] & \text{for } x \geq 0, \ \alpha > 0, \end{cases}$$

$$\bar{\Re}_3(x) = \exp[-\exp(x)] \quad \text{for } x \in (-\infty, \infty),$$

for a series homogeneous system and $\Re_i(x) = 1 - \bar{\Re}_i(-x)$, $i = 1, 2, 3$, for a parallel homogeneous system. In Kołowrocki (1993), it is stated that depending on the relationship between k_n and l_n the only possible limit reliability functions are:

$$\Re_1(x) = \begin{cases} 1 & \text{for } x \leq 0, \\ 1 - \exp[-x^{-\alpha}] & \text{for } x > 0, \ \alpha > 0, \end{cases}$$

$$\Re_2(x) = \begin{cases} 1 - \exp[-(-x)^\alpha] & \text{for } x \leq 0, \ \alpha > 0, \\ 0 & \text{for } x < 0, \ \alpha > 0, \end{cases}$$

$$\Re_3(x) = 1 - \exp[-\exp(-x)] \quad \text{for } x \in (-\infty, \infty),$$

if $k_n = n$, $|l_n - c \log n| >> s$, $s > 0$, $c > 0$;

$$\Re_4(x) = \begin{cases} 1 & \text{for } x < 0, \\ 1 - \exp[-\exp[-x^\alpha - s/c]] & \text{for } x \geq 0, \ \alpha > 0, \end{cases}$$

$$\Re_5(x) = \begin{cases} 1 - \exp[-\exp[(-x)^\alpha - s/c]] & \text{for } x < 0, \\ 0 & \text{for } x \geq 0, \ \alpha > 0, \end{cases}$$

$$\Re_6(x) = \begin{cases} 1 - \exp[-\exp[(-x)^\alpha - s/c]] & \text{for } x < 0, \\ 1 - \exp[-\exp[-x^\alpha - s/c]] & \text{for } x \geq 0, \ \alpha > 0, \ \beta > 0, \end{cases}$$

$$\Re_7(x) = \begin{cases} 1 & \text{for } x < x_1, \\ 1 - \exp[-\exp[-s/c]] & \text{for } x_1 \leq x < x_2, \\ 0 & \text{for } x \geq x_2, \ x_1 < x_2, \end{cases}$$

if $k_n = n$, $l_n - c \log n \sim s$, $s \in (-\infty, \infty)$, $c > 0$;

$$\Re_8(x) = \begin{cases} 1 - [1 - \exp[-(-x)^\alpha]]^k & \text{for } x < 0, \ \alpha > 0, \\ 0 & \text{for } x \geq 0, \ \alpha > 0, \end{cases}$$

$$\Re_9(x) = \begin{cases} 1 & \text{for } x < 0, \\ 1-[1-\exp[-x^{\alpha}]]^k \text{for } x \geq 0, \ \alpha > 0, \end{cases}$$

$$\Re_{10}(x) = 1-[1-\exp[-\exp x]]^k \quad \text{for } x \in (-\infty, \infty);$$

if $k_n \to k$, $l_n \to \infty$; for a series-parallel homogeneous system and $\bar{\Re}_i(x) = 1 - \Re_i(-x)$, $i = 1, 2, \ldots, 10$, for a parallel- series homogeneous system. In proving facts on limit reliability functions of homogeneous series and parallel systems we use the following lemmas [Barlow and Proschan (1975)].

Lemma 11.2.1 *A nondegenerate reliability function* $\bar{\Re}(x) = \exp[-\bar{V}(x)]$ *is the limit reliability function of the homogeneous series system, if and only if,* $\lim_{n\to\infty} nF(a_n x + b_n) = \bar{V}(x)$ *for* $x \in C_{\bar{V}}$*, where* $C_{\bar{V}}$ *is the set of continuity points of* $\bar{V}(x)$ *and points such that* $\bar{V}(x) = \infty$.

Lemma 11.2.2 *A nondegenerate reliability function* $\Re(x) = 1 - \exp[-V(x)]$ *is the limit reliability function of the homogeneous parallel system if and only if* $\lim_{n\to\infty} nR(a_n x + b_n) = V(x)$ *for* $x \in C_V$*, where* C_V *is the set of continuity points of* $V(x)$ *and points such that* $V(x) = \infty$.

Fact 11.2.1 If system components have triangle reliability functions given by $R(x) = 1-(x-a)^2/(b-a)(c-a)$ for $a < x \leq c$ and $R(x) = (b-x)^2/(b-a)(b-c)$ for $c < x \leq b$, where $a \leq c \leq b$, and $a_n = ((b-a)(c-a)/n)^{1/2}$, $b_n = a$, then $\bar{\Re}_2(x) = \exp[-x^2]$ for $x > 0$, is the limit reliability function of the series homogeneous system.

JUSTIFICATION. Since for sufficiently large n

$$a_n x + b_n \leq a \text{ for } x \leq 0 \text{ and } a_n x + b_n \in (a, c) \text{ for } x > 0,$$

then for sufficiently large n we have

$$F(a_n x + b_n) = 0 \qquad \text{for } x \leq 0$$

and

$$F(a_n x + b_n) = \frac{(a_n x + b_n - a)^2}{(b-a)(c-a)} = \frac{x^2}{n} \qquad \text{for } x > 0.$$

Hence

$$\lim_{n\to\infty} nF(a_n x + b_n) = 0 \qquad \text{for } x \leq 0$$

and

$$\lim_{n\to\infty} nF(a_n x + b_n) = x^2 \qquad \text{for } x > 0,$$

which, by Lemma 11.2.1, means that $\bar{\Re}_2(x)$ is the limit reliability of the system. ∎

Example 11.2.1 If system components have triangle reliability functions given by $R(x) = 1 - x^2/2$ for $0 < x < 1$ and $R(x) = 2 - 2x + x^2/2$ for $1 < x < 2$, then $\bar{R}_n(x) = [1 - x^2/2]^n$ for $0 < x < 1$ and $\bar{R}_n(x) = [2 - 2x + x^2/2]^n$ for $1 < x < 2$, is the exact reliability function of the homogeneous series system. Assuming norming constants $a_n = (2/n)^{1/2}$, $b_n = 0$, by Fact 11.2.1, we get that the limit reliability function of the system is given by $\bar{\Re}_2(x) = \exp[-x^2]$ for $x > 0$. Hence, the approximate formula takes form $\bar{R}_n(x) \simeq \bar{\Re}_2((x - b_n)/a_n) = \bar{\Re}_2((n/2)^{1/2}x) = \exp[-nx^2/2]$ for $x > 0$. In Table 11.1 the speed of convergence of the exact system reliability functions sequence to its limit reliability function is illustrated. The component mean life time is $E[X_i] = 1$. The expected values of the system lifetimes for different numbers of system components are given in Table 11.1.

Table 11.1: The speed of convergence of the reliability functions sequence

	$n = 10$		$n = 20$		$n = 40$		$n = 80$	
x	R_n	$\Re_2$	R_n	$\Re_2$	R_n	$\Re_2$	R_n	$\Re_2$
0.0	1.000	1.000	1.000	1.000	1.000	1.000	1.000	1.000
0.1	0.951	0.951	0.904	0.905	0.818	0.819	0.670	0.670
0.2	0.817	0.819	0.668	0.670	0.446	0.449	0.197	0.197
0.3	0.631	0.638	0.398	0.407	0.159	0.165	0.025	0.025
0.4	0.434	0.449	0.189	0.202	0.036	0.041	0.001	0.002
0.5	0.263	0.287	0.069	0.082	0.005	0.007	0.000	0.000
0.6	0.137	0.165	0.019	0.027	0.000	0.001		
0.7	0.060	0.086	0.004	0.007				
0.8	0.021	0.041	0.000	0.002				
0.9	0.006	0.017						
1.0	0.001	0.007						
1.1	0.000	0.002						
$E[X] \simeq$	0.40		0.28		0.20		0.14	
$\max\lvert\Delta\rvert \simeq$	0.028		0.013		0.006		0.001	

In proving facts on limit reliability functions of homogeneous series-parallel systems we use the following lemmas [Kołowrocki (1993)].

Lemma 11.2.3 *If $k_n \to \infty$, then a nondegenerate reliability function $\Re(x) = 1 - \exp[-V(x)]$ is the limit reliability function of the homogeneous series-parallel system if and only if $\lim_{n\to\infty} k_n[R(a_n x + b_n)]^{l_n} = V(x)$ for $x \in C_V$.*

Lemma 11.2.4 *If $k_n \to k$, $l_n \to \infty$, then a nondegenerate reliability function $\Re(x)$ is the limit reliability function of the homogeneous series-parallel system if and only if $\lim_{n\to\infty}[R(a_n x + b_n)]^{l_n} = \Re_0(x)$ for $x \in C_{\Re_0}$, and moreover $\Re(x) = 1 - [1 - \Re_0(x)]^k$ for $x \in (-\infty, \infty)$.*

Fact 11.2.2 If system components have Weibull reliability functions given by $R(x) = \exp[-\beta x^{\alpha}]$ for $x \geq 0$, $\alpha > 0$, $\beta > 0$ and $k_n \to k$, $l_n > 0$, $a_n = 1/(\beta l_n)^{1/\alpha}$, $b_n = 0$, $\Re_9(x) = 1 - [1 - \exp[-x^{\alpha}]]^k$ for $x \geq 0$, is the limit reliability function of the homogeneous series-parallel system. [The proof is given in Kołowrocki (1998)].

Example 11.2.2 Let us consider the gas piping composed of $k_n = 2$ lines each of them composed of $l_n = 1000$ segment pipes with the Weibull reliability function $R(x) = \exp[-0.0002x^3]$ for $x \geq 0$. We may consider it as the homogeneous series-parallel system. Its exact reliability function is given by $R_n(x) = 1 - [1 - \exp[-0.2x^3]]^2$ for $x \geq 0$. Assuming norming constants $a_n = (0.0002 \cdot 1000)^{1/3} = 1.71$, $b_n = 0$, by Fact 11.2.2, we get that the limit reliability function of the system is given by $\Re_9(x) = 1 - [1 - \exp[-x^3]]^2$ for $x \geq 0$. Hence the exact reliability function of the system is approximately (in this case the formula is exact) given by $R_n(x) \simeq \Re_9((x - b_n)/a_n) = 1 - [1 - \exp[-0.2x^3]]^2$ for $x \geq 0$. The component (the pipe) mean lifetimes are $E(X_i) = \Gamma(4/3)(0.0002)^{-1/3} \simeq 15.3$ years. The expected value of the system lifetime is $E(X) = 2\Gamma(4/3)(0.2)^{-1/3} - \Gamma(4/3)(0.04)^{-1/3} = 2.4$ years.

The behaviour of the gas piping reliability function is illustrated in Figure 11.9.

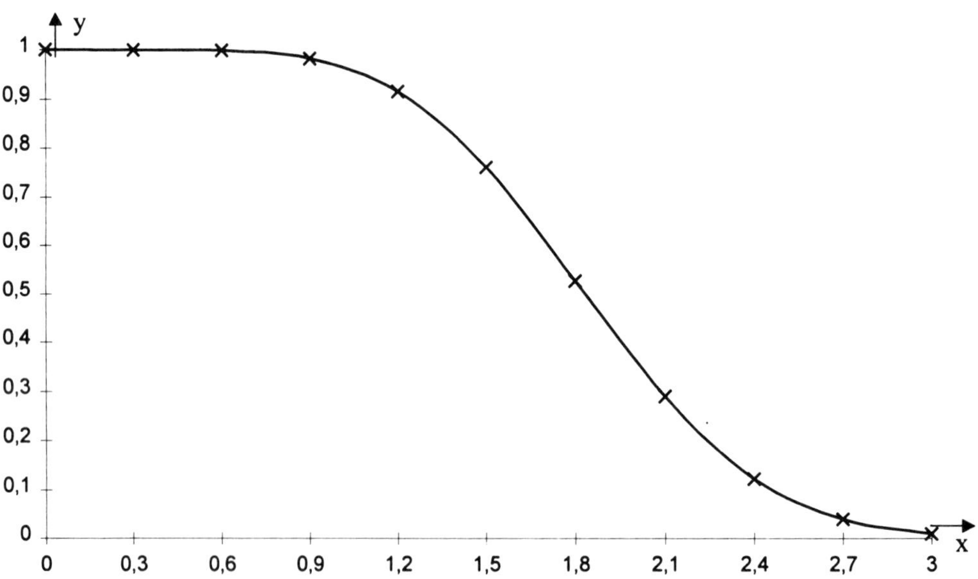

Figure 11.9: Exact and limit reliability functions of the gas piping system

In proving facts on limit reliability functions of homogeneous parallel-series systems we use the following lemmas [Kołowrocki (1993)].

Lemma 11.2.5 *If $k_n \to \infty$, then a nondegenerate reliability function $\bar{\Re}(x) = \exp[-\bar{V}(x)]$ is the limit reliability function of the homogeneous parallel-series system if and only if $\lim_{n\to\infty} k_n[F(a_n x + b_n)]^{l_n} = \bar{V}(x)$ for $x \in C_{\bar{V}}$.*

Lemma 11.2.6 *If $k_n \to k$, $l_n \to \infty$, then a nondegenerate reliability function $\bar{\Re}(x)$ is the limit reliability function of the homogeneous parallel-series system if and only if $\lim_{n\to\infty}[F(a_n x + b_n)]^{l_n} = F_0(x)$ for $x \in C_{F_0}$, and moreover $\bar{\Re}(x) = [1 - F_0(x)]^k$ for $x \in (-\infty, \infty)$.*

Fact 11.2.3 If system components have Weibull reliability functions given by $R(x) = \exp[-\beta x^\alpha]$ for $x \geq 0$, $\alpha > 0$, $\beta > 0$ and $k_n = n$, $l_n - c\log n >> s$, $c > 0$, $s > 0$, $a_n = b_n/\alpha\beta(b_n)^\alpha \log n$, $b_n = 1/\beta[\log(l_n/\log n)]^{1/\alpha}$, then $\bar{\Re}_3(x) = \exp[-\exp[x]]$ for $x \in (-\infty, \infty)$, is the limit reliability function of the homogeneous parallel-series system. [The proof is based on Lemma 11.2.5 and given in Kołowrocki (1998)].

Example 11.2.3 Let us consider the line composed of $k_n = 100$ segments each of them composed of $l_n = 20$ wires with the Weibull distribution function of their strengths (the times up to the wire break) given by $F(x) = 1 - \exp[-0.004x^2]$ for $x \geq 0$. We may consider this line as the homogeneous parallel-series system. Its exact reliability function is given by $\bar{R}_n(x) = [1 - [1 - \exp[-0.004x^2]]^{20}]^{100}$. Assuming norming constants $a_n = 1/(2 \cdot 0.004 \cdot b_n \cdot \log 100) = 1.4$, $b_n = (1/0.004) \cdot \log(20/\log 100) = 19.2$, by Fact 11.2.3, we get that the limit reliability function of the line is $\bar{\Re}_3(x) = \exp[-\exp[x]]$ for $x \in (-\infty, \infty)$. Therefore the exact reliability function of the line is approximately given by $\bar{\Re}_n(x) \simeq \bar{\Re}_3((x-b_n)/a_n) = \exp[-\exp[(x-19.2)/1.4]]$ for $x \in (-\infty, \infty)$. The component (the wire) mean lifetimes are $E(X_i) = \alpha(3/2)(0.004)^{-1/2} \simeq 14$ months. The expected value of the system lifetime evaluated from the limit reliability function is $E(X) \geq 17.5$ months.

The differences between exact and asymptotic reliability functions of the line are illustrated in Figure 11.10.

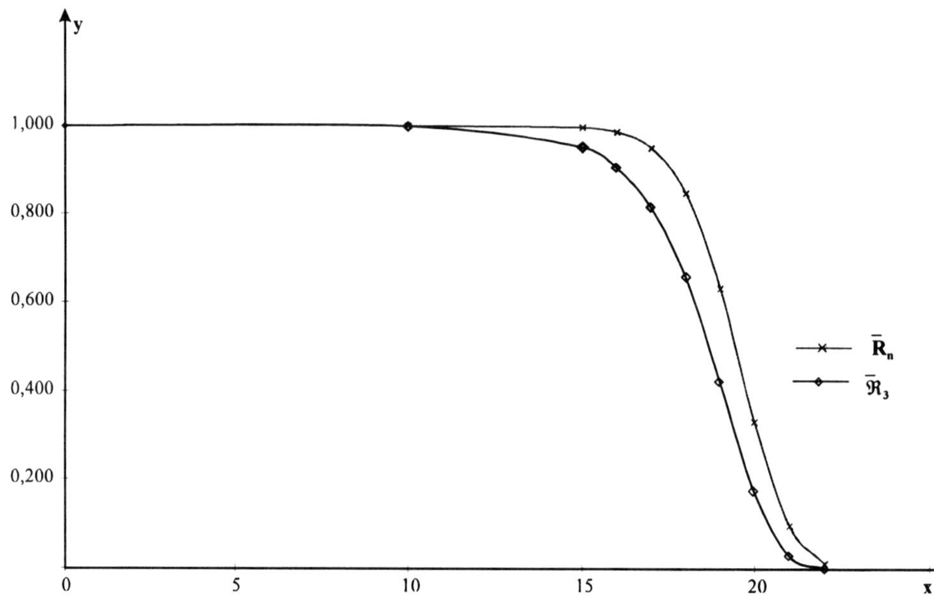

Figure 11.10: Exact and limit reliability functions of the line

11.3 Limit Reliability Functions of Nonhomogeneous Systems

From the results given in [Kołowrocki (1994)] it follows that the only possible limit reliability functions are:

$$\Re'_1(x) = \begin{cases} 1 & \text{for } x \leq 0, \\ 1 - \exp[-d(x)x^{-\alpha}] & \text{for } x > 0, \ \alpha > 0, \end{cases}$$

$$\Re'_2(x) = \begin{cases} 1 - \exp[-d(x)(-x)^{\alpha}] & \text{for } x \geq 0, \ \alpha > 0, \\ 0 & \text{for } x > 0, \end{cases}$$

$$\Re'_3(x) = 1 - \exp[-d(x)\exp(-x)] \quad \text{for } x \in (-\infty, \infty),$$

for a parallel nonhomogeneous system, where $d(x)$ is a nonincreasing function depending on the reliability functions of particular components and on their fractions in the system, and $\bar{\Re}'_i(x) = 1 - \Re'_i(-x)$, $i = 1, 2, 3$, for a series non-homogeneous system. In [Kołowrocki (1994)] it is stated that depending on the relationship between the number of series subsystems k_n and the number of the components in the series subsystems l_n the only possible limit reliability functions of a nonhomogeneous series-parallel system are:

$$\Re'_1(x) = \begin{cases} 1 & \text{for } x \geq 0, \\ 1 - \exp[-d(x)x^{-\alpha}] & \text{for } x > 0,\ \alpha > 0, \end{cases}$$

$$\Re'_2(x) = \begin{cases} 1 - \exp[-d(x)(-x)^{\alpha}] & \text{for } x \leq 0,\ \alpha > 0, \\ 0 & \text{for } x < 0, \end{cases}$$

$$\Re'_3(x) = 1 - \exp[-d(x)\exp(-x)] \quad \text{for } x \in (-\infty, \infty),$$

if $k_n = n$, $|l_n - c\log n| >> s$, where $s > 0$, $c > 0$;

$$\Re'_4(x) = \begin{cases} 1 & \text{for } x < 0, \\ \exp[-d(x)\exp[-x^{\alpha} - s/c]] & \text{for } x \geq 0,\ \alpha > 0, \end{cases}$$

$$\Re'_5(x) = \begin{cases} 1 - \exp[-d(x)\exp[(-x)^{\alpha} - s/c]] & \text{for } x < 0, \\ 0 & \text{for } x \geq 0,\ \alpha > 0, \end{cases}$$

$$\Re'_6(x) = \begin{cases} 1 - \exp[-d(x)\exp[\beta(-x)^{\alpha} - s/c]], & \text{for } x < 0, \\ 1 - \exp[-d(x)\exp[-x^{\alpha} - s/c]] & \text{for } x \geq 0,\ \alpha > 0, \end{cases}$$

$$\Re'_7(x) = \begin{cases} 1 & \text{for } x < x_1, \\ 1 - \exp[-d(x)\exp[-s/c]] & \text{for } x_1 \leq x < x_2, \\ 0 & \text{for } x \geq x_2,\ x_1 < x_2, \end{cases}$$

if $k_n = n$, $l_n - c\log n \sim s$, $s \in (-\infty, \infty)$, $c > 0$;

$$\Re'_8(x) = \begin{cases} 1 - \prod_{i=1}^{a}[1 - d_i(x)\exp[-(-x)^{\alpha}]]^{q_i k} & \text{for } x < 0,\ \alpha > 0, \\ 0 & \text{for } x \geq 0, \end{cases}$$

$$\Re' 9(x) = \begin{cases} 1 & \text{for } x < 0, \\ 1 - \prod_{i=1}^{a}[1 - d_i(x)\exp[-x^{\alpha}]]^{q_i k} & \text{for } x \geq 0,\ \alpha > 0, \end{cases}$$

$$\Re'_{10}(x) = 1 - \prod_{i=1}^{a}[1 - d_i(x)\exp[-\exp x]]^{q_i k} \quad \text{for } x \in (-\infty, \infty),$$

if $k_n \to k$, $l_n \to \infty$; where $d(x)$ and $d_i(x)$ are nonincreasing functions depending on the reliability functions of particular components and on their fractions in the system. In [Kołowrocki (1994)] it is also stated that depending on the relationship between k_n and l_n the only possible limit reliability functions of a nonhomogeneous parallel-series system are: $\bar{\Re}'_i(x) = 1 - \Re'_i(-x)$, $i = 1, 2, \ldots, 10$. In proving facts on limit reliability functions of nonhomogeneous series and parallel systems we use the following lemmas, which immediately follow from lemmas given in [Kołowrocki (1994)].

Lemma 11.3.1 *If*

(i) $\bar{\mathfrak{R}}'(x) = \exp[-\bar{V}'(x)]$ *is a nondegenerate reliability function,*

(ii) $\bar{R}'_n(x)$ *is given by (11.3),*

(iii) $F(x)$ *is one of the distribution functions* $F^{(1)}(x), F^{(2)}(x), \ldots, F^{(a)}(x)$ *such that*

$$\lim_{n\to\infty} \frac{F^{(i)}(a_n x + b_n)}{F(a_n x + b_n)} \le 1 \quad \text{for } x \ge x_0,\ i = 1, 2, \ldots, a,$$

and moreover there exists

$$d(x) = \begin{cases} 0 & \text{for } x < x_0 \\ \lim_{n\to\infty} \sum_{i=1}^{a} q_i d_i(a_n x + b_n) & \text{for } x \ge x_0. \end{cases}$$

where

$$d_i(a_n x + b_n) = \frac{F^{(i)}(a_n x + b_n)}{F(a_n x + b_n)}$$

and $x_0 \in [-\infty, \infty)$ *is such a point that* $\exists\ N\ \forall\ n > N$

$$F(a_n x + b_n) = 0 \text{ for } x < x_0 \text{ and } F(a_n x + b_n) \ne 0 \text{ for } x \ge x_0,$$

then

$$\bar{R}'_n(a_n x + b_n) = \bar{\mathfrak{R}}'(x) \qquad \text{for } x \in C_{\bar{\mathfrak{R}}'} \text{ iff}$$
$$\lim_{n\to\infty} nF(a_n x + b_n) d(x) = \bar{V}'(x) \qquad \text{for } x \in C_{\bar{V}'}.$$

Lemma 11.3.2 *If*

(i) $\mathfrak{R}'(x) = 1 - \exp[-V'(x)]$ *is a nondegenerate reliability function,*

(ii) $R_n(x)$ *is given by (11.3),*

(iii) $R(x)$ *is one of the reliability functions* $R^{(1)}(x), R^{(2)}(x), \ldots, R^{(a)}(x)$ *such that*

$$\lim_{n\to\infty} \frac{R^{(i)}(a_n x + b_n)}{R(a_n x + b_n)} \le 1 \quad \text{for } x < x_0,\ i = 1, 2, \ldots, a,$$

and moreover there exists

$$d(x) = \begin{cases} \lim_{n\to\infty} \sum_{i=1}^{a} q_i d_i(a_n x + b_n) & \text{for } x < x_0 \\ 0 & \text{for } x \ge x_0, \end{cases}$$

where

$$d_i(a_n x + b_n) = \frac{R^{(i)}(a_n x + b_n)}{R(a_n x + b_n)}$$

and $x_0 \in (-\infty, \infty]$ *is such a point that* $\exists\ N, \forall\ n > N$

$$R(a_n x + b_n) \ne 0 \text{ for } x < x_0 \text{ and } R(a_n x + b_n) = 0 \text{ for } x \ge x_0,$$

then

$$\lim_{n\to\infty} R'_n(a_nx+b_n) = \Re'(x) \qquad \text{for } x \in C_{\Re'} \text{ iff}$$
$$\lim_{n\to\infty} nR(a_nx+b_n)d(x) = V'(x) \qquad \text{for } x \in C_{V'}.$$

In proving facts on limit reliability functions of nonhomogeneous series-parallel systems we use the following modifications of lemmas given in [Kołowrocki (1994)].

Lemma 11.3.3 *If*

(i) $\Re'(x) = 1 - \exp[-V'(x)]$ *is a nondegenerate reliability function,*

(ii) $R'_n(x)$ *is given by (11.4),*

(iii) $\lim_{n\to\infty} k_n = \infty$,

(iv) $R(x)$ *is one of the reliability functions* $R^{(1)}(x), R^{(2)}(x), \ldots, R^{(a)}(x)$ *given by (11.4) such that*

$$\lim_{n\to\infty} \frac{R^{(i)}(a_nx+b_n)}{R(a_nx+b_n)} \le 1 \qquad \text{for } x < x_0,\ i = 1, 2, \ldots, a,$$

and moreover there exists

$$d(x) = \begin{cases} \lim_{n\to\infty} \sum_{i=1}^{a} q_i d_i(a_nx+b_n) & \text{for } x < x_0 \\ 0 & \text{for } x \ge x_0, \end{cases}$$

where

$$d_i(a_nx+b_n) = \left(\frac{R^{(i)}(a_nx+b_n)}{R(a_nx+b_n)}\right)^{l_n}$$

and $x_0 \in (-\infty, \infty\rangle$ *is such a point that* $\exists\ N, \forall\ n > N$

$$R(a_nx+b_n) \ne 0 \text{ for } x < x_0 \text{ and } R(a_nx+b_n) = 0 \text{ for } x \ge x_0,$$

then

$$\lim_{n\to\infty} R'_n(a_nx+b_n) = \Re'(x) \qquad \text{for } x \in C_{\Re'} \text{ iff}$$
$$\lim_{n\to\infty} k_n(R(a_nx+b_n))^{l_n} d(x) = V'(x) \qquad \text{for } x \in C_{V'}.$$

Lemma 11.3.4 *If*

(i) $\Re'(x)$ *is a nondegenerate reliability function,*

(ii) $R'_n(x)$ *is given by (11.4),*

(iii) $\lim_{n\to\infty} k_n = k$, $\lim_{n\to\infty} l_n = \infty$,

(iv) $R(x)$ *is one of the reliability functions* $R^{(1)}(x), R^{(2)}(x), \ldots, R^{(a)}(x)$ *given by (11.4) such that*

$$\lim_{n\to\infty} \frac{R^{(i)}(a_n x + b_n)}{R(a_n x + b_n)} \leq 1 \quad \textit{for } x < x_0, \; i = 1, 2, \ldots, a,$$

and moreover there exist

$$d_i(x) = \begin{cases} \lim_{n\to\infty} d_i(a_n x + b_n) & \textit{for } x < x_0 \\ 0 & \textit{for } x \geq x_0, \end{cases}$$

where

$$d_i(a_n x + b_n) = \left(\frac{R^{(i)}(a_n x + b_n)}{R(a_n x + b_n)} \right)^{l_n}$$

and $x_0 \in (-\infty, \infty]$ *is such a point that* $\exists\, N, \forall\, n > N$

$$R(a_n x + b_n) \neq 0 \textit{ for } x < x_0 \textit{ and } R(a_n x + b_n) = 0 \textit{ for } x \geq x_0,$$

then

$$\lim_{n\to\infty} R'_n(a_n x + b_n) = \Re'(x) \qquad \textit{for } x \in C_{\Re'} \textit{ iff}$$
$$\lim_{n\to\infty} (R(a_n x + b_n))^{l_n} = \Re_0(x) \qquad \textit{for } x \in C_{\Re_0},$$

where $\Re_0(x)$ *is a nondegenerate reliability function and moreover*

$$\Re'(x) = 1 - \prod_{i=1}^{a} [1 - d_i(x)\Re_0(x)]^{q_i k}, \qquad x \in (-\infty, \infty).$$

Fact 11.3.1 If system components have Weibull reliability functions given by $R^{(i,j)}(x) = \exp[-\beta_{ij} x^{\alpha_{ij}}]$ for $x \geq 0$, $\alpha_{ij} > 0$, $\beta_{ij} > 0$, for $i = 1, 2, \ldots, a$, $j = 1, 2, \ldots, e_i$, and $k_n \to k$, $l_n \to \infty$, $a_n = 1/(\beta l_n)^{1/\alpha}$, $b_n = 0$, where $\alpha_i = \min_{1 \leq j \leq e_i}\{\alpha_{ij}\}$, $\beta_i = \sum_{[j:\alpha_{ij}=\alpha_i]} p_{ij}\beta_{ij}$, $\alpha = \max_{1 \leq i \leq a}\{\alpha_i\}$, $\beta = \min\{\beta_i : \alpha_i = \alpha\}$, then

$$\Re'_9(x) = 1 - \prod_{[i:\alpha_i=\alpha]} \left[1 - \exp\left[-\frac{\beta_i}{\beta}\, x^{\alpha}\right]\right]^{q_i k} \qquad \text{for } x \geq 0,$$

is the limit reliability function of the nonhomogeneous series-parallel system. (The proof is based on Lemma 11.3.4 and given in [Kołowrocki (1998)]).

Example 11.3.1 Let us consider the water supply composed of $k_n = 3$ lines each of them composed of $l_n = 100$ segment pipes of two kinds. In two of the lines there are 40 segment pipes with the reliability functions given by $R^{(i,j)}(x) = \exp[-0.05x]$ for $x \geq 0$, $i = 1, 2$, $j = 1, 2, \ldots, 40$, and 60 segment pipes with the reliability functions given by $R^{(i,j)}(x) = \exp[-0.0015x^2]$ for $x \geq 0$, $i = 1, 2$, $j = 41, 42, \ldots, 100$. In the third line there are 50 segment pipes with the reliability functions given by $R^{(i,j)}(x) = \exp[-0.0007x^3]$ for $x \geq 0$, $i = 3$, $j = 1, 2, \ldots, 50$ and 50 segment pipes with the reliability functions given by $R^{(i,j)}(x) = \exp[-0.2\sqrt{x}]$ for $x \geq 0$, $i = 3$, $j = 51, 52, \ldots, 100$. We may consider this supply as the nonhomogeneous series-parallel system. In order to evaluate the reliability of this system we shall apply Fact 11.3.1. According to Definition 11.1.4, we have $k_n = k = 3$, $l_n = 100$, $a = 2$, $q_1 = 2/3$, $q_2 = 1/3$. Therefore, by (11.4) we get

$$R_n(x) = 1 - \prod_{i=1}^{2}[1 - (R^{(i)}(x))^{100}]^{q_i 3} = 1 - [-(R^{(1)}(x))^{100}]^2[1 - (R^{(2)}(x))^{100}],$$

where substituting: $e_1 = 2$, $p_{11} = 0.4$, $p_{12} = 0.6$, $\alpha_{11} = 1$, $\beta_{11} = 0.05$, $\alpha_{12} = 2$, $\beta_{12} = 0.0015$, by (11.4)

$$\begin{aligned} R^{(1)}(x) &= \prod_{j=1}^{e_1}(R^{(1,j)}(x))^{p_{1j}} = (R^{(1,1)}(x))^{0.4}(R^{(1,2)}(x))^{0.6} \\ &= \exp[-0.02x - 0.0009x^2] \end{aligned}$$

and substituting: $e_2 = 2$, $p_{21} = 0.5$, $p_{22} = 0.5$, $\alpha_{21} = 3$, $\beta_{21} = 0.0007$, $\alpha_{22} = 0.5$, $\beta_{22} = 0.2$, by (11.4)

$$\begin{aligned} R^{(2)}(x) &= \prod_{j=1}^{e_2}(R^{(2,j)}(x))^{p_{2j}} = (R^{(2,1)}(x))^{0.5}(R^{(2,2)}(x))^{0.5} \\ &= \exp[-0.00035x^3 - 0.1\sqrt{x}]. \end{aligned}$$

Hence, the exact reliability function of the system is given by $R_n(x) = 1 - [1 - \exp[-2x - 0.09x^2]]^2 \cdot [1 - \exp[-0.035x^3 - 10\sqrt{x}]$ for $x \geq 0$. Further, according to Fact 11.2.3, we get $\alpha_1 = \min\{\alpha_{11}, \alpha_{12}\} = \min\{1, 2\} = 1$, $\beta_1 = p_{11}\beta_{11} = 0.4 \cdot 0.05 = 0.02$, $\alpha_2 = \min\{\alpha_{21}, \alpha_{22}\} = \min\{3, 0.5\} = 0.5$, $\beta_2 = p_{22}\beta_{22} = 0.5 \cdot 0.2 = 0.1$, $\alpha = \max\{\alpha_1, \alpha_2\} = \max\{1, 0.5\} = 1$, $\beta = \min\{\beta_1\} = \min\{0.02\} = 0.02$, $a_n = 0.002 \cdot 100 = 0.5$, $b_n = 0$, and the limit reliability function of the system is $\Re_9(x) = 1 - [1 - \exp[-x]]^2$ for $x \geq 0$. Hence, the exact reliability function of the system is approximately given by $R_n(x) \simeq \Re_9((x - b_n)/a_n) = 1 - [1 - \exp[-2x]]^2$ for $x \geq 0$. The data is close to the practice. According to the experts opinion the mean lifetimes of the pipes change in the range from 10 to 50 years. The component mean lifetimes considered in the example are: $E(X_{11}) = 1/0.05 = 20$ years, $E(X_{12}) = \Gamma(3/2)(0.0015)^{-1/2} \simeq 23$ years, $E(X_{21}) = \Gamma(4/3)(0.0007)^{-1/3} \simeq 10$ years, $E(X_{22}) = \Gamma(3)(0.2)^{-2} \simeq 50$ years. The expected value of the system lifetime calculated from the limit reliability function is $E(X) = 0.75$ year.

The differences between exact and asymptotic reliability functions of the water supply system are shown in Figure 11.11.

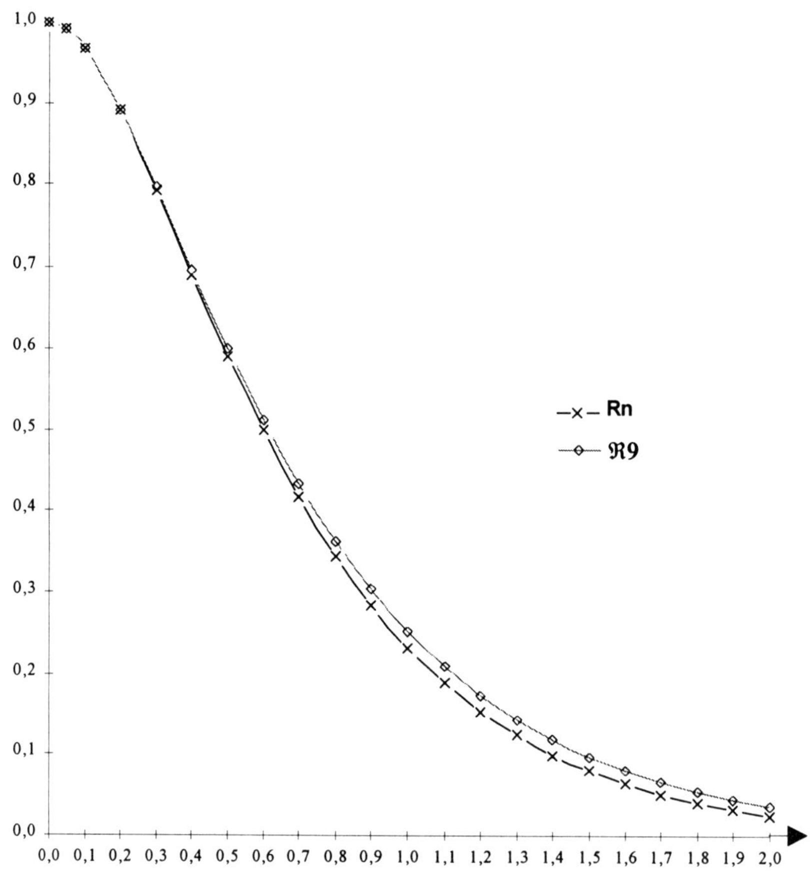

Figure 11.11: Exact and limit reliability functions of the water supply system

In proving facts on limit reliability functions of nonhomogeneous parallel-series systems we use the following modifications of lemmas given in [Kołowrocki (1994)].

Lemma 11.3.5 *If*

(i) $\Re'(x) = \exp[-\bar{V}'(x)]$ *is a nondegenerate reliability function,*

(ii) $R'_n(x)$ *is given by (11.5),*

(iii) $\lim_{n\to\infty} k_n = \infty$,

(iv) $F(x)$ *is one of the distribution functions* $F^{(1)}(x), F^{(2)}(x), \ldots, F^{(a)}(x)$ *given by (11.5) such that*

$$\lim_{n\to\infty} \frac{F^{(i)}(a_n x + b_n)}{F(a_n x + b_n)} \leq 1 \quad \text{for } x \geq x_0,\ i = 1, 2, \ldots, a,$$

and moreover there exists

$$d(x) = \begin{cases} 0 & \text{for } x < x_0 \\ \lim_{n\to\infty} \sum_{i=1}^{a} q_i d_i(a_n x + b_n) & \text{for } x \geq x_0, \end{cases}$$

where

$$d_i(a_n x + b_n) = \left(\frac{F^{(i)}(a_n x + b_n)}{F(a_n x + b_n)} \right)^{l_n}$$

and $x_0 \in [-\infty, \infty)$ *is such a point that* $\exists\ N, \forall\ n > N$

$$F(a_n x + b_n) = 0 \text{ for } x < x_0 \text{ and } F(a_n x + b_n) \neq 0 \text{ for } x \geq x_0,$$

then

$$\lim_{n\to\infty} \bar{R}'_n(a_n x + b_n) = \bar{\Re}'(x) \qquad \text{for } x \in C_{\bar{\Re}'} \text{ iff}$$

$$\lim_{n\to\infty} k_n (F(a_n x + b_n))^{l_n} d(x) = \bar{V}'(x) \qquad \text{for } x \in C_{\bar{V}'}.$$

Lemma 11.3.6 *If*

(i) $\bar{\Re}'(x)$ *is a nondegenerate reliability function,*

(ii) $\bar{R}'_n(x)$ *is given by (11.5),*

(iii) $\lim_{n\to\infty} k_n = k$, $\lim_{n\to\infty} l_n = \infty$,

(iv) $F(x)$ *is one of the distribution functions* $F^{(1)}(x), F^{(2)}(x), \ldots, F^{(a)}(x)$ *given by (11.5) such that*

$$\lim_{n\to\infty} \frac{F^{(i)}(a_n x + b_n)}{F(a_n x + b_n)} \leq 1 \quad \text{for } x \geq x_0,\ i = 1, 2, \ldots, a,$$

and moreover there exist

$$d_i(x) = \begin{cases} 0 & \text{for } x < x_0 \\ \lim_{n\to\infty} d_i(a_n x + b_n) & \text{for } x \geq x_0, \end{cases}$$

where

$$d_i(a_n x + b_n) = \left(\frac{F^{(i)}(a_n x + b_n)}{F(a_n x + b_n)} \right)^{l_n}$$

and $x_0 \in [-\infty, \infty)$ *is such a point that* $\exists\ N, \forall\ n > N$

$$F(a_n x + b_n) = 0 \text{ for } x < x_0 \text{ and } F(a_n x + b_n) \neq 0 \text{ for } x \geq x_0,$$

then

$$\lim_{n\to\infty} \bar{R}'_n(a_n x + b_n) = \bar{\Re}'(x) \qquad \textit{for } x \in C_{\bar{\Re}'} \textit{ iff}$$
$$\lim_{n\to\infty} (F(a_n x + b_n))^{l_n} = \mathcal{T}_0(x) \qquad \textit{for } x \in C_{\mathcal{T}_0},$$

where $\mathcal{T}_0(x)$ *is a nondegenerate distribution function and moreover*

$$\bar{\Re}'(x) = \prod_{i=1}^{a} [1 - d_i(x)\mathcal{T}_0(x)]^{q_i k}, \quad x \in (-\infty, \infty).$$

11.4 Remarks on Limit Reliability Functions of Multi-State Systems

We suppose that:

- E_i, $i = 1, 2, \ldots, n$, are components of a system,
- all components and a system under consideration have the state set $\{0, 1, \ldots, m\}$,
- $X_i(k)$ are independent random variables representing the lifetimes of components E_i in the state subset $\{k, k+1, \ldots, m\}$,
- $X(k)$ is a random variable representing the lifetime of a system in the state subset $\{k, k+1, \ldots, m\}$,
- the state indexes are ordered, the state 0 is the worst and the state m is the best,
- the system state degrades with time x without repair,
- $e_i(x)$ is a component E_i state at the moment x, $x \in (-\infty, \infty)$.

Definition 11.4.1 [Xue and Yang (1995)] A vector

$$R_i(x) = [R_i(x, 0), R_i(x, 1), \ldots, R_i(x, m)], \quad x \in (-\infty, \infty), \; i = 1, 2, \ldots, n,$$

where

$$R_i(x, k) = P(e_i(x) \geq k) = P(X_i(k) > x), \quad x \in (-\infty, \infty), \; k = 0, 1, \ldots, m,$$

is the probability that the component E_i is in the state subset $\{k, k+1, \ldots, m\}$ at the moment x, $x \in (-\infty, \infty)$, is called the multi-state reliability function of a component E_i.

Under the above definition the following property is obvious

$$1 = R_i(x,0) \geq R_i(x,1) \geq \cdots \geq R_i(x,m), \quad x \in (-\infty,\infty),\ i = 1,2,\ldots,n.$$

Definition 11.4.2 We call the system homogeneous if $X_i(k)$ for a fixed k have an identical distribution function

$$F_i(x,k) = F(x,k), \quad x \in (-\infty,\infty),\ i = 1,2,\ldots,n,$$

i.e., if

$$R_i(x,k) = R(x,k) = 1 - F(x,k), \quad x \in (-\infty,\infty),\ i = 1,2,\ldots,n.$$

We are interested in limit distributions of a standardized random variable

$$(X(k) - b_n(k))/a_n(k), \qquad k = 1,2,\ldots,m,$$

where $X(k)$ is the lifetime of the system in the state subset $\{k, k+1, \ldots, m\}$ and

$$a_n(k) > 0,\ b_n(k) \in (-\infty,\infty), \quad k = 1,2,\ldots,m,$$

are some suitably chosen numbers. And, since

$$\begin{aligned} P((X(k) - b_n(k))/a_n((k) > x) &= P(X(k) > a_n(k)x + b_n(k)) \\ &= R_n(a_n(k)x + b_n(k), k), \quad k = 1,2,\ldots,m, \end{aligned}$$

where

$$R_n(x) = [R_n(x,0), R_n(x,1), \ldots, R_n(x,m)], \quad x \in (-\infty,\infty),$$

is the multi-state reliability function of the system, then we assume the following definition.

Definition 11.4.3 A vector

$$\Re(x) = [\Re(x,0), \Re(x,1), \ldots, \Re(x,m)], \quad x \in (-\infty,\infty),$$

is called the limit multi-state reliability function of the system if there exist norming constants $a_n(k) > 0$, $b_n(k) \in (-\infty,\infty)$ such that

$$\lim_{n\to\infty} R_n(a_n(k)x + b_n(k), k) = \Re(x,k) \text{ for } x \in C_{\Re(k)}, \quad k = 1,2,\ldots,m,$$

where $C_{\Re(k)}$ is the set of continuity points of $\Re(x,k)$.

Hence, for sufficiently large n, we get the following approximate formula

$$[R_n(x,0), R_n(x,1), \ldots, R_n(x,m)] \simeq \left[1, \Re\left(\frac{x-b_n(1)}{a_n(1)}, 1\right), \ldots, \Re\left(\frac{x-b_n(m)}{a_n(m)}, m\right)\right], \quad x \in (-\infty, \infty). \tag{11.6}$$

The simplest reliability structures are series and parallel systems with lifetimes of systems in the state subset $\{k, k+1, \ldots, m\}$ respectively given by

$$X(k) = \min_{1 \le i \le n} \{X_i(k)\}, \quad k = 1, 2, \ldots, m,$$

and

$$X(k) = \max_{1 \le i \le n} \{X_i(k)\}, \quad k = 1, 2, \ldots, m.$$

It is easy to notice that the reliability functions of homogeneous series and parallel multi-state systems are respectively given by

$$\bar{R}_n(x) = [\bar{R}_n(x,0), \bar{R}_n(x,1), \ldots, \bar{R}_n(x,m)],$$

where

$$\bar{R}_n(x,k) = [R(x,k)]^n, \qquad k = 0, 1, \ldots, m,$$

and

$$R_n(x) = [R_n(x,0), R_n(x,1), \ldots, R_n(x,m)],$$

where

$$R_n(x,k) = 1 - [F(x,k)]^n, \qquad k = 0, 1, \ldots, m,$$

where n is the number of system components.

The next simple systems are series-parallel and parallel-series homogeneous systems with lifetimes of systems in the state subset $\{k, k+1, \ldots, m\}$ respectively given by

$$X(k) = \max_{1 \le i \le k_n} \left\{ \min_{1 \le j \le l_n} \{X_{ij}(k)\} \right\}, \quad k = 1, 2, \ldots, m,$$

and

$$X(k) = \min_{l \le i \le k_n} \left\{ \max_{1 \le j \le l_n} \{X_{ij}(k)\} \right\}, \quad k = 1, 2, \ldots, m.$$

The reliability functions of these systems are respectively given by

$$R_n(x) = [R_n(x,0), R_n(x,1), \ldots, R_n(x,m)],$$

where

$$R_n(x,k) = 1 - [1 - [R(x,k)]^{l_n}]^{k_n}, \quad k = 0, 1, \ldots, m,$$

and

$$\bar{R}_n(x) = [\bar{R}_n(x,0), \bar{R}_n(x,1), \ldots, \bar{R}_n(x,m)],$$

where

$$\bar{R}_n(x,k) = [1 - [F(x,k)]^{l_n}]^{k_n}, \quad k = 0, 1, \ldots, m.$$

Definition 11.4.4 We call the homogeneous multi-state system an exp-system if

$$R(x,k) = \begin{cases} 1 & \text{for } x < 0, \\ \exp[-\lambda(k)x] & \text{for } x \geq 0, \end{cases} \qquad \lambda(k) > 0,\ k = 1,2,\ldots,m.$$

In proving facts on limit reliability functions of homogeneous series and parallel systems we use the following slight extensions of Lemmas 11.2.1 and 11.2.2.

Lemma 11.4.1 *A nondegenerate multi-state reliability function*

$$\bar{\mathfrak{R}}(x) = [\bar{\mathfrak{R}}(x,0), \bar{\mathfrak{R}}(x,1), \ldots, \bar{\mathfrak{R}}(x,m)], \quad x \in (-\infty, \infty),$$

where

$$\bar{\mathfrak{R}}(x,k) = \exp[-\bar{V}(x,k)], \quad k = 1,2,\ldots,m,$$

is the limit multi-state reliability function of the homogeneous series system if and only if

$$\lim_{n\to\infty} nF(a_n(k)x + b_n(k), k) = \bar{V}(x,k) \text{ for } x \in C_{\bar{V}(k)},\ k = 1,2,\ldots,m,$$

where $C_{\bar{V}(k)}$ is the set of continuity points of $\bar{V}(x,k)$ and points such that $\bar{V}(x,k) = \infty$.

Lemma 11.4.2 *A nondegenerate multi-state reliability function*

$$\mathfrak{R}(x) = [\mathfrak{R}(x,0), \mathfrak{R}(x,1), \ldots, \mathfrak{R}(x,m)], \quad x \in (-\infty, \infty),$$

where

$$\mathfrak{R}(x,k) = 1 - \exp[-V(x,k)],\ k = 1,2,\ldots,m,$$

is the limit multi-state reliability function of the homogeneous parallel system if and only if

$$\lim_{n\to\infty} nR(a_n(k)x + b_n(k), k) = V(x,k) \text{ for } x \in C_{V(k)},\ k = 1,2,\ldots,m,$$

where $C_{V(k)}$ is the set of continuity points of $V(x,k)$ and points such that $V(x,k) = \infty$.

Proposition 11.4.1 *If*

$$a_n(k) = \frac{1}{\lambda(k)n}, \quad b_n(k) = 0, \quad k = 1,2,\ldots,m,$$

then

$$\bar{\mathfrak{R}}_2(x) = [\bar{\mathfrak{R}}_2(x,0), \bar{\mathfrak{R}}_2(x,1), \ldots, \bar{\mathfrak{R}}_2(x,m)], \quad x \in (-\infty, \infty),$$

where

$$\bar{\mathfrak{R}}_2(x,k) = \begin{cases} 1 & \text{for } x < 0, \\ \exp[-x] & \text{for } x \geq 0, \end{cases} \qquad k = 1,2,\ldots,m,$$

is the limit multi-state reliability function of the series exp-system.

Proposition 11.4.2 *If*

$$a_n(k) = \frac{1}{\lambda(k)}, \quad b_n(k) = \frac{1}{\lambda(k)} \log n, \quad k = 1, 2, \ldots, m,$$

then

$$\bar{\mathfrak{R}}_3(x) = [\bar{\mathfrak{R}}_3(x,0), \bar{\mathfrak{R}}_3(x,1), \ldots, \bar{\mathfrak{R}}_3(x,m)], \quad x \in (-\infty, \infty),$$

where

$$\bar{\mathfrak{R}}_3(x,k) = 1 - \exp[-\exp[-x]] \text{ for } x \in (-\infty, \infty), \ k = 1, 2, \ldots, m,$$

is the limit multi-state reliability function of the parallel exp-system.

In proving facts on limit reliability functions of homogeneous series-parallel systems we use the following extensions of Lemmas 11.2.3 and 11.2.4.

Lemma 11.4.3 *If*

$$k_n \to \infty,$$

then a nondegenerate multi-state reliability function

$$\mathfrak{R}(x) = [\mathfrak{R}(x,0), \mathfrak{R}(x,1), \ldots, \mathfrak{R}(x,m)], \quad x \in (-\infty, \infty),$$

where

$$\mathfrak{R}(x,k) = 1 - \exp[-V(x,k)], \quad k = 1, 2, \ldots, m,$$

is the limit multi-state reliability function of the homogeneous series-parallel system if and only if

$$\lim_{n\to\infty} k_n[R(a_n(k)x + b_n(k), k)]^{l_n} = V(x,k) \text{ for } x \in C_{V(k)}, \ k = 1, 2, \ldots, m,$$

where $C_{V(k)}$ is the set of continuity points of $V(x,k)$ and points such that $V(x,k) = \infty$.

Lemma 11.4.4 *If*

$$k_n \to c, \ c > 0, \qquad l_n \to \infty,$$

then a nondegenerate multi-state reliability function

$$\mathfrak{R}(t) = [\mathfrak{R}(x,0), \mathfrak{R}(x,1), \ldots, \mathfrak{R}(x,m)], \quad x \in (-\infty, \infty),$$

is the limit multi-state reliability function of the homogeneous series-parallel system if and only if

$$\lim_{n\to\infty} [R(a_n(k)x + b_n(k), k)]^{l_n} = \mathfrak{R}_0(x,k) \text{ for } x \in C_{\mathfrak{R}_0(k)}, \ k = 1, 2, \ldots, m,$$

where $\mathfrak{R}_0(x,k)$ is a nondegenerate reliability function, and moreover

$$\mathfrak{R}(x,k) = 1 - [1 - \mathfrak{R}_0(x,k)]^c \text{ for } x \in (-\infty, \infty), \ k = 1, 2, \ldots, m.$$

Proposition 11.4.3 *If*

$$k_n = n, \quad l_n > 0,$$
$$a_n(k) = \frac{1}{\lambda(k)l_n}, \quad b_n(k) = \frac{1}{\lambda(k)l_n}\log n, \quad k = 1, 2, \ldots, m,$$

then

$$\Re_3(x) = [\Re_3(x,0), \Re_3(x,1), \ldots, \Re_3(x,m)], \quad x \in (-\infty, \infty),$$

where

$$\Re_3(x,k) = 1 - \exp[-\exp[-x]] \text{ for } x \in (-\infty, \infty), \; k = 1, 2, \ldots, m,$$

is the limit multi-state reliability function of the series-parallel exp-system.

Proposition 11.4.4 *If*

$$k_n \to c, \quad c > 0, \quad l_n > 0,$$
$$a_n(k) = \frac{1}{\lambda(k)l_n}, \quad b_n(k) = 0, \quad k = 1, 2, \ldots, m,$$

then

$$\Re_9(x) = [\Re_9(x,0), \Re_9(x,1), \ldots, \Re_9(x,m)], \quad x \in (-\infty, \infty),$$

where

$$\Re_9(x,k) = \begin{cases} 1 & \text{for } x < 0, \\ 1 - [1 - \exp[-x]]^c & \text{for } x \geq 0, \end{cases} \quad k = 1, 2, \ldots, m,$$

is the limit multi-state reliability function of the homogeneous series-parallel exp-system.

In proving facts on limit reliability functions of homogeneous parallel-series systems we use the following extensions of Lemmas 11.2.5 and 11.2.6.

Lemma 11.4.5 *If*

$$k_n \to \infty,$$

then a nondegenerate multi-state reliability function

$$\bar{\Re}(x) = [\bar{\Re}(x,0), \bar{\Re}(x,1), \ldots, \bar{\Re}(x,m)], \quad x \in (-\infty, \infty),$$

where

$$\bar{\Re}(x,k) = \exp[-\bar{V}(x,k)], \quad k = 1, 2, \ldots, m,$$

is the limit multi-state reliability function of the homogeneous parallel-series system if and only if

$$\lim_{n\to\infty} k_n[F(a_n(k)x + b_n(k), k)]^{l_n} = \bar{V}(x,k) \text{ for } x \in C_{\bar{V}(k)}, \; k = 1, 2, \ldots, m,$$

where $C_{\bar{V}(k)}$ is the set of continuity points of $\bar{V}(x,k)$ and points such that $\bar{V}(x,k) = \infty$.

Lemma 11.4.6 *If*

$$k_n \to c, \quad c > 0, \qquad l_n \to \infty,$$

then a nondegenerate multi-state reliability function

$$\bar{\Re}(x) = [\bar{\Re}(x,0), \bar{\Re}(x,1), \ldots, \bar{\Re}(x,m)], \quad x \in (-\infty, \infty),$$

is the limit multi-state reliability function of the homogeneous parallel-series system if and only if

$$\lim_{n \to \infty} [F(a_n(k)x + b_n(k), k)]^{l_n} = F_0(x,k) \text{ for } x \in C_{F_0(k)}, \; k = 1, 2, \ldots, m,$$

where $F_0(x,k)$ is a nondegenerate distribution function, and moreover

$$\bar{\Re}(x,k) = [1 - F_0(x,k)]^c \text{ for } x \in (-\infty, \infty), \; k = 1, 2, \ldots, m.$$

Proposition 11.4.5 *If*

$$\begin{aligned} k_n &= n, \quad l_n \to c, \quad c > 0, \\ a_n(k) &= \frac{1}{\lambda(k) n^{\frac{1}{l_n}}}, \qquad b_n(k) = 0, \quad k = 1, 2, \ldots, m, \end{aligned}$$

then

$$\bar{\Re}_2(x) = [\bar{\Re}_2(x,0), \bar{\Re}_2(x,1), \ldots, \bar{\Re}_2(x,m)], \quad x \in (-\infty, \infty),$$

where

$$\bar{\Re}_2(x,k) = \begin{cases} 1 & \text{for } x < 0, \\ \exp[-x^c] & \text{for } x \geq 0, \end{cases} \qquad k = 1, 2, \ldots, m,$$

is the limit multi-state reliability function of the homogeneous parallel-series exp-system.

Proposition 11.4.6 *If*

$$\begin{aligned} k_n &= n, \quad c << l_n, \quad c \log n - l_n >> s, \quad c > 0, \; s > 0, \\ a_n(k) &= \frac{1}{(n^{\frac{1}{l_n}} - 1)\lambda(k) l_n}, \\ \exp[-\lambda(k) b_n(k)] &= 1 - n^{-\frac{1}{l_n}}, \qquad k = 1, 2, \ldots, m, \end{aligned}$$

then

$$\bar{\Re}_3(x) = [\bar{\Re}_3(x,0), \bar{\Re}_3(x,1), \ldots, \bar{\Re}_3(x,m)], \quad x \in (-\infty, \infty),$$

where

$$\bar{\Re}_3(x,k) = \exp[-\exp[x]] \text{ for } x \in (-\infty, \infty), \quad k = 1, 2, \ldots, m,$$

is the limit reliability function of the homogeneous parallel-series exp-system.

Proposition 11.4.7 *If*

$$\begin{aligned} k_n &= n, \quad l_n - c\log n \sim s, \quad c > 0,\ s \in (-\infty, \infty), \\ a_n(k) &= \frac{1}{(e^{\frac{1}{c}} - 1)\lambda(k) l_n}, \\ b_n(k) &= -\frac{1}{\lambda(k)} \log(1 - e^{-\frac{1}{c}}), \quad k = 1, 2, \ldots, m, \end{aligned}$$

then

$$\bar{\mathfrak{R}}_3(x) = [\bar{\mathfrak{R}}_3(x,0), \bar{\mathfrak{R}}_3(x,1), \ldots, \bar{\mathfrak{R}}_3(x,m)], \quad x \in (-\infty, \infty),$$

where

$$\bar{\mathfrak{R}}_3(x,k) = \exp\left[-\exp\left[x - \frac{s}{c}\right]\right] \text{ for } x \in (-\infty, \infty),\ k = 1, 2, \ldots, m,$$

is the limit multi-state reliability function of the homogeneous parallel-series exp-system.

Proposition 11.4.8 *If*

$$\begin{aligned} k_n &= n, \quad l_n - c\log n >> s, \quad c > 0,\ s > 0, \\ a_n(k) &= \frac{1}{\lambda(k)\log n}, \\ b_n(k) &= \frac{1}{\lambda(k)} \log\left(\frac{l_n}{\log n}\right), \quad k = 1, 2, \ldots, m, \end{aligned}$$

then

$$\bar{\mathfrak{R}}_3(x) = [\bar{\mathfrak{R}}_3(x,0), \bar{\mathfrak{R}}_3(x,1), \ldots, \bar{\mathfrak{R}}_3(x,m)], \quad x \in (-\infty, \infty),$$

where

$$\bar{\mathfrak{R}}_3(x,k) = \exp[-\exp[x]] \text{ for } x \in (-\infty, \infty), \quad k = 1, 2, \ldots, m,$$

is the limit multi-state reliability function of the homogeneous parallel-series exp-system.

Proposition 11.4.9 *If*

$$\begin{aligned} k_n &\to c, \quad c > 0, \quad l_n \to \infty, \\ a_n(k) &= \frac{1}{\lambda(k)}, \\ b_n(k) &= \frac{1}{\lambda(k)} \log l_n, \quad k = 1, 2, \ldots, m, \end{aligned}$$

then

$$\bar{\mathfrak{R}}_{10}(t) = [\bar{\mathfrak{R}}_{10}(x,0), \bar{\mathfrak{R}}_{10}(x,1), \ldots, \bar{\mathfrak{R}}_{10}(x,m)], \quad x \in (-\infty, \infty),$$

where

$$\bar{\Re}_{10}(x,k) = [1 - \exp[-exp[-x]]^c \text{ for } x \in (-\infty, \infty), \quad k = 1, 2, \ldots, m,$$

is the limit multi-state reliability function of the homogeneous parallel-series exp-system.

Proofs of all propositions are given in [Kołowrocki (1998)].

Example 11.4.1 If a parallel-series system is such that $k_n = 30$, $l_n = 60$, $m = 5$, $\lambda(k) = 10^{k-6}$, $k = 1, 2, 3, 4, 5$, then according to Proposition 11.4.8, assuming $a_n(k) = 1/(10^{k-6} \cdot \log 30) = 10^{6-k} \cdot 0.294$, $b_n(k) = (1/10^{6-k}) \cdot \log(60/log30)$, $k = 1, 2, 3, 4, 5$, by (11.6), we get the approximate formula for its exact multi-state reliability function given by

$$\begin{aligned}
&[\bar{R}_{30}(x,0), \bar{R}_{30}(x,1), \bar{R}_{30}(x,2), \bar{R}_{30}(x,3), \bar{R}_{30}(x,4), \bar{R}_{30}(x,5)] \\
&\simeq \left[1, \exp\left[-\exp\frac{x-287000}{29400}\right], \exp\left[-\exp\frac{x-28700}{2940}\right],\right. \\
&\quad \left.\exp\left[-\exp\frac{x-2870}{294}\right], \exp\left[-\exp\frac{x-287}{29.4}\right], \exp\left[-\exp\frac{x-28.7}{2.94}\right]\right] \\
&= [1, \exp[-\exp[0.000034x - 9.76]], \exp[-\exp[0.00034x - 9.76]], \\
&\quad \exp[-\exp[0.0034x - 9.76]], \exp[-\exp[0.034x - 9.76]], \\
&\quad \exp[-\exp[0.34x - 9.76]]] \qquad \text{for } x \in (-\infty, \infty).
\end{aligned}$$

Hypothesis. The results presented in Sections 11.2 and 11.3 allow us to state that the classes of possible limit multi-state reliability functions for series and parallel systems are composed of 3^m members. Their forms for a series system are

$$\bar{\Re}(x) = [\bar{\Re}(x,0), \bar{\Re}(x,1), \ldots, \bar{\Re}(x,m)], \quad x \in (-\infty, \infty),$$

where

$$\bar{\Re}(x,k) \in \{\bar{\Re}_2(x), \bar{\Re}_2(x), \bar{\Re}_3(x)\}, \quad k = 1, 2, \ldots, m,$$

if a system is homogeneous and

$$\bar{\Re}'(x) = [\bar{\Re}'(x,0), \bar{\Re}'(x,1), \ldots, \bar{\Re}'(x,m)], \quad x \in (-\infty, \infty),$$

where

$$\bar{\Re}'(x,k) \in \{\bar{\Re}'_2(x), \bar{\Re}'_2(x), \bar{\Re}'_3(x)\}, \quad k = 1, 2, \ldots, m,$$

if a system is nonhomogeneous.

Their forms for a parallel system are

$$\Re(x) = [\Re(x,0), \Re(x,1), \ldots, \Re(x,m)], \quad x \in (-\infty, \infty),$$

where

$$\Re(x,k) \in \{\Re_2(x), \Re_2(x), \Re_3(x)\}, \quad k = 1, 2, \ldots, m,$$

if a system is homogeneous and

$$\Re'(x) = [\Re'(x,0), \Re'(x,1), \ldots, \Re'(x,m)], \quad x \in (-\infty, \infty),$$

where

$$\Re'(x,k) \in \{\Re_2'(x), \Re_2'(x), \Re_3'(x)\}, \quad k = 1, 2, \ldots, m,$$

if a system is nonhomogeneous.

Moreover, under the assumptions on the regularity of changing of k_n and l_n given in [Kołowrocki (1993)] and [Kołowrocki (1994)], the classes of possible limit multi-state reliability functions for series-parallel and parallel-series systems are composed of $3^m + 4^m + 3^m$ members.

Their forms for a series-parallel system are

$$\Re(x) = [\Re(x,0), \Re(x,1), \ldots, \Re(x,m)], \quad x \in (-\infty, \infty),$$

where

$$\begin{aligned}
\Re(x,k) &\in \{\Re_2(x), \Re_2(x), \Re_3(x)\}, \\
&\quad k = 1, 2, \ldots, m, \text{ if } k_n = n, \ |l_n - c\log n| >> s, \ s > 0, \ c > 0; \\
\Re(x,k) &\in \{\Re_4(x), \Re_5(x), \Re_6(x), \Re_7(x)\}, \\
&\quad k = 1, 2, \ldots, m, \text{ if } k_n = n, \ l_n - c\log n \sim s, \ s \in (-\infty, \infty), \ c > 0; \\
\Re(x,k) &\in \{\Re_8(x), \Re_9(x), \Re_{10}(x)\}, \\
&\quad k = 1, 2, \ldots, m, \text{ if } k_n \to k, \ l_n \to \infty;
\end{aligned}$$

if a system is homogeneous and

$$\Re'(x) = [\Re'(x,0), \Re'(x,1), \ldots, \Re'(x,m)], \quad x \in (-\infty, \infty),$$

where

$$\begin{aligned}
\Re'(x,k) &\in \{\Re_2'(x), \Re_2'(x), \Re_3'(x)\}, \\
&\quad k = 1, 2, \ldots, m, \text{ if } k_n = n, \ |l_n - c\log n| >> s, \ s > 0, \ c > 0; \\
\Re'(x,k) &\in \{\Re_4'(x), \Re_5'(x), \Re_6'(x), \Re_7'(x)\}, \\
&\quad k = 1, 2, \ldots, m, \text{ if } k_n = n, \ l_n - c\log n \sim s, \ s \in (-\infty, \infty), \ c > 0; \\
\Re'(x,k) &\in \{\Re_8'(x), \Re_9'(x), \Re_{10}'(x)\}, \\
&\quad k = 1, 2, \ldots, m, \text{ if } k_n \to k, \ l_n \to \infty;
\end{aligned}$$

if a system is nonhomogeneous.

Their forms for a parallel-series system are

$$\bar{\Re}(x) = [\bar{\Re}(x,0), \bar{\Re}(x,1), \ldots, \bar{\Re}(x,m)], \quad x \in (-\infty, \infty),$$

where

$$\begin{aligned}
\bar{\Re}(x,k) &\in \{\bar{\Re}_2(x), \bar{\Re}_2(x), \bar{\Re}_3(x)\}, \\
&\quad k = 1,2,\ldots,m, \text{ if } k_n = n, \ |l_n - c\log n| >> s, \ s > 0, \ c > 0; \\
\bar{\Re}(x,k) &\in \{\bar{\Re}_4(x), \bar{\Re}_5(x), \bar{\Re}_6(x), \bar{\Re}_7(x)\}, \\
&\quad k = 1,2,\ldots,m, \text{ if } k_n = n, \ l_n - c\log n \sim s, \ s \in (-\infty,\infty), \ c > 0; \\
\bar{\Re}(x,k) &\in \{\bar{\Re}_8(x), \bar{\Re}_9(x), \bar{\Re}_{10}(x)\}, \\
&\quad k = 1,2,\ldots,m, \text{ if } k_n \to k, \ l_n \to \infty;
\end{aligned}$$

if a system is homogeneous and

$$\bar{\Re}'(x) = [\bar{\Re}'(x,0), \bar{\Re}'(x,1), \ldots, \bar{\Re}'(x,m)], \quad x \in (-\infty,\infty),$$

where

$$\begin{aligned}
\bar{\Re}'(x,k) &\in \{\bar{\Re}'_2(x), \bar{\Re}'_2(x), \bar{\Re}'_3(x)\}, \\
&\quad k = 1,2,\ldots,m, \text{ if } k_n = n, \ |l_n - c\log n| >> s, \ s > 0, \ c > 0; \\
\bar{\Re}'(x,k) &\in \{\bar{\Re}'_4(x), \bar{\Re}'_5(x), \bar{\Re}'_6(x), \bar{\Re}'_7(x)\}, \\
&\quad k = 1,2,\ldots,m, \text{ if } k_n = n, \ l_n - c\log n \sim s, \ s \in (-\infty,\infty), \ c > 0; \\
\bar{\Re}'(x,k) &\in \{\bar{\Re}'_8(x), \bar{\Re}'_9(x), \bar{\Re}'_{10}(x)\}, \\
&\quad k = 1,2,\ldots,m, \text{ if } k_n \to k, \ l_n \to \infty;
\end{aligned}$$

if a system is nonhomogeneous.

11.5 Summary

In the paper series-parallel and parallel-series systems with regular reliability structures are considered, i.e., all subsystems have equal numbers of components. The regularity of these systems is formal because assuming suitably chosen reliability functions of the particular components (unfailing components in series subsystems and failed components in parallel subsystems) it is possible to get any nonregularity of the system structures. Therefore the results are applicable to nonregular homogeneous and nonhomogeneous systems as well. In practical applications of the presented results the speed of convergence of the system reliability functions sequences to the limit reliability functions is very important. The general solution of this problem for "k out of n" system is solved in [Dziubdziela (1977)] and it seems to be possible to transfer this result to series-parallel and parallel-series systems.

References

1. Barlow, R. E. and Proschan F. (1975). *Statistical Theory of Reliability and Life Testing, Probability Models*, New York: Holt, Rinehart and Winston.

2. Kołowrocki, K. (1993). On asymptotic reliability functions of series-parallel and parallel-series systems with identical components, *Reliability Engineering and System Safety*, **41**, 251–257.

3. Kołowrocki, K. (1994). The classes of asymptotic reliability functions for series-parallel and parallel-series systems, *Reliability Engineering and System Safety*, **46**, 179–188.

4. Kołowrocki, K. (1998). Limit reliability functions and their application to the evaluation of large Weibull systems, *Reliability Engineering and System Safety* (submitted).

5. Xue J. and Yang K. (1995). Dynamic reliability analysis of coherent multistate systems, *IEEE Transactions on Reliability*, **44**, 683–688.

6. Kołowrocki, K. (1998). Reliability analysis of large multistate systems, *Proceedings of the International Conference ESREL'98* (submitted).

7. Dziubdziela, W. (1977). Rozk ady graniczne ekstremalnych statystyk pozycyjnych, *Applied Mathematics*, **9**, 45–71.

12

On Limit Reliability Functions of Large Systems. Part II

Adam Cichocki, Dorota Kurowicka and Beata Milczek

Gdynia Maritime Academy, Gdynia, Poland

Abstract: Reliability functions of particular components have an influence on the form of limit reliability functions of the system. This problem is strictly related to the domains of attraction. In the paper the theorems about domains of attraction of limit reliability functions of series-parallel systems are given. Moreover, limit reliability functions of series-"k out of n" systems are discussed, as a generalization of previously solved problem for series and "k out of n" systems separately. The class of possible limit reliability functions for these systems is given. Next, the class of limit reliability functions for the systems being a generalization of series-parallel and parallel-series systems discussed in part 1 is presented. Namely, the partial solution of this problem for hierarchical series-parallel homogeneous regular systems is given.

Keywords and phrases: Large scale systems, asymptotic reliability, "k out of n" systems, hierarchical systems

12.1 Domains of Attraction of Limit Reliability Functions

There are known ten types of limit reliability functions for series-parallel and ten types for parallel- series systems given in [Kołowrocki (1998)]. We can formulate conditions which the reliability functions of particular components should satisfy so that the limit reliability function of the system is one of the possible types. This way the reliability functions of particular components will be divided into the groups, called domains of attraction.

We assume the following definition.

Definition 12.1.1 A reliability functions $\Re$ is called a limit reliability function of the regular homogeneous series-parallel system if there exist functions $a_t = a(n) > 0$ and $b_n = b(n) \in (-\infty, \infty)$ such that

$$\lim_{n\to\infty} R_n(a_n x + b_n) = \Re(x) \quad \text{for } x \in C_{\Re}$$

where $C_{\Re}$ is the set of continuity points of $\Re$. Moreover, we say that a reliability functions R belongs to the domain of attraction of $\Re$ and write $R \in D_{\Re}$. The pair (a_n, b_n) is called a norming function pair.

In [Kołowrocki (1998)] it is solved that in the case when $k_n = n$, $l_n << c(\ln n)^{\rho(n)}$, $c > 0$, $0 \le \rho(n) < 1$, the only possible nondegenerate limit reliability functions are $\Re_1(x), \Re_2(x), \Re_3(x)$.

For this relationship between the number of series l_n and the number of parallel k_n components the following theorem holds.

Theorem 12.1.1 *Let $k_n = n$, $l_n << c(\ln n)^{\rho(n)}$, $c > 0$, $0 \le \rho(n) < 1$.*

(i) If $R \in D_{\Re_1}$ then $\exists\ a > 0$

$$\lim_{r\to\infty} \left(\frac{R(r)}{R(rx)}\right)^{l_r} = x^{a\alpha} \quad \textit{for } x > 0.$$

If

$$\lim_{r\to\infty} \left(\frac{R(r)}{R(rx)}\right)^{l_r} = x^{a\alpha} \quad \textit{for } x > 0,$$

where

$$a = \frac{l_{a_n}}{l_n} \textit{ and } a_n = \inf\left\{x : R(x(1+0)) \le \frac{1}{k_n^{\frac{1}{l_n}}} \le R(x(1-0))\right\}$$

then $R \in D_{\Re_1}$.

(ii) If $R \in D_{\Re_2}$ then

(a) there exists y such that $R(y) = 0$ and $R(y-\varepsilon) > 0$ for all $\varepsilon > 0$,

(b) $\exists\ a > 0$ $\lim_{r\to 0^-} \left(\frac{R(rx+y)}{R(r+y)}\right)^{l_r} = x^{a\alpha}$ for $x > 0$.

If

(a′) there exists y such that $R(y) = 0$ and $R(y-\varepsilon) > 0$ for all $\varepsilon > 0$,

(b′) $\lim_{r\to 0^-} \left(\frac{R(rx+y)}{R(r+y)}\right)^{l_r} = x^{a\alpha}$ for $x > 0$,

where

$$a = \frac{l_{a_n}}{l_n} \text{ and } a_n = \inf\left\{x : R(-x(1-0)+y) \le \frac{1}{k_n^{\frac{1}{l_n}}} \le R(-x(1+0)+y)\right\}$$

then $R \in D_{\Re_2}$.

(iii) $R \in D_{\Re_2}$ *if and only if the condition*

$$\lim_{n\to\infty} k_n(R(a_n x + b_n))^{l_n} = e^{-x}$$

is satisfied for all x, *where functions* b_n *and* a_n *are defined as*

$$\begin{aligned} b_n &= \inf\left\{x : R(x(1+0)+b_n) \le \frac{1}{(k_n e)^{\frac{1}{l_n}}} \le R(x(1-0)+b_n)\right\}, \\ a_n &= \inf\left\{x : R(x+0) \le \frac{1}{k_n^{\frac{1}{l_n}}} \le R(x-0)\right\}. \end{aligned}$$

The proof is given in Kurowicka (1997).

If $k_n = n$, $l_n - c\ln k_n \sim s$, where $s \in (-\infty, \infty)$, $c > 0$ then from Kołowrocki (1998) we learn that the only possible nondegenerate limit reliability functions are $\Re_4(x), \Re_5(x), \Re_6(x), \Re_7(x)$ and next theorem is valid.

Theorem 12.1.2 *Let* $k_n = n$, $l_n - c\ln k_n \sim s$, *where* $s \in (-\infty, \infty)$, $c > 0$.

(i) $R \in D_{\Re_4}$ *if and only if*

(a) there exists y *such that* $R(y) = e^{-\frac{1}{c}}$ *and* $R(y-0) > e^{-\frac{1}{c}}$,

(b) $\lim_{r\to 0^+} \frac{1+c\cdot\ln R(rx+y)}{1+c\cdot\ln R(r+y)} = x^\alpha$ *for* $x > 0$.

(ii) $R \in D_{\Re_5}$ *if and only if*

(a) there exists y *such that* $R(y) < e^{-\frac{1}{c}}$ *and* $R(y-0) = e^{-\frac{1}{c}}$,

(b) $\lim_{r\to 0^-} \frac{1+c\cdot\ln R(rx+y)}{1+c\cdot\ln R(r+y)} = x^\alpha$ *for* $x > 0$.

(iii) $R \in D_{\Re_6}$ *if and only if*

(a) there exists y *such that* $R(y) = e^{-\frac{1}{c}}$,

(b) $\lim_{r\to 0} \frac{1+c\cdot\ln R(rx+y)}{1+c\cdot\ln R(r+y)} = x^\alpha$ *for* $x > 0$.

(iv) $R \in D_{\Re}$ *if and only if* $\exists\ b\ \forall\ x \in [x_1 + b,\ x_2 + b)$

$$R(x) = e^{-\frac{1}{c}},\ R(x_1 + b - 0) > e^{-\frac{1}{c}},\ R(x_2 + b) < e^{-\frac{1}{c}}.$$

The proof is given in Kurowicka (1998).

From Kołowrocki (1998) , we also get that if $\lim_{n\to\infty} k_n = k$, $\lim_{n\to\infty} l_n = \infty$ then the only possible nondegenerate limit reliability functions are $\Re_8(x), \Re_9(x)$, $\Re_{10}(x)$. In this case the theorem about domain of attraction takes the form.

Theorem 12.1.3 *Let* $\lim_{n\to\infty} k_n = k$, $\lim_{n\to\infty} = \infty$.

(i) $R \in D_{\Re_8}$ *if and only if*

$$\lim_{r\to-\infty} \frac{1-R(r)}{1-R(rx)} = x^\alpha \qquad \textit{for } x > 0.$$

(ii) $R \in D_{\Re_9}$ *if and only if*

(a) there exists y *such that* $R(y) = 1$ *and* $R(y+\varepsilon) < 1$ *for all* $\varepsilon > 0$,

(b) $\lim_{r\to 0^+} \frac{1-R(rx+y)}{1-R(r+y)} = x^\alpha$ *for* $x > 0$.

(iii) $R \in D_{\Re_{10}}$ *if and only if the condition*

$$\lim_{n\to\infty} l_n(R(a_n x + b_n)) = e^x$$

is satisfied for all x, *where functions* b_n *and* a_n *are defined as*

$$\begin{aligned} b_n &= \inf\left\{x : R(x+0) \le 1 - \frac{1}{l_n} \le R(x-0)\right\}, \\ a_n &= \inf\left\{x : R(x(1+0)+b_n) \le 1 - \frac{e}{l_n} \le R(x(1-0)+b_n)\right\}. \end{aligned}$$

The proof is given in Kurowicka (1994).

Example 12.1.1 If the regular homogeneous series-parallel system is such that the reliability of the particular components is given by

$$R(x) = \begin{cases} 1 & x \le 1 \\ \exp[-\sqrt[3]{\ln x}] & x > 1 \end{cases}$$

and the pair (k_n, l_n) satisfy the condition

$$k_n = n, \quad l_n = \sqrt[3]{(\ln k_n)^2}$$

then $R \in D_{\Re_1}$.

JUSTIFICATION. We have to verify that

$$\begin{aligned}\lim_{r\to\infty}\left(\frac{\exp[-\sqrt[3]{\ln r}]}{\exp-\sqrt[3]{\ln rx}]}\right)^{\sqrt[3]{(\ln r)^2}} &= \lim_{r\to\infty}\exp[(\sqrt[3]{\ln r+\ln x}-\sqrt[3]{\ln r})\cdot\sqrt[3]{(\ln r)^2}]\\ &= \lim_{r\to\infty}\exp\left[\ln r\left(\sqrt[3]{1+\frac{\ln x}{\ln r}}-1\right)\right]\\ &= \lim_{r\to\infty}\exp\left[\ln r\left(1+\frac{1}{3}\frac{\ln x}{\ln r}-1\right)\right]=x^{\frac{1}{3}}.\end{aligned}$$

We also get that $a_n = n$ and $a = 1$. ■

12.2 Asymptotic Reliability Functions of a Regular Homogeneous Series–"k out of n" System

Let $E_{11},\ldots,E_{1m},E_{21},\ldots,E_{nm}$ be components of a homogeneous, regular system and

$$\begin{pmatrix} X_{11} & X_{12} & \cdots & X_{1m}\\ X_{21} & X_{22} & \cdots & X_{2m}\\ \cdots & \cdots & \cdots & \cdots\\ X_{n1} & X_{n2} & \cdots & X_{nm}\end{pmatrix}$$

be the matrix of independent random variables representing their lifetimes.

We define random variables

$$X_i = \min_{1\le j\le m}\{X_{ij}\}, \qquad i=1,2,\ldots,n. \tag{12.1}$$

Definition 12.2.1 A system S is called series–"k out of n", if its lifetime X is given by

$$X = X_{(n-k+1)}, \qquad k=1,2,\ldots,n,$$

where $X_{(n-k+1)}$ is the k-th maximal order statistics in a sample of random variables $X_1,X_2,\ldots,X_n$ defined by (12.1).

Under the above definition the reliability function of series–"k out of n", homogeneous, regular system is given by

$$\Re^{(k)}(x) = 1-\sum_{r=0}^{k-1}\binom{n}{r}(R^m(x))^r(1-R^m(x))^{n-r}, \qquad x\in(-\infty,\infty). \tag{12.2}$$

Assuming $m = m_t$, $n = n_t$, where t, n_t, m_t are positive real numbers and if $t\to\infty$, we obtain the family of series–"k out of n" systems. According to

(12.2) for this family there exists the family of reliability functions given by

$$\Re_t^{(k)}(x) = 1 - \sum_{r=0}^{k-1} \binom{n_t}{r} (R^{m_t}(x))^r (1 - R^{m_t}(x))^{n_t - r}. \tag{12.3}$$

We are looking for the class of possible limit distributions of the standardized lifetime $\frac{X_{(n-k+1)}-b_t}{a_t}$ what is related to considering limits of reliability functions family $R_t^{(k)}(x) = (a_t x + b_t)$, where $a_t > 0$, $b_t \in (-\infty, \infty)$.

We shall search for non-degenerate asymptotic reliability functions of the form

$$\Re^{(k)}(x) = 1 - \sum_{r=0}^{k-1} \exp[-L(x)] \frac{L^r(x)}{r!}, \quad x \in (-\infty, \infty),\ k \in N_+, \tag{12.4}$$

where $L(x)$ is a nonnegative nonincreasing right-continuous function, $L(-\infty) = \infty$, $L(\infty) = 0$ and besides $L(x)$ may be identically equal to ∞ in an interval. In order to solve this problem we need the following generalization of Lemma 12.2.1 given in Kołowrocki (1993a).

Lemma 12.2.1 *If*

(i) the family $R_t^{(k)}(x)$ is given by (12.3),

(ii) the non-degenerate reliability function $\Re^{(k)}(x)$ is given by (12.4),

(iii) $\lim_{t\to\infty} n_t = \infty$,

then the assertion

$$\lim_{t\to\infty} R_t^{(k)}(a_t x + b_t) = \Re^{(k)}(x), \qquad x \in C_{\Re^{(k)}}, \tag{12.5}$$

is equivalent to the assertion

$$\lim_{t\to\infty} n_t R^{m_t}(a_t x + b_t) = L(x), \qquad x \in C_L. \tag{12.6}$$

The proof is given in Milczek (1998).

Applying the results given in Kołowrocki (1993a), we may formulate the next lemma.

Lemma 12.2.2 *The only possible forms of a non-degenerate function $L(x)$ are:*

$$L_1(x) = \begin{cases} \infty, & x < 0 \\ x^{-\alpha}, & x > 0, \end{cases}$$

$$L_2(x) = \begin{cases} (-x)^{\alpha}, & x < 0 \\ 0, & x > 0, \end{cases}$$

$$L_3(x) = e^{-x}, \qquad x \in (-\infty, \infty)$$

if $n_t = t$, $|m_t - c \cdot \ln t| >> s$, $s > 0$, $c > 0$,

$$L_4(x) = \begin{cases} \infty, & x < 0 \\ \exp(-x^\alpha), & x > 0, \end{cases}$$

$$L_5(x) = \begin{cases} \exp[(-x)^\alpha], & x < 0 \\ 0, & x > 0, \end{cases}$$

$$L_6(x) = \begin{cases} \exp[\beta(-x)^\alpha], & x < 0 \\ \exp(-x^\alpha), & x > 0, \end{cases}$$

$$L_7(x) = \begin{cases} \infty, & x < x_1 \\ 1, & x_1 < x < x_2 \\ 0, & x > x_2, \end{cases}$$

if $n_t = t$, $m_t - c \cdot \ln t \sim s$, $s \in (-\infty, \infty)$, $c > 0$.

The above lemma and Gnedenko's results [Gnedenko (1943)] given also in [Barlow and Proschan (1975)] implies the theorem:

Theorem 12.2.1 *The only possible forms of a non-degenerate limit reliability functions of the series–"k out of n" systems are:*

$$\Re_1^{(k)}(x) = \begin{cases} 1, & x < 0 \\ 1 - \sum_{r=0}^{k-1} \exp(-x^{-\alpha}) \ \frac{x^{-r\alpha}}{r!}, & x \geq 0, \end{cases}$$

$$\Re_2^{(k)}(x) = \begin{cases} 1 - \sum_{r=0}^{k-1} \exp[-(-x^\alpha)] \ \frac{(-x)^{-r\alpha}}{r!}, & x < 0 \\ 0, & x \geq 0, \end{cases}$$

$$\Re_3^{(k)}(x) = 1 - \sum_{r=0}^{k-1} \exp[-\exp(-x)] \ \frac{e^{-rx}}{r!}, \quad x \in (-\infty, \infty)$$

if $n_t = t$, $|m_t - c \cdot \ln t| >> s$, $s > 0$, $c > 0$,

$$\Re_4^{(k)}(x) = \begin{cases} 1, & x < 0 \\ 1 - \sum_{r=0}^{k-1} \exp\left[-\exp\left(-x^\alpha - \frac{s}{c}\right)\right] \ \frac{\left[\exp\left(-x^\alpha - \frac{s}{c}\right)\right]^r}{r!}, & x \geq 0, \end{cases}$$

$$\Re_5^{(k)}(x) = \begin{cases} 1 - \sum_{r=0}^{k-1} \exp\left[-\exp\left((-x)^\alpha - \frac{s}{c}\right)\right] \ \frac{\left[\exp\left((-x)^\alpha - \frac{s}{c}\right)\right]^r}{r!}, & x < 0 \\ 0, & x \geq 0, \end{cases}$$

$$\Re_6^{(k)}(x) = \begin{cases} 1 - \sum_{r=0}^{k-1} \exp\left[-\exp\left[\beta(-x)^\alpha - \frac{s}{c}\right]\right] \ \frac{\left[\exp\left[\beta(-x)^\alpha - \frac{s}{c}\right]\right]^r}{r!}, & x < 0, \\ 1 - \sum_{r=0}^{k-1} \exp\left[-\exp\left(-x^\alpha - \frac{s}{c}\right)\right] \ \frac{\left[\exp\left(-x^\alpha - \frac{s}{c}\right)\right]^r}{r!}, & x \geq 0, \end{cases}$$

$$\Re_7^{(k)}(x) = \begin{cases} 1, & x < x_1 \\ 1 - \sum_{r=0}^{k-1} \exp\left[-\exp\left(-\frac{s}{c}\right)\right] \ \frac{\left[\exp\left(-\frac{s}{c}\right)\right]^r}{r!}, & x_1 \leq x < x_2 \\ 0, & x \geq x_2, \end{cases}$$

if $n_t = t,\ m_t - c\ln t \sim s,\ s \in (-\infty,\infty),\ c > 0,$

$$\Re_8^{(k)}(x) = \begin{cases} 1 - \sum_{r=0}^{k-1}\binom{n}{r}[\exp(-(-x)^{-\alpha})]^r[1-\exp(-(-x)^{-\alpha})]^{n-r}, & x < 0 \\ 0, & x \geq 0, \end{cases}$$

$$\Re_9^{(k)}(x) = \begin{cases} 1, & x < 0 \\ 1 - \sum_{r=0}^{k-1}\binom{n}{r}[\exp(-x^{-\alpha})]^r[1-\exp(-x^{-\alpha})]^{n-r}, & x \geq 0, \end{cases}$$

$$\Re_{10}^{(k)}(x) = 1 - \sum_{r=0}^{k-1}\binom{n}{r}[\exp(-\exp(x))]^r[1-\exp(-\exp(x))]^{n-r}, \quad x \in (-\infty,\infty),$$

if $n_t \to n \geq k,\ m_t \to \infty.$

Naturally the problem of possible limit reliability functions of "k out of n"–series systems appears. It is currently under consideration.

12.3 Limit Reliability Functions of Homogeneous Regular Series-Parallel Systems of Higher Order

Definition 12.3.1 System S is called the series-parallel system of order n if its lifetime X is given by

$$X = \max_{1\leq i_1\leq k}\left\{\min_{1\leq j_1\leq l_{i_1}}\left[\max_{1\leq i_2\leq k^{(i_1j_1)}}\left(\min_{1\leq j_2\leq l_{i_2}^{(i_1j_1)}}\cdots\left(\max_{1\leq i_n\leq k^{(i_1j_1\cdots i_{n-1}j_{n-1})}}\left(\min_{1\leq j_n\leq l_{i_n}^{(i_1j_1\cdots i_{n-1}j_{n-1})}} X_{i_1j_1\ldots i_nj_n}\right)\right)\right)\right]\right\},$$

where

$$\mathcal{X} = \{X_{i_1j_1\ldots i_nj_n},\ i_1 = 1,2,\ldots,k;\ j_1 = 1,2,\ldots l_{i_1};\ i_2 = 1,2,\ldots,k^{(i_1j_1)};\ j_2 = 1,2,\ldots,l_{i_2}^{(i_1j_1)};\ \cdots\ i_n = 1,2,\ldots,k^{(i_1j_1\ldots i_{n-1}j_{n-1})};\ j_n = 1,2,\ldots,l_{i_n}^{(i_1j_1\ldots i_{n-1}j_{n-1})}\}$$

is the family of independent random variables corresponding to the lifetimes of particular components.

Definition 12.3.2 A series-parallel system of order n is regular if

$$l = l_{i_1} = l_{i_2}^{(i_1j_1)} = \cdots = l^{(i_1j_1\ldots i_{n-1}j_{n-1})} \quad \text{and } k =^{(i_1j_1)}= \cdots = k^{(k_1j_1\ldots i_{n-1}j_{n-1})},$$

where k and l are natural numbers.

Definition 12.3.3 A regular series-parallel system of order n is homogeneous if random variables from Definition 12.1.1 have the same distribution function $F(x) = P(X_{i_1 j_1 \dots i_n j_n} \le x)$, $x \in (-\infty, +\infty)$, i.e., if components of the system have the same reliability function

$$R(x) = P(X_{i_1 j_1 \dots i_n j_n} > x) = 1 - F(x), \quad x \in (-\infty, +\infty).$$

Now assuming in Definition 12.3.2, $k = k_m$ and $l = l_m$, where m is a natural number and k_m, l_m are sequences of natural numbers such that at least one of them tends to infinity, we obtain a sequence of regular homogeneous series-parallel systems of order n corresponding to a sequence (k_m, l_m). For this sequence of systems there exists a sequence of their reliability functions given by

$$R_m^{(k)}(x) = 1 - \left\{1 - \left[R_m^{(k-1)}(x)\right]^{l_m}\right\}^{k_m}, \quad k = 2, 3, \dots, n \tag{12.7}$$

and

$$R_m^{(1)}(x) = 1 - \left\{1 - [R(x)]^{l_m}\right\}^{k_m}. \tag{12.8}$$

Next replacing m by a positive real number t and assuming that k_t and l_t are positive real numbers we obtain a family of systems corresponding to a pair (k_t, l_t). For this family of systems there exist a family of their reliability functions given by

$$R_t^{(k)}(x) = 1 - \left\{1 - \left[R_t^{(k-1}(x)\right]^{l_t}\right\}^{k_t}, \qquad \text{for } k = 2, 3, \dots, n \tag{12.9}$$

and

$$R_t^{(1)}(x) = 1 - \left\{1 - [R(x)]^{l_t}\right\}^{k_t}. \tag{12.10}$$

A problem of assignation of a closed class of possible nondegenerate limit reliability functions for the regular, homogeneous, series-parallel system of order n we shall resolve analogically as in Kołowrocki (1993a,b).

The proof of the next lemma is based on induction and approximate formulae for functions $\exp x$ and $\ln(1 \pm x)$.

Lemma 12.3.1 *If*

(i) $\mathfrak{R}(x) = 1 - \exp[-V(x)]$,

(ii) $R_t^{(n)}(x)$ *is given by Equations (12.9) and (12.10),*

(iii)

(case I) $\lim_{t\to\infty} l_t^{1-n} k_t^{\frac{1+l_t+\cdots+l_t^{n-1}}{l_t^{n-2}}} = \infty$ *for* $n \geq 2$ *and* $\lim_{t\to\infty} k_t = \infty$ *for* $n = 1$,

(case II) $\lim_{t\to\infty} k_t = \infty$, $\lim_{t\to\infty} l_t^{n-1}[\ln k_t - cl_t - \ln(-\ln(1-e^{-c}))] = A$,

where $c > 0$ *and* A *are some real numbers,*

(iv) $a_t > 0$, $b_t \in (-\infty, +\infty)$, *are some functions,*

then the assertion

$$\lim_{t\to\infty} R_t^{(n)}(a_t x + b_t) = \Re(x), \qquad x \in C_{\Re}$$

is equivalent to the assertion

$$\lim_{t\to\infty} \bar{k}_t [R(a_t x + b_t)]^{\bar{l}_t} = V(x), \qquad x \in C_V$$

where

(caseI) $\bar{k}_t = k^{1+l_1+\cdots+l_t^{n-1}}$ *and* $\bar{l}_t = l_t^n$

(caseII)
$$\bar{k}_t = -\ln(1-e^{-c}) \left[\frac{k_t}{-\ln(1-e^{-c})}\right]^{l_t^{n-1}(e^c-1)^{n-1}[-\ln(1-e^{-c})]^{n-1}}$$

and

$$\bar{l}_t = l_t^n (e^c - 1)^{n-1} [-\ln(1-e^{-c})]^{n-1}.$$

Lemma 12.3.2 *[Kołowrocki (1993a)] If*

(i) $\lim_{t\to\infty} \bar{k}_t = \infty$,

(ii) there exist functions $a_t > 0$, $b_t \in (-\infty, +\infty)$ *such that* $\lim_{t\to\infty} \bar{k}_t [R(a_t x + b_t)]^{\bar{l}_t} = V(x)$, $x \in C_V$,

(iii) for every natural $\nu \geq 2$, $\lim_{t\to\infty} \bar{k}_t [R(a_{t_\nu} x + b_{t_\nu})]^{\bar{l}_t} = V_0(x)$, $x \in C_{V_0}$, *where,* $\tau_\nu = \tau_\nu(t)$, $t \in (-\infty, +\infty)$, $\tau_\nu \in (0, \infty)$,

(iv) $R(x)$ *is a reliability function of a particular component of the regular, homogeneous, series-parallel system of order* n,

(v) $V(x)$ *and* $V_0(x)$ *are nondegenerate functions;*

then the functions $V(x)$ *and* $V_0(x)$ *are of the same type; i.e., there exist* $\alpha_\nu > 0$, $\beta_\nu \in (-\infty, +\infty)$ *such that for every natural* $\nu \geq 2$, $V_0(x) = V(\alpha_\nu x + \beta_\nu)$, $x \in (-\infty, +\infty)$.

Lemma 12.3.3 *[Kołowrocki (1993a)] If*

(i) $\lim_{t\to\infty} \bar{k}_t = \infty$,

(ii) $a_t > 0$, $b_t \in (-\infty, +\infty)$ *are some functions;*

then from the assertion

$$\lim_{t\to\infty} R_t^{(n)}(a_t x + b_t) = \Re(x), \qquad x \in C_{\Re},$$

where $\Re(x) = 1 - \exp[-V(x)]$ *is a reliability function and* $R_t^n(x)$ *is the family given by (12.9) and (12.10), follows that for every natural* $\nu \geq 2$

$$\lim_{t\to\infty} \bar{k}_{\tau_\nu} \bar{k}_t^{-\frac{\bar{l}_{\tau_\nu}}{\bar{l}_t}} \left\{\bar{k}_t[R(a_{\tau_\nu}x + b_{\tau_\nu})]^{\bar{l}_t}\right\}^{\frac{\bar{l}_{\tau_\nu}}{\bar{l}_t}} = V(x), \qquad x \in C_V \tag{12.11}$$

where $\tau_\nu = \tau_\nu(t)$, $t \in (-\infty, +\infty)$, $\tau_\nu \in (0, \infty)$ *and* $\lim_{t\to\infty} \tau_\nu = \infty$.

The result (12.11) in Lemma 12.3.3 is a necessary condition for a reliability function $\Re(x) = 1 - \exp[-V(x)]$ to be a limit reliability function of the system for a suitably chosen functions a_t and b_t. From this condition, we intend to get some information on the possible types of functions $\Re(x)$ by the assignation of the possible types of functions $V(x)$ which fulfill it. To this end we need the next two lemmas.

Lemma 12.3.4 *[Cichocki (1997)] If*

(i) $k_t = t$, $l_t = c(\ln t)^{\rho(t)}$, $t \in (0, \infty)$, $c > 0$, $\lim_{t\to\infty} l_t^{l-n} k_t^{\frac{1+l_t+\cdots+l_t^{n-2}}{l_t^{n-1}}} = \infty$ *for* $n = 2, 3, \ldots$,

(ii) for every natural $\nu \geq 2$, $\tau_\nu = \tau_\nu(t)$ *is given by* $\tau_\nu = t\nu^{\frac{1}{[1-\rho(t)]l_t^{n-1}}}$, $t \in (0, \infty)$,

(iii) $|\rho(\tau_\nu) - \rho(t)| << \frac{1}{(\ln t)^n \ln\ln t}$,

(iv) a nondegenerate function $\Re(x) = 1 - \exp[-V(x)]$ *is a limit reliability function of the regular, homogeneous, series-parallel system of order* n $(n \geq 2)$*;*

then for every natural $\nu \geq 2$ *there exist* $\alpha_\nu > 0$, $\beta_\nu \in (-\infty, +\infty)$ *and* $\mu \neq 0$ *such that*

$$\nu^\mu V(\alpha_\nu x + \beta_\nu) = V(x), \qquad x \in (-\infty, +\infty). \tag{12.12}$$

Lemma 12.3.5 *[Cichocki (1997)] If*

(i) $\lim_{t\to\infty} k_t = \infty$, $\lim_{t\to\infty} l_t^{n-1}[\ln k_t - cl_t - \ln(-\ln(1-e^{-c}))] = A$, $c > 0$, $A \in (-\infty, +\infty)$,

(ii) a nondegenerate function $\Re(x) = 1 - \exp[-V(x)]$ *is a limit reliability function of the regular, homogeneous, series-parallel system of order* n $(n \geq 2)$,

then for every natural $\nu \geq 2$ *there exist* $\alpha_\nu > 0$, $\beta_\nu \in (-\infty, +\infty)$ *such that*

$$[V(\alpha_\nu x + \beta_\nu)]^\nu = V(x), \qquad x \in (-\infty, +\infty). \tag{12.13}$$

The functional equations (12.12) and (12.13) will allow us to fix a closed class of nondegenerate functions $V(x)$ for the considered cases of pairs (k_t, l_t). Having this class of functions we shall immediately obtain a class of limit reliability functions of the regular, homogeneous, series-parallel system of order n we are looking for. (The classes are the same as in Kołowrocki (1993a,b) because functional equations (12.12) and (12.13) are the same as in mentioned papers). The result is as follows.

Theorem 12.3.1 *If*

(i) $k_t = t$, $l_t = c(\ln t)^{\rho(t)}$, $t \in (0, \infty)$, $c > 0$, $\lim_{t\to\infty} l_t^{l-n} k_t^{\frac{l_t^1 + l_t^2 + \cdots + l_t^{n-2}}{l_t^{n-1}}} = \infty$ *for* $n = 2, 3, \ldots$,

(ii) $\tau_\nu = t\nu^{\frac{1}{[1-\rho(t)]l_t^{n-1}}}$, $|\rho(\tau_\nu) - \rho(t)| << \frac{1}{(\ln t)^n \ln\ln t}$, *for every natural* $\nu \geq 2$,

then the only possible nondegenerate limit reliability functions of the regular, homogeneous, series-parallel system of order n $(n \geq 2)$ *are the same type functions as* $\Re_1(x), \Re_2(x), \Re_3(x)$, *given in Kołowrocki (1998).*

Theorem 12.3.2 *If*

$$\lim_{t\to\infty} k_t = \infty, \qquad \lim_{t\to\infty} l_t^{n-1}[\ln k_t - cl_t - \ln(-\ln(1-e^{-c}))] = A,$$

then the only possible nondegenerate limit reliability functions of the regular, homogeneous, series- parallel system of order n $(n \geq 2)$ *are the same type functions as* $\Re_4(x), \Re_5(x), \Re_6(x), \Re_7(x)$, *given in Kołowrocki (1998).*

References

1. Kołowrocki K. (1998). On limit reliability functions of large systems. Part I, In *Statistical and Probabilistic Models in Reliability* (Eds., D. C. Ionescu and N. Limnios) pp. 153–184, Boston, MA: Birkhaüser.

2. Barlow R. E. and Proschan F. (1975). *Statistical Theory of Reliability and Life Testing, Probability Models*, New York: Holt, Rinehart and Winston.

3. Cichocki A. (1997), *Report*, Materials of Gdynia Maritime Academy.

4. Chernoff H. and Teicher H. (1965). Limit distribution of minimax of independent identically distributed random variables, *Transactions of the American Mathematical Society*, **116**, 474–491.

5. Gnedenko B. W. (1943). Sur la distribution limite du terme maximum d'une serie aleatorie, *Annals of Mathematics*, **44**, 432–453.

6. Kołowrocki K. (1993a). On a class of limit reliability functions of some regular homogeneous series-parallel and parallel-series systems, *Reliability Engineering and System Safety*, **39**, 11–23.

7. Kołowrocki K. (1993b). On a class of limit reliability functions of some regular homogeneous series-parallel systems, *Applied Mathematics*, **36**, 55–69.

8. Kurowicka D. (1994). Domains of attraction of asymptotic reliability functions for series-parallel and parallel-series systems with bonded numbers of parallel components, *Advances in Modeling and Analysis, C*, **44**, 55–64.

9. Kurowicka D. (1997). Domains of attraction of asymptotic reliability functions of some homogeneous series-parallel systems, In *Applied Mathematics and Computations*, New York, NY: Elsevier Science (in press).

10. Kurowicka D. (1998). Domains of attraction of asymptotic reliability functions for homogeneous series-parallel and parallel-series systems, *Applied Mathematics* (submitted).

11. Milczek B. (1998). Asymptotic reliability functions of a regular homogeneous series–"k out of n" systems, *Proceedings of the International Conference ESREL'98*, June 16–19, 1998, Trondheim, Norway.

13

Error Bounds for a Stiff Markov Chain Approximation Technique and an Application

Olivier Pourret, Jérôme Collet and Jean-Louis Bon

Direction des Etudes et Recherches, Clamart, France

Abstract: A classical stiff Markov chain solution technique is adapted to analysis of dependability models, and given a new interpretation. This allows the derivation of bounds for the approximation error. A numerical example illustrates the practical use of these error bounds.

Keywords and phrases: Markov Chain, dependability, transient solution, approximation, error bounds

13.1 Introduction

Continuous-time Markov chains (CTMC) are widely used models in the fields of dependability and performance evaluation. The problem of computing the steady-state probability vector has been thoroughly studied, but few methods are available for the transient analysis of a Markov model [Stewart (1994)].

The computation of the transient probabilities is especially difficult for large and *stiff* models. The most commonly accepted definition of stiffness is the presence of "fast states", i.e., states with average sojourn time much smaller than observation time t [Dunkel and Stahl (1993), Malhotra et al. (1994)].

Several authors have suggested techniques to reduce the size of the graph, and to avoid (or tolerate) stiffness. In 1986, Bobbio and Trivedi [Bobbio and Trivedi (1986)] proposed an approximation technique based on a classification of the state space, generalizing previous work on the subject [Courtois and Semal (1984), McGough et al. (1985), White (1991)]. The method has been further analyzed and developed in [Bobbio and Trivedi (1990)] and [Reibman et al. (1990)].

We have adapted and used this method to analyze stiff dependability models. Although the results seem generally accurate, the general validity of the approximation had to be demonstrated through the derivation of error bounds. We indeed show with a counter-example that the computation of a system unavailability using this method may not be conservative; yet possibly underestimated risk values are not acceptable in the field of dependability.

To our knowledge the problem of bounding the approximation error has not yet been solved, except in the more specific context of "instantaneous coverage approximation" [White (1991)].

The paper is organized as follows. Section 13.2 briefly introduces the main notations used in the article. In Section 13.3, we suggest a "path-based" method to compute the state vector. In Section 13.4, this method is connected to Bobbio and Trivedi's algorithm. This enables us to bound the approximation error of this classical algorithm (Theorem 13.4.1). Section 13.5 presents a numerical example. The appendix contains the proofs of our main results.

13.2 Notations

We consider a continuous-time Markov chain $\{X(t); t \geq 0\}$ on a finite state space $\mathcal{E}$. Let A denote its transition rate matrix (or infinitesimal generator). Entry a_{ij} $(i \neq j)$ of matrix A is the transition rate from i to j. The exit rate of state i is defined as

$$a_i = \sum_{j \in \mathcal{E};\, j \neq i} a_{ij},$$

and the diagonal entry of A is $a_{ii} = -a_i$.

The subject of this paper is the computation of the state vector $P(t)$ at time t, given its initial value $P(0)$.

13.3 Approximation Techniques

13.3.1 A path-based technique

It is well known that $P(t)$ may be computed as the solution of the Chapman-Kolmogorov differential system, or from the exponential of matrix tA. We will though focus our attention on another expression of $P(t)$.

Given a state j in $\mathcal{E}$, let us consider all the sequences s of states that may have been visited before the process hit j for the last time. These sequences are called "paths". We denote by $\mathcal{P}_j$ the set of all these paths:

$$\mathcal{P}_j = \bigcup_{n \in \mathbb{N}^*} \{(i_1, \ldots, i_n) \in \mathcal{E}^n \mid i_n = j;\ \forall k, i_k \neq i_{k+1}\}.$$

We associate to any path s its *length* n, and its *duration* T_s, defined as the sum of $n-1$ independent random variables, exponentially distributed with rates a_{i_k} $(k = 1, \ldots, n-1)$.

Let $p_j(t)$ denote the probability that the process is in state j at time t. The starting point of our approach is the expression of $p_j(t)$ given below.

Proposition 13.3.1 *The following expression of $p_j(t)$ holds:*

$$p_j(t) = \sum_{s \in \mathcal{P}_j} p_s \int_0^t f_s(t-u) e^{-a_j u}\, du, \tag{13.1}$$

where, for any path $s = (i_1, \ldots, i_n)$ (with $i_n = j$),

- $p_s = p_{i_1}(0) \prod_{k=1}^{n-1} \frac{a_{i_k i_{k+1}}}{a_{i_k}}$,
- *f_s is the density of probability of T_s.*

A proof of Proposition 13.3.1 is given in the Appendix.

If state j is absorbing, then $a_j = 0$, and Equation (13.1) becomes

$$p_j(t) = \sum_{s \in \mathcal{P}_j} p_s F_s(t), \tag{13.2}$$

where F_s is the cumulative distribution function of T_s.

Assuming that j is not an absorbing state allows a different simplification of equation (13.1):

$$p_j(t) = \frac{1}{a_j} \sum_{s \in \mathcal{P}_j} p_s g_s(t). \tag{13.3}$$

Function $g_s(t)$ represents the probability density function of the sum of T_s and the sojourn time in state j, i.e., the sum of n independent random variables, exponentially distributed with rates a_{i_k} $(k = 1, \ldots, n)$.

Probability $p_j(t)$ appears then as an infinite weighted sum of density functions $g_s(t)$.

Assuming that exit rates can be classified into "slow" and "fast" rates, a natural approximation principle is to compute the $g_s(t)$'s (or the $F_s(t)$'s, in the case of an absorbing state) by taking into account only the "greatest" (in a stochastic sense) exponential random variables, e.g., those which rate is lower than a given threshold θ.

Our path-based algorithm for the computation of $p_j(t)$ can be summed up as follows:

- Partition state space $\mathcal{E}$ into two subsets: subset $\mathcal{F}$ of "fast states", i.e., with exit rate greater than threshold θ, and subset $\mathcal{S} = \mathcal{E} - \mathcal{F}$ of "slow states",
- Choose a finite subset $\mathcal{P}_j^*$ of $\mathcal{P}_j$,
- Enumerate the paths s of $\mathcal{P}_j^*$, and
  - Compute p_s,
  - Compute approximate value $\tilde{g}_s(t)$ of $g_s(t)$ by taking into account only the slow rate exponential random variables.
- Addition the $p_s\tilde{g}_s(t)$,
- Estimate $p_j(t)$ using (13.3), and a truncation error bound.

Let us denote this principle as "BCP's algorithm".

When used for the computation of an absorbing state probability, BCP's algorithm is clearly equivalent to "instantaneous coverage" approximation.

13.3.2 Bobbio and Trivedi's algorithm

The approximation algorithm proposed in [Bobbio and Trivedi (1986)] is also based on the existence of fast and slow exit rates in the Markov graph, but involves matrix computations instead of path enumeration. We have adapted it to the field of dependability as follows:

- Partition space state $\mathcal{E}$ into subsets $\mathcal{F}$ and $\mathcal{S}$, similarly to BCP's algorithm,
- Renumber the states to get a block-decomposition of A such as:

$$A = \begin{pmatrix} A_S & A_{SF} \\ A_{FS} & A_F \end{pmatrix}. \tag{13.4}$$

- Compute approximate state vector P_S of $\mathcal{S}$ using

$$P_S \simeq P_S(0) \exp\left[(A_S - A_{SR}(A_F)^{-1}A_{FS})t\right]. \tag{13.5}$$

- Compute approximate state vector P_F of $\mathcal{F}$ using

$$P_F \simeq P_S A_{SF}(-A_F^{-1}). \tag{13.6}$$

This algorithm which has proved efficient in many applications, will be referred to as "BT's algorithm". For the sake of historical accuracy, it is worth mentioning the two points of Bobbio and Trivedi's approach that we do not take into account.

1. *Possible existence of "fast-recurrent" states.* In a dependability model, fast transitions are mostly repair transitions. Since a sequence of repairs cannot bring a system back to its initial state, a plausible assumption is that the graph of fast transitions is acyclic.

2. *Normalization of the approximate state vector.* Normalization lets the $\tilde{p}_j(t)$ sum to one, and yields asymptotically exact results [Bobbio and Trivedi (1986)]. However, normalization was not given any natural probabilistic interpretation. In the field of dependability, the sum of the $\tilde{p}_j(t)$ is not important, because we rather focus on unavailability or failure state probabilities, than on a global process analysis. Moreover, reliability estimations should generally be conservative, whereas normalization decreases the probabilities. Finally, as $t \to +\infty$, the system generally reaches its absorbing states, so the asymptotical values of the probabilities are trivial.

13.4 Main Results

13.4.1 Equivalence

Our first result is the following:

Proposition 13.4.1 *BCP's and BT's algorithms are equivalent.*

Therefore we shall define $\tilde{p}_j(t)$, for any $j \in \mathcal{F}$, as the approximate value of $p_j(t)$ computed with either method. Similarly, we no longer need to specify whether "BT's" or "BCP's" is referred to.

13.4.2 A non-conservative case

This approximation often leads to overestimates of the state probabilities. As mentioned above, conservative assessments are especially important when a dependability level is to be guaranteed.

Absorbing state probabilities are overestimated at any time t, which has been shown in [McGough et al. (1985)] and [White (1991)] in the semi-Markov case. In [McGough et al. (1985)], the proof uses matrix computations, whereas that of [White (1991)] is based on path analysis. The conservativeness result does not extend to non-homogeneous Markov processes [Geist et al. (1989)].

In the Markovian case, it does not hold for non-absorbing states, as the following counter-example shows. Let us examine the graph of Figure 13.1.

This graph models a system which can be repaired only once (e.g., with one spare component).

Figure 13.1: State graph of a dependability model

Assuming that the system is initially in state 1, a standard calculus yields:

$$p_3(t) = \frac{\mu\lambda}{(\mu-\lambda)^2}\left(e^{-\lambda t}(\mu t - \lambda t - 1) + e^{-\mu t}\right).$$

We assume $\lambda \ll \mu$, and therefore $\mathcal{S} = \{1,3,4\}$, $\mathcal{F} = \{2\}$. Applying BT's or BCP's algorithm, we get a simpler expression

$$\tilde{p}_3(t) = \lambda t e^{-\lambda t}.$$

As t approaches infinity, we have

$$\frac{\tilde{p}_3(t)}{p_3(t)} \to 1 - \frac{\lambda}{\mu}. \tag{13.7}$$

This result proves that the approximation is non-conservative after a certain time. Note that in this case, normalizing $\tilde{p}_3(t)$ would produce a larger approximation error.

13.4.3 Error bounds

Our main result is the derivation of error bounds for the approximation. These bounds are presented in Theorem 13.4.1. It must be emphasized that application of this theorem does not require any particular assumption about the structure of the state graph.

Theorem 13.4.1 (Approximation error bounds) *Let Λ the maximum of the exit rates a_i for $i \in \mathcal{S}$. We assume that the duration of a visit to $\mathcal{F}$ is stochastically lower than a random variable $T_{\mathcal{F}}$ with mean $\tau_0 = \mathbb{E}(T_{\mathcal{F}})$ and Laplace transform $L(s) = \mathbb{E}(\exp(sT_{\mathcal{F}}))$. Then for any $j \in \mathcal{E}$, the following double-sided estimation of the approximation error holds for any $n \in \mathbb{N}^*$, and at any time $t > 0$:*

$$-a_n\tilde{p}_j(t) - P_n(\Lambda t) \le \tilde{p}_j(t) - p_j(t) \le b_n\tilde{p}_j(t) + P_n(\Lambda t),$$

where

$$\begin{cases} a_n = L(\Lambda)^n - 1 \\ b_n = n(n - \mathbf{1}_{j\in\mathcal{F}})\dfrac{\tau_0}{t} \end{cases}$$

and $P_n(x)$ is the probability that a Poisson's random variable with rate x is greater than or equal to n.

This theorem is proved in the appendix.

13.5 Numerical Example

13.5.1 Model used

In order to illustrate the use of our error bounds theorem, we present a numerical example. The graph of Figure 13.2 is a Markov dependability model of a 4-component system that we had to study.

We assume the system is initially in state 1. The numerical values of the rates are the following:

$$\alpha = 10^{-7};\ \lambda = 10^{-6};\ \mu = 1.$$

We want to estimate the probability of state 4, known as a dominant failure state, at time $t = 10^4$.

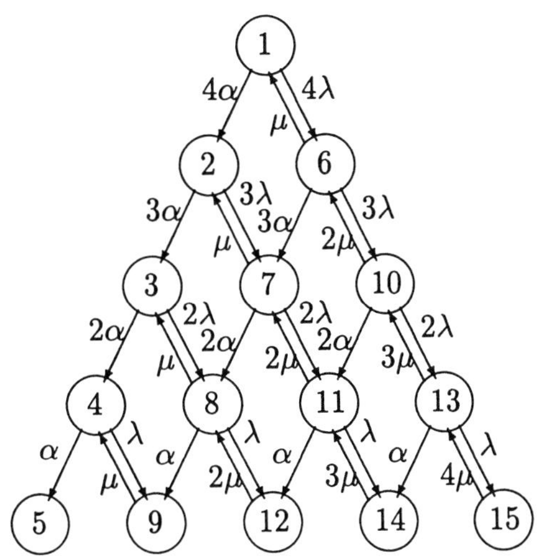

Figure 13.2: A 15-state Markov graph

13.5.2 Results

We may emphasize that because of the *stiffness* of the graph, some standard methods for computing $\exp(At)$ (e.g., based on the computation of the eigenvalues of A) fail in this case. Though we mainly focus on the accuracy of our bounds.

The exit rates considered as high are naturally those of states $6 \leq k \leq 15$. Applying BT's algorithm (with the Maple package), we get an estimate of the probability of state 4 at time t:

$$\tilde{p}_4(t) = 3.99 \cdot 10^{-9}.$$

In order to apply Theorem 13.4.1, we have to find a stochastic upper bound of the duration of any visit to subset $\mathcal{F}$. We can then use the model given in Figure 13.3. Let $\lambda_0 = 3\lambda$ and $\mu_0 = \mu$. Then it seems clear (and has been proved) that the total time $T_{\mathcal{F}}$ spent in states 1,2,3 and 4 (with initial state 4) in the graph of Figure 13.3 is stochastically greater than the duration of any visit to $\mathcal{F}$. The Laplace transform of $T_{\mathcal{F}}$ is:

$$L(s) = \frac{\mu_0^4(s - \lambda_0)}{s(\mu_0 + \lambda_0 - s)^4 - \lambda_0\mu_0^4},$$

and its mean value is:

$$\tau_0 = L'(0) = \frac{1}{\lambda_0}\left(\left(1 + \frac{\lambda_0}{\mu_0}\right)^4 - 1\right)$$

The model of Figure 13.3 may appear over-pessimistic, but this method for bounding the duration of a sojourn in $\mathcal{F}$ is generic to any dependability model, whenever the graph of fast transitions is acyclic.

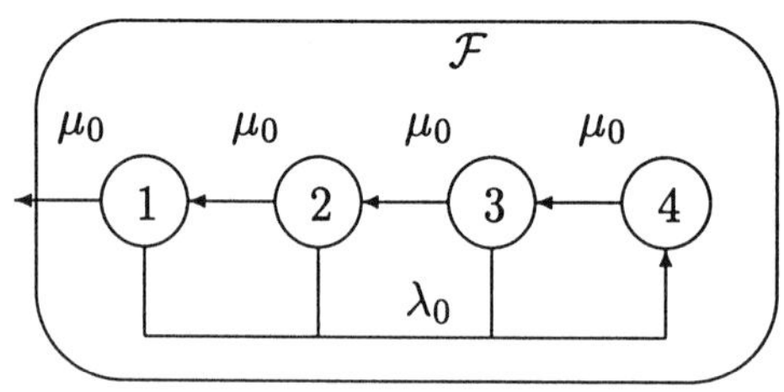

Figure 13.3: Pessimistic model of transitions inside $\mathcal{F}$

The greatest slow state exit rate is $\Lambda = 4(\lambda + \alpha)$. Applying Theorem 13.4.1, we prove that the approximation error is greater than

$$-4 \cdot 10^{-9}((1 + 1.8 \cdot 10^{-5})^n - 1) - P_n(0.044)$$

and lower than

$$1.6 \cdot 10^{-11} n^2 + P_n(0.044).$$

By choosing the optimal values of n, which are in this case 6 (upper bound) and 7 (lower bound), we prove that the approximation error is in

$$[-5.53 \cdot 10^{-13}, 6.72 \cdot 10^{-11}].$$

In the field of dependability, the most important bound is the lower one. It indeed proves that the true value of $p_4(t)$ is lower than

$$3.99 \cdot 10^{-9} + 5.53 \cdot 10^{-13}.$$

In spite of stiffness, randomization is successful in analyzing the 15-state Markov model of Figure 13.2. This allows us to compute the actual approximation error:

$$\delta_4(t) = \tilde{p}_4(t) - p_4(t) = 4 \cdot 10^{-14}.$$

The bounds of Theorem 13.4.1 are correct. We observe that the right bound is much larger than the approximation error (but small compared to the approximate value $\tilde{p}_4(t)$).

In order to test the general accuracy of our bounds, we have also computed $\tilde{p}_4(t)$, $p_4(t)$ and applied Theorem 13.4.1 with a wide range of t values.

Table 13.1: Instantaneous probability estimates and error bounds

t	$\tilde{p}_4(t)$	$\delta_4(t)$	lower b.	upper b.
1.00e-1	4.00e-24	2.78e-27	-3.52e-28	2.56e-21
1.00e0	4.00e-21	9.09e-26	-3.52e-25	2.56e-19
1.00e1	4.00e-18	4.26e-23	-3.53e-22	2.58e-17
1.00e2	4.00e-15	4.02e-20	-4.22e-19	4.00e-15
1.00e3	4.00e-12	4.00e-17	-4.32e-16	4.14e-13
1.00e4	3.99e-09	3.99e-14	-5.53e-13	6.72e-11
1.00e5	3.90e-06	3.87e-11	-7.37e-10	1.38e-08
1.00e6	3.12e-03	2.89e-08	-1.15e-06	4.72e-06
1.00e7	3.72e-01	1.37e-06	-5.04e-04	8.32e-04
1.00e8	1.82e-04	-2.18e-09	-1.82e-06	2.30e-06
1.00e9	1.49e-43	-1.52e-47	-1.48e-44	1.73e-44

Table 13.1 displays the estimates $\tilde{p}_4(t)$, the approximation error $\delta_4(t)$, and the lower and upper bounds provided by Theorem 13.4.1. All lower bounds represent less than 1% of the approximate value, except for $t = 10^9$. The upper bounds are less tight, especially for very small and very large values of t.

We also note that the approximation is non-conservative for large values of t, as in our counter-example of Section 13.4.2.

Application of Theorem 13.4.1 to state 13 produced similar results for the same range of t values.

13.6 Conclusion

A path-based expression of the instantaneous state probability has allowed us to show that the two steps of BT's algorithm (i.e., computation of the slow states, and then fast states probabilities) come under the same approximation principle, and to derive easily computable error bounds. Their accuracy has been checked in several numerical examples.

A disadvantage of Theorem 13.4.1 is that one needs to choose a value for the integer parameter n, in order to get the best possible bounds. Helping error bounds "users" to simply get acceptable values of n is still the subject of our research. Experiments show that optimal values of n mainly depend on the order of magnitude of the slow transitions rates, and of time t.

Possible extension of the approximation error theorem to the computation of "fast-recurrent" states probabilities is also being investigated.

Appendix

A.1 Proof of Proposition 13.3.1

Using the formula of total probability, we get

$$p_j(t) = \sum_{s \in \mathcal{P}_j} \mathbb{P}\left(\{X(t) = j\} \cap s\right)$$

Given a path s of length n, we condition the probability $\mathbb{P}\left(\{X(t) = j\} \cap s\right)$ on the first n states visited by the process X. The probability that the first n states are those of s is p_s, i.e., the probability of the path in the embedded Markov chain. Furthermore, X being a Markov process, the sojourn time in any state $k \in \mathcal{E}$ is exponentially distributed with rate a_k, and is independent of the exit transition. Finally, expression (13.1) is obtained by conditioning on the total time $t - u$ elapsed before hitting j for the last time.

A.2 Proof of Proposition 13.4.1

We prefer to give only a sketch of this proof. A complete proof of Proposition 13.4.1 presents no difficulty, but requires many additional notations.

As far as the computation of *slow* states probabilities is concerned, equivalence of BT's and BCP's algorithm is quite obvious. The probabilistic interpretation of the slow states approximation given in [Bobbio and Trivedi (1986)] indeed shows that the very nature of BT's approximation is to neglect the time spent in the fast subset.

Equivalence for the computation of *fast* states probabilities is shown as follows. We use BCP's approximation of $p_j(t)$, with $j \in \mathcal{F}$. Given a path with last state j, let i denote the (possible) last slow state visited. With a little manipulation, we get equation (13.6), i.e., the expression of $\tilde{p}_j(t)$ in terms of the slow states approximate probabilities $\tilde{p}_i(t)$, $i \in \mathcal{S}$ of BT's algorithm.

A.3 Proof of Theorem 13.4.1

Let us assume first that j is not an absorbing state, i.e., $a_j \neq 0$. We shall use expression (13.3) of $p_j(t)$ to prove Theorem 13.4.1. BCP's algorithm consists in replacing $g_s(t)$ with $\tilde{g}_s(t)$ in this equation. Therefore we have to bound the ratio

$$\frac{\tilde{g}_s(t)}{g_s(t)}.$$

Let us consider a path s, with last state j. Let k_s be the total number of slow states in path s (possibly including j). Using BCP's algorithm, we approximate function $g_s(t)$ by the density $\tilde{g}_s(t)$ of the sum of k_s exponentially distributed random variables, with rates $\lambda_1, \ldots, \lambda_{k_s}$, i.e., the slow rates of the path.

Density $\tilde{g}_s(t)$ may be expressed as follows:

$$\tilde{g}_s(t) = \int_{t_1+\cdots+t_{k_s}=t} \prod_{i=1}^{k_s} \left(\lambda_i \, e^{-\lambda_i t_i} dt_i\right).$$

Let $u < t$. The change of variable $t_i \to t_i \cdot (t-u)/t$ yields after a little manipulation

$$\left(\frac{t-u}{t}\right)^{k_s-1} \tilde{g}_s(t) \leq \tilde{g}_s(t-u). \leq \left(\frac{t-u}{t}\right)^{k_s-1} e^{\Lambda u} \tilde{g}_s(t). \tag{13.8}$$

This may be simplified using

$$1-(k_s-1)\frac{u}{t} \leq \left(\frac{t-u}{t}\right)^{k_s-1} \leq 1.$$

We then have

$$1-k_s\frac{u}{t} \leq \frac{\tilde{g}_s(t-u)}{\tilde{g}_s(t)} \leq e^{\Lambda u}. \tag{13.9}$$

Denoting by h_s the density function of the cumulated time spent in subset $\mathcal{F}$, we know that

$$g_s(t) = \int_0^t \tilde{g}_s(t-u) h_s(u) du. \tag{13.10}$$

Let ${k'}_s$ be the number of entrances into subset $\mathcal{F}$ in path s. From Bienaymé-Chebyshev's inequality, applied to the cumulated time spent in subset $\mathcal{F}$, we have:

$$\int_t^{+\infty} h_s(u) du \leq k'_s \frac{\tau_0}{t}.$$

It follows from (13.9) and (13.10)

$$-k_s k'_s \frac{\tau_0}{t} \tilde{g}_s(t) \leq g_s(t) \leq L(\Lambda)^{k'_s} \tilde{g}_s(t). \tag{13.11}$$

Let n_s be the number of slow states in s apart from j. Then we obviously have:

$$k_s = n_s + \mathbf{1}_{j\in\mathcal{S}} \tag{13.12}$$

and

$$k'_s \leq k_s. \tag{13.13}$$

Furthermore the probability of having visited and left n_s or more slow states at time t is lower than the probability that a Poisson process with rate Λ has n_s or more occurrences in $[0, t]$.

Theorem 13.4.1 is finally obtained by combining equations (13.11) for all paths s which number n_s is lower than a given integer $n \geq 1$, and by using (13.12) and (13.13).

Extension of Theorem 13.4.1 to absorbing states follows the same lines. It may also be obtained by letting the exit rates approach 0, using the continuity of the matrix exponential.

References

1. Bobbio, A. and Trivedi, K. S. (1986). An aggregation technique for the transient analysis of stiff Markov chains, *IEEE Transactions on Computers*, **35**, 803–814.

2. Bobbio, A. and Trivedi, K. S. (1990). Computing cumulative measures of stiff Markov chains using aggregation, *IEEE Transactions on Computers*, **39**, 1291–1298.

3. Courtois, P.-J. and Semal, P. (1984). Bounds for positive eigenvectors of nonnegative matrices and for their approximation by decomposition, *J. Ass. Comput. Mach.*, **31**, 804–825.

4. Dunkel, J. and Stahl, H. (1993). On the transient analysis of stiff Markov chains, *Proceedings of 3rd IFIP Working Conference on Dependable Computing for Critical Applications*, Mondello, Italy, June 1993, pp. 137-160, IFIP, Wien, Austria: Springer-Verlag.

5. Geist R., Smotherman, M. and Brown, M. (1989). Ultrahigh reliability estimates for systems exhibiting globally time-dependent failure processes, *The Nineteenth International Symposium on Fault-Tolerant Computing*, pp. 152–158, Chicago, June 1989.

6. Malhotra, M., Muppala, J. K. and Trivedi, K. S. (1994). Stiffness-tolerant methods for transient analysis of stiff Markov chains, *Microelectronics and Reliability*, **34**, 1825–1841.

7. McGough, J., Smotherman, M. and Trivedi, K. S. (1985). The conservativeness of reliability estimates based on instantaneous coverage, *IEEE Transactions on Computers*, **34**, 602–609.

8. Reibman, A., Trivedi, K. S., Kumar, S. and Ciardo, G. (1990). Analysis of stiff Markov chains, *ORSA Journal of Computing*, **1**, 126–133.

9. Stewart, W. J. (1994). *Introduction to the Numerical Solution of Markov Chains*, Princeton, NJ: Princeton University Press.

10. White, A. L. (1991). An error bound for instantaneous coverage, In *Annual Reliability and Maintenability Symposium 1991 Proceedings*, pp. 587–591, Orlando, USA.

14

On the Failure Rate of Components Subjected to a Diffuse Stress Environment

A. Le Breton and J.-L. Soler

LMC–IMAG, Grenoble, France

Abstract: The aim of this paper is to investigate the failure rate behaviour of a component subjected to a random diffuse stress environment, that is a stress whose intensity is an extremely variable continuous random function. This intensity is taken as the square of an Ornstein-Uhlenbeck random process path. Moreover, a concept of memory of the component with respect to the stress is introduced, which allows to consider different kinds of stress action (instantaneous, cumulative and others). Explicit and implicit formulas are derived for the resulting failure rate of the component from which it is shown that the stress contribution part rapidly converges to some finite constant value at least in the case of an instantaneous effect of the stress.

Keywords and phrases: reliability, failure rate, diffuse stress, Ornstein-Uhlenbeck process

14.1 Introduction

In many practical situations systems must operate in a dynamic environment generating *stresses* of randomly variable intensity which could affect the life-lengths of the involved components. Stresses may occur as a series of events or continuously in time and their effects can be either *instantaneous* or *cumulative*. This suggests that the components could eventually keep some memory of the stresses and that the action of the environment results into a combination of the stresses and of the component memory. Here, this action is assumed to have an additive effect on the proper failure rate of the components, conditionally to the stress process. The main objective is to investigate the resulting behaviour of the failure rate in a rather general model allowing to take into account various kinds of stresses and memory. The model is described in Section 14.2. Some general mathematical results are established in Section 14.3. Several interesting particular cases are investigated in Section 14.4 and a shot-noise model with diffuse intensity is discussed in Section 14.5.

14.2 The Mathematical Model

According to the model introduced in [Gaudoin and Soler (1997)], the stress, which is assumed to be randomly distributed over the time, will be represented by a random Stieltjes measure on the positive real half-line $\mathbb{R}_+$, with a cumulative distribution function $S(t)$, and then its intensity is the derivative $dS(t)$ in the Stieltjes sense. This allows to consider both a point process occurrence of the stress or a continuous one whose intensity is some continuous random function.

A stress that occurs at some time ($t = 0$, for example) may influence in the future the component which may have kept some memory of it. This memory which is distributed over the time will be represented by a deterministic Stieltjes measure on $\mathbb{R}_+$, with cumulative distribution function $M(t)$. The case of lack of memory, which stands for an instantaneous effect, is then represented by the Dirac measure at 0, that is $M(t) = 1$ for all $t \geq 0$, whereas the case of total memory, which stands for a cumulative effect, is represented by the Lebesgue measure, that is $M(t) = t$. Other cases, such as a decreasing intensity memory or an intensity vanishing after some time, may be investigated by choosing appropriate functions M.

Let us now define the action of the stress on a component as the random Stieltjes measure, whose cumulative function A is given, for all $t \geq 0$, by:

$$A(t) = \int_0^t M(t-u)dS(u).$$

The quantity $A(t)$ represents the total memory kept by the component of all the stresses that occurred up to time t.

Finally, the model is based on the assumption that, conditionally to the stress process, the cumulative hazard function Λ of the component is given, for all $t \geq 0$, by $\Lambda(t) = \alpha A(t) + B(t)$, that is:

$$\Lambda(t) = \alpha \int_0^t M(t-u)dS(u) + B(t),$$

where B is its proper cumulative hazard function, and α is a positive sensitivity coefficient. Then, the resulting failure rate of the component subject to the stress is obtained by calculating:

$$\lambda(t) = -\frac{d}{dt}\Big[\log E\Big[\exp\{-\alpha \int_0^t M(t-u)dS(u)\}\Big] \exp\{-B(t)\}\Big].$$

Since $B(t)$ is assumed to be known, we shall neglect it in the following and focus on the stress contribution part which will be added to the proper failure rate.

Some results have been obtained in [Gaudoin and Soler (1997)] in several cases of stress modelled by point processes or processes with alternating intensity. It was shown there that the failure rate generally increases up to some constant value, even when the effect of the stress is cumulative.

Here we will investigate the case of a stress whose intensity is related to an extremely variable random continuous function representing a *diffuse* environment. One can suppose that this function results at each time from numerous elementary stresses of diverse origins (electrical, mechanical, etc. ...). So this function will be represented by means of a continuous Gaussian process X, and here in fact, we choose an Ornstein-Uhlenbeck process (see e.g. [Breiman (1968)]), which is a particular case of diffusion process. Since the stress intensity must be a *positive* function, we assume that it is connected to the diffuse environment in terms of an energy. Then we set:

$$dS(t) = X_t^2 dt,$$

where, for some drift and diffusion coefficients $\beta \in \mathbb{R}$ and $\sigma > 0$ and some Gaussian initial condition X_0, the process $X = (X_t; t \geq 0)$ is the solution of the stochastic differential equation:

$$dX_t = -\beta X_t dt + \sigma dW_t, \quad t \geq 0, \tag{14.1}$$

where $W = (W_t; t \geq 0)$ is a standard Brownian motion independent of X_0. Recall that when $\beta > 0$, then the process X is stationary if and only if X_0 has the Gaussian distribution $\mathcal{N}(0, \frac{\sigma^2}{2\beta})$.

14.3 General Results

For the case of a general memory $M(t)$, to compute the reliability we need to evaluate the following Laplace transform of a quadratic functional of a process X satisfying (14.1):

$$R^M(t) = E\left[\exp\left\{-\alpha \int_0^t M(t-u) X_u^2 du\right\}\right].$$

Then of course the corresponding failure rate will be derived as

$$\lambda^M(t) = -\frac{\dot{R}^M(t)}{R^M(t)}.$$

14.3.1 The case of a stress starting from a fixed level

Here we consider the case where the process X starts from some fixed point $x_0 \in \mathbb{R}$ at time 0. The computations will be based on the following key technical lemma:

Lemma 14.3.1 *Let θ and x be arbitrary fixed real numbers. Let $\xi^x = (\xi^x_t, t \geq 0)$ be the solution process of the stochastic differential equation $d\xi_t = \theta\xi_t dt + dW_t$, $t \geq 0$, starting from x at time 0 i.e. such that $\xi_0 = x$. Let $Q = (Q_t, t \geq 0)$ be any continuous nonnegative real-valued function on $\mathbb{R}_+$. Define for $a \geq 0$ and $t \geq 0$ the quantity $\Psi^Q_x(\theta, a, t)$ by:*

$$\Psi^Q_x(\theta, a, t) = E\left[\exp\left\{-a\int_0^t Q(u)(\xi^x_u)^2 du\right\}\right]. \tag{14.2}$$

Then the following equality holds:

$$\Psi^Q_x(\theta, a, t) = \delta_t^{-\frac{1}{2}} \exp\Big(\frac{1}{2}\Big[\gamma_0 - \gamma_t\varphi_t^2\delta_t^{-1}\Big]x^2 + \frac{1}{2}\int_0^t \gamma_u du\Big), \tag{14.3}$$

where $\gamma = (\gamma_u, u \geq 0)$ is the unique nonnegative solution of the Riccati differential equation

$$\dot{\gamma}_u = -2\theta\gamma_u - \gamma_u^2 + 2aQ(u);\ u \geq 0, \tag{14.4}$$

for some initial condition $\gamma_0 \geq 0$ and

$$\varphi_s = \exp\,(\theta s + \int_0^s \gamma_u du),\ 0 \leq s \leq t; \quad \delta_t = 1 + \gamma_t\varphi_t^2 \int_0^t \varphi_s^{-2} ds. \tag{14.5}$$

PROOF. See Appendix. ∎

From Equation (14.1) we may write $d\frac{X_t}{\sigma} = -\beta\frac{X_t}{\sigma}dt + dW_t$. Therefore, for a deterministic initial condition $X_0 = x_0 \in \mathbb{R}$, the process $\frac{X}{\sigma} = (\frac{X_t}{\sigma}; t \geq 0)$ can be taken as the process ξ in Lemma 14.3.1 with $\theta = -\beta$ and $x = \frac{x_0}{\sigma}$. Moreover we have

$$\alpha\int_0^t M(t-u)X_u^2 du = \alpha\sigma^2\int_0^t Q(u)(\frac{X_u}{\sigma})^2 du,$$

for Q defined by $Q(u) = M(t-u), 0 \leq u \leq t$. Hence, choosing that function Q and $a = \alpha\sigma^2$ we may compute the corresponding reliability

$$R^M_{x_0}(t) = \Psi^Q_{\frac{x_0}{\sigma}}(-\beta, \alpha\sigma^2, t),$$

by Equations (14.3), (14.4) and (14.5). Then for instance, in the particular case where $x_0 = 0$, it is readily seen that the following statement holds:

Proposition 14.3.1 *Consider a component subject to a stress with diffuse intensity starting from* 0 *and whose memory is the continuous function M. Then the stress contribution parts to its reliability and failure rate are given by:*

$$R_0^M(t) = \Delta_t^{-\frac{1}{2}} \exp\Big(\frac{1}{2}\int_0^t \Gamma(t,u)du\Big), \tag{14.6}$$

and

$$\lambda_0^M(t) = \frac{1}{2}\left[\Delta_t^{-1}\dot{\Delta}_t - \frac{\partial}{\partial t}\left(\int_0^t \Gamma(t,u)du\right)\right], \tag{14.7}$$

respectively, where $\Gamma(t,.) = (\Gamma(t,u), 0 \le u \le t)$ is the unique nonnegative solution of the differential equation

$$\frac{\partial}{\partial u}\Gamma(t,u) = 2\beta\Gamma(t,u) - \Gamma^2(t,u) + 2\alpha\sigma^2 M(t-u);\ 0 \le u \le t, \tag{14.8}$$

for some initial condition $\Gamma(t,0) = \Gamma_0 \ge 0$ and

$$\Delta_t = 1 + \Gamma(t,t)\int_0^t \Phi^2(t,s)ds, \tag{14.9}$$

with

$$\Phi(t,s) = \exp\big(-\beta(t-s) + \int_s^t \Gamma(t,u)du\big),\ 0 \le s \le t. \tag{14.10}$$

14.3.2 The case of a stationary stress process

Here we consider the case where in Equation (14.1) the parameter β is positive and the process X starts from $X_0 \sim \mathcal{N}(0, \frac{\sigma^2}{2\beta})$ and consequently X is stationary. We will make use of the following result:

Lemma 14.3.2 *Let θ be an arbitrary fixed negative real number. Let $\bar{\xi} = (\bar{\xi}_t, t \ge 0)$ be the solution process of the stochastic differential equation $d\xi_t = \theta\xi_t dt + dW_t,\ t \ge 0$, with initial condition $\xi_0 = \bar{\xi}_0 \sim \mathcal{N}(0, (2|\theta|)^{-1})$. Let $Q = (Q_t, t \ge 0)$ be any continuous nonnegative real-valued function on $\mathbb{R}_+$. Define for $a \ge 0$ and $t \ge 0$ the quantity $\Psi_{st}^Q(\theta, a, t)$ by:*

$$\Psi_{st}^Q(\theta,a,t) = E\left[\exp\left\{-a\int_0^t Q(u)\bar{\xi}_u^2 du\right\}\right]. \tag{14.11}$$

Then the following equality holds:

$$\Psi_{st}^Q(\theta,a,t) = (2|\theta|)^{\frac{1}{2}}\Psi_0^Q(\theta,a,t)\left[\gamma_t\varphi_t^2\delta_t^{-1} + 2|\theta| - \gamma_0\right]^{-\frac{1}{2}}, \tag{14.12}$$

where $\Psi_0^Q(\theta,a,t)$ is given by (14.3), $\gamma = (\gamma_u, u \ge 0)$ satisfies (14.4) for some initial condition $\gamma_0 \ge 0$ and φ_t and δ_t are both given by (14.5).

PROOF. First notice that since for all $x \in \mathbb{R}$ the conditional distribution of the process $\bar{\xi}$ given $\bar{\xi}_0 = x$ coincides with that of the process ξ^x appearing in Lemma 14.3.1, from (14.11) we get

$$\Psi_{st}^Q(\theta, a, t) = \int_{-\infty}^{+\infty} \Psi_x^Q(\theta, a, t) \, \frac{|\theta|}{\pi} \, \exp\{-|\theta| x^2\} dx.$$

Now due to (14.3) we have also

$$\Psi_x^Q(\theta, a, t) = \Psi_0^Q(\theta, a, t) \exp\left(\frac{1}{2}\left[\gamma_0 - \gamma_t \varphi_t^2 \delta_t^{-1}\right] x^2\right).$$

Then, combining these two equalities and computing the integral, the result is readily obtained. ∎

Here, again from Equation (14.1), for $\beta > 0$ and for the initial condition $X_0 \sim \mathcal{N}(0, \frac{\sigma^2}{2\beta})$, choosing Q defined by $Q(u) = M(t-u), 0 \le u \le t$ and $a = \alpha\sigma^2$ we may compute the corresponding reliability

$$R_{st}^M(t) = \Psi_{st}^Q(-\beta, \alpha\sigma^2, t),$$

by Equations (14.12), (14.3), (14.4) and (14.5). Then we get the following statement:

Proposition 14.3.2 *Consider a component subject to a stress with stationary diffuse intensity and whose memory is the continuous function M. Then the stress contribution parts to its reliability and failure rate are given by:*

$$R_{st}^M(t) = (2\beta)^{\frac{1}{2}} R_0^M(t) [\Gamma(t,t)\Phi^2(t,0)\Delta_t^{-1} + 2\beta - \Gamma_0]^{-\frac{1}{2}}, \tag{14.13}$$

and

$$\lambda_{st}^M(t) = \lambda_0^M(t) + \frac{\frac{\partial}{\partial t}[\Gamma(t,t)\Phi^2(t,0)\Delta_t^{-1}]}{2[\Gamma(t,t)\Phi^2(t,0)\Delta_t^{-1} + 2\beta - \Gamma_0]}, \tag{14.14}$$

respectively, where $R_0^M(t)$ and $\lambda_0^M(t)$ are given by (14.6) and (14.7) respectively while $\Gamma(t,.) = (\Gamma(t,u), 0 \le u \le t)$ satisfies (14.8) for some initial condition $\Gamma(t,0) = \Gamma_0 \ge 0$ and Δ_t and $\Phi(t,.) = (\Phi(t,s), 0 \le s \le t)$ are given by (14.9) and (14.10) respectively.

14.4 Particular Cases of Interest

14.4.1 Instantaneous action of the stress

Here we assume that the effect of the stress on the component is instantaneous. This is expressed by setting $M(t) = 1$ for all $t \ge 0$. Then the results of

Propositions 14.3.1 and 14.3.2 can be explicit. Equation (14.8) becomes

$$\frac{\partial}{\partial u}\Gamma(t,u) = 2\beta\Gamma(t,u) - \Gamma^2(t,u) + 2\alpha\sigma^2;\ 0 \le u \le t.$$

Since the initial condition $\Gamma(t,0) = \Gamma_0$ can be arbitrarily chosen we can take it as the positive solution

$$\omega = \beta + \sqrt{\beta^2 + 2\alpha\sigma^2}, \tag{14.15}$$

of the Riccati algebraic equation

$$2\beta\Gamma - \Gamma^2 + 2\alpha\sigma^2 = 0.$$

Then of course we have $\Gamma(t,.) \equiv \omega$. Inserting this into Equations (14.9) and (14.10) we get:

$$\Phi(t,s) = \exp[\omega(t-s)],\ 0 \le s \le t; \quad \Delta_t = \frac{(\omega-\beta) + (\omega+\beta)e^{2\omega t}}{2\omega}.$$

Instantaneous action of a stress starting from zero

Here we take $x_0 = 0$ in Equation (14.1). Due to the above discussion, the following statement is an immediate consequence of Proposition 14.3.1:

Corollary 14.4.1 *Consider a component subject to a stress with instantaneous action and diffuse intensity starting from* 0. *Then the stress contribution parts to its reliability and failure rate are given by:*

$$R_0(t) = e^{\frac{\beta}{2}t}\Big[\frac{(\omega-\beta)e^{-\omega t} + (\omega+\beta)e^{\omega t}}{2\omega}\Big]^{-\frac{1}{2}}, \tag{14.16}$$

and

$$\lambda_0(t) = \frac{1}{2}\Big[\frac{\omega(\beta-\omega)e^{-\omega t} + \omega(\omega+\beta)e^{\omega t}}{(\omega-\beta)e^{-\omega t} + (\omega+\beta)e^{\omega t}} - \beta\Big], \tag{14.17}$$

respectively, where ω *is defined by (14.15).*

Notice that the failure rate $\lambda_0(t)$ is a function of t which increases from the value 0 at $t = 0$ and converges to the limit $\frac{1}{2}(\omega - \beta) = \frac{1}{2}(\sqrt{\beta^2 + 2\alpha\sigma^2} - \beta)$ as t tends to infinity. Moreover it is an increasing function of the diffusion coefficient σ, what is rather natural.

Observe also that when $\beta = 0$ (and then the density of the stress intensity is the square of a Brownian motion), the reliability is:

$$R_0(t) = [\cosh(\sigma\sqrt{2\alpha}t)]^{-\frac{1}{2}},$$

and the failure rate is:

$$\lambda_0(t) = \sigma\sqrt{\frac{\alpha}{2}}\tanh(\sigma\sqrt{2\alpha}t).$$

Instantaneous action of a stationary stress

Here we consider the case where in Equation (14.1) the parameter β is positive and the process X is stationary. The following statement is an immediate consequence of Proposition 14.3.2:

Corollary 14.4.2 *Consider a component subject to a stress with instantaneous action and stationary diffuse intensity. Then the stress contribution parts to its reliability and failure rate are given by:*

$$R_{st}(t) = e^{\frac{\beta}{2}t}\Big[\frac{(\omega+\beta)^2 e^{\omega t} - (\omega-\beta)^2 e^{-\omega t}}{4\omega\beta}\Big]^{-\frac{1}{2}}, \tag{14.18}$$

and

$$\lambda_{st}(t) = \frac{1}{2}\Big[\frac{\omega(\omega+\beta)^2 e^{\omega t} + \omega(\omega-\beta)^2 e^{-\omega t}}{(\omega+\beta)^2 e^{\omega t} - (\omega-\beta)^2 e^{-\omega t}} - \beta\Big], \tag{14.19}$$

respectively, where ω is defined by (14.15).

Here the failure rate $\lambda_0(t)$ is a function of t which decreases from the value $\frac{\alpha\sigma^2}{2\beta}$ at $t = 0$ and converges to the same limit as in the non-stationary case as t tends to infinity. Again it is an increasing function of the diffusion coefficient σ.

14.4.2 Cumulative action of the stress

The cumulative effect of the stress is expressed by setting $M(t) = t$ for all $t \geq 0$. Unfortunately, in this case the results of Propositions 14.3.1 and 14.3.2 do not lead to formulas for the reliability and the failure rate more explicit than (14.6)–(14.7) and (14.18)–(14.19). Nevertheless the corresponding functions can be drawn by means of symbolic calculus systems.

14.5 A Shot-Noise Model With Diffuse Stress

As an extension of the previous results, one may study a shot-noise stress process itself generated by a diffuse environment.

Let us first recall the result obtained in [Gaudoin and Soler (1997)] about the shot-noise model. Suppose that a component with memory $M(t)$ is subject to a stress occurring at times produced by a Poisson process with intensity $\eta(t)$ and associated counting process $N = (N_t, t \geq 0)$. Then, neglecting the proper failure rate, the action of such a stress is expressed at any time t as:

$$A(t) = \alpha \int_0^t M(t-u)dN_u,$$

and the reliability is obtained by calculating:

$$R(t) = E\Big[\exp\{-\alpha \int_0^t M(t-u)dN_u\}\Big].$$

Hence, thanks to the well-known result about the Laplace transform of a Poisson process (see, e.g., [Snyder, 1975]), one gets:

$$R(t) = \exp\left\{-\int_0^t [1 - e^{-\alpha M(t-u)}]\eta(u)du\right\}.$$

In the case of a homogeneous Poisson process with (constant) intensity η, the reliability reduces to:

$$R(t) = \exp\left\{-\eta \int_0^t [1 - e^{-\alpha M(t-u)}]du\right\}.$$

This shows that if the effect of the stress is instantaneous $(M(t) = 1)$ then the lifelength distribution of the component is exponential with a constant failure rate equal to the value $\eta(1-e^{-\alpha})$. Similarly if the stress has a cumulative effect $(M(t) = t)$, then the reliability is

$$R(t) = \exp\left\{-\eta \left[t - \frac{1-e^{-\alpha t}}{\alpha}\right]\right\},$$

and the corresponding failure rate

$$\lambda(t) = \eta(1 - e^{-\alpha t}),$$

is increasing and bounded.

Now, let us consider a doubly stochastic non homogeneous Poisson process N with diffuse intensity $\eta(t) = X_t^2$, where X satisfies Equation (14.1), and suppose that a component with memory $M(t)$ is affected by the realization of the point process. Then, the reliability is obtained by calculating:

$$R(t) = E\Big[E\Big(\exp\{-\alpha \int_0^t M(t-u)dN_u\}/X\Big)\Big].$$

Since the conditional expectation inside is:

$$E\left(\exp\left\{-\alpha \int_0^t M(t-u)dN_u\right\}\Big/X\right) = \exp\left\{-\int_0^t [1 - e^{-\alpha M(t-u)}]X_u^2 du\right\},$$

the reliability is still of the form:

$$R(t) = E\left[\exp\{-\int_0^t Q(u)X_u^2 du\}\right].$$

Then the results of Lemmas 14.3.1 and 14.3.2 are applicable and again the above mentioned methods for drawing the failure rate can be used.

Actually, as above, explicit formulas can be obtained in the case of an instantaneous action of the stress since then

$$R(t) = E\Big[\exp\{-(1-e^{-\alpha})\int_0^t X_u^2 du\}\Big].$$

Then the failure rate is still a bounded increasing function or a decreasing function (with the same positive limit) depending on whether the stress starts from 0 or is stationary.

14.6 Conclusion

We have proposed a quite general model for representing the influence of a stress environment on the lifelength of a component. This model can take into account diverse kinds of stress process and effects (instantaneous, cumulative and others). Here we focused on a continuous diffuse stress and have obtained the most general results we could derive from stochastic calculus methods. In the particular case of an instantaneous effect of the stress (lack of memory of the component), explicit formulas show that the additive contribution part of the stress on the failure rate is an increasing (or decreasing) bounded function. In more general cases, the results do not lead to quite explicit formulas but they allow some investigations by means of symbolic calculus techniques. Finally we have presented some extension of the results to a new shot-noise model in diffuse environment.

Appendix (Proof of Lemma 14.3.1)

The proof is similar to that of Lemma 2.2.2 in [Le Breton and Musiela, 1985]. At first we notice that the Riccati differential Equation (14.4) can be seen as the equation governing the variance error in a linear filtering problem [see, e.g., Liptser and Shiryaev (1978)]. Hence the existence of a unique nonnegative solution $\gamma = (\gamma_t, t \geq 0)$ to this equation is ensured.

Now recall that the process $\xi^x = (\xi_s^x, 0 \leq s \leq t)$ satisfies the stochastic differential equation:

$$d\xi_s^x = \theta\xi_s^x ds + dW_s, \ 0 \leq s \leq t; \ \xi_0^x = x.$$

Let us introduce the process $\zeta^x = (\zeta_s^x, 0 \leq s \leq t)$ which is the solution of the auxiliary stochastic differential equation:

$$d\zeta_s^x = (\theta + \gamma_s)\zeta_s^x ds + dW_s, \ 0 \leq s \leq t; \ \zeta_0^x = x,$$

where γ is the solution of (14.4). Now let us define the random variable

$$Z_t = \exp\left\{-\int_0^t \gamma_s \zeta_s^x dW_s - \frac{1}{2}\int_0^t \gamma_s^2 (\zeta_s^x)^2 ds\right\},$$

and the process $\bar{W} = (\bar{W}_s, 0 \leq s \leq t)$ by

$$\bar{W}_s = W_s + \int_0^s \gamma_u \zeta_u^x du; \; 0 \leq s \leq t.$$

Applying the Girsanov Theorem [see, e.g., Liptser and Shiryaev (1978)] we get that $E(Z_t) = 1$ and moreover, defining the probability $\bar{P}$ as the probability with density Z_t with respect to the reference probability P, then the process $\bar{W}$ is a Brownian motion under $\bar{P}$. In others words, with respect to the probability $\bar{P}$, we may write

$$d\zeta_s^x = \theta \zeta_s^x ds + d\bar{W}_s, \; 0 \leq s \leq t; \; \zeta_0^x = x,$$

where $\bar{W}$ is a Brownian motion. This means that the distribution of ζ^x under $\bar{P}$ coincides with that of ξ^x under P. Therefore the quantity $\Psi_x^Q(\theta, a, t)$ defined by (14.2) can be evaluated by

$$\Psi_x^Q(\theta, a, t) = \bar{E}\left[\exp\left\{-a\int_0^t Q(s)(\zeta_s^x)^2 ds\right\}\right],$$

where the symbol $\bar{E}$ stands for the expectation with respect to $\bar{P}$. Making use of $\bar{P} = Z_t.P$, it follows that

$$\Psi_x^Q(\theta, a, t) = E\left[\exp\left\{-\int_0^t \left(aQ(s) + \frac{\gamma^2}{2}\right)(\zeta_s^x)^2 ds - \int_0^t \gamma_s \zeta_s^x dW_s\right\}\right].$$

Evaluating the stochastic integral above as

$$\int_0^t \gamma_s \zeta_s^x dW_s = \int_0^t \gamma_s \zeta_s^x d\zeta_s^x - \int_0^t \gamma_s(\theta + \gamma_s)(\zeta_s^x)^2 ds,$$

we obtain

$$\Psi_x^Q(\theta, a, t) = E\left[\exp\left\{-\int_0^t \gamma_s \zeta_s^x d\zeta_s^x + \int_0^t \left(\theta\gamma_s + \frac{\gamma^2}{2} - aQ(s)\right)(\zeta_s^x)^2 ds\right\}\right].$$

Now making use of the Ito formula to compute the stochastic integral and taking Equation (14.4) into account in the ordinary integral, the quantity appearing in the exponential above can be written as

$$\frac{1}{2}\left[\int_0^t \gamma_s ds - \int_0^t \gamma_s d(\zeta_s^x)^2 - \int_0^t \dot{\gamma}_s (\zeta_s^x)^2 ds\right] = \frac{1}{2}\left[\gamma_0 x^2 - \gamma_t(\zeta_t^x)^2 + \int_0^t \gamma_s ds\right].$$

Hence we get

$$\Psi_x^Q(\theta, a, t) = \exp\left[\frac{1}{2}\left\{\gamma_0 x^2 + \int_0^t \gamma_s ds\right\}\right] E\left[\exp\left\{-\frac{1}{2}\gamma_t(\zeta_t^x)^2\right\}\right].$$

Let us observe that from the equation generating ζ^x we have

$$\zeta_t^x = \varphi_t \left\{x + \int_0^t \varphi_s^{-1} dW_s\right\},$$

where φ satisfies Equation (14.5). Therefore the random variable ζ_t^x has the Gaussian distribution with mean φ_t and variance $\varphi_t^2 \int_0^t \varphi_s^{-2} ds$. But if $Y \sim \mathcal{N}(m, \tau^2)$ then for any positive number r the following equality holds:

$$E(e^{-rY^2}) = (1 + 2r\tau^2)^{-\frac{1}{2}} \exp\{-r(1 + 2r\tau^2)^{-1} m^2\}.$$

Finally, making use of this to evaluate the expectation in the last expression of $\Psi_x^Q(\theta, a, t)$, it is easy to check that Equation (14.3) is valid. ■

References

1. Breiman, L. (1968). *Probability*, Reading: Addison-Wesley.

2. Gaudoin, O. and Soler, J.-L. (1997). Failure rate behaviour of components subject to random stresses, *Reliability Engineering and System Safety*, **58**, 19–30.

3. Le Breton, A. and Musiela, M. (1985). Some parameter estimation problems for hypoelliptic homogeneous Gaussian diffusions, In *Sequential Methods in Statistics, Banach Center Publications, Vol. 16*, pp. 337–356, Warsaw: PWN-Polish Scientific Publications.

4. Liptser, R. S. and Shiryaev, A. N. (1978). *Statistics of Random Processes*, Berlin: Springer-Verlag.

5. Snyder, D. L. (1975). *Random Point Processes*, New-York: John Wiley & Sons.

15

Modelling the Reliability of a Complex System Under Stress Environment

Cristina Zahalca and Mohamed Chardi

Laboratoire IMAG-LMC, Grenoble, France

Abstract: The aim of the paper is to study the reliability of a complex system, all components of which are subjected to a common point stress process. The point stress process may represent impulsions, shocks, radioactive emissions. We generalize the results obtained for one component, using the model introduced in [Gaudoin and Soler (1997)]. We investigate the stress consequences with respect to the system architecture.

Keywords and phrases: Reliability, failure rate, mean time to failure, survival function, point stress process, homogeneous Poisson process, multivariate exponential distribution

15.1 Introduction

The main objective of this work is to study the lifetime of a system made up of components. Failure renders a component non-operational due to damage caused by failure mechanism, actuated generally by external and/or internal conditions. External conditions that affect the state of a component are broadly termed **stresses**.

The nature of the stresses which trigger the failure mechanism can be electrical, mechanical, thermal, chemical or radiative. The types of stresses, which represent diverse physical phenomena are point stresses, diffused stresses, alternating stresses.

According to the model proposed by [Gaudoin and Soler (1997)], a component of the system which undergoes a stress at a given time can possibly keep a memory of it. Our stress model described in Section 15.2 study a system when the components keep no memory of a point stress process. Section 15.3 and

15.4 generalize the model for a system of n identical components subjected to an homogeneous Poisson stress process. The main objective is to investigate the dependence between the system architecture and the stress influence. In Section 15.5 an indicator of the stress influence is proposed and some interesting results are derived. Section 15.6 presents the particular case of a system of two identical components system subjected to a common homogeneous Poisson stress process.

15.2 Modelling the Stress

According to the model proposed by [Gaudoin and Soler (1997)], the stress is characterized by a random distribution of mass among the time, which can be discrete or continuous. We represent the stress by a random *Stieltjes measure* m_S *on* $(\mathcal{R}^+, \mathcal{B}(\mathcal{R}^+))$. A Stieltjes measure is a measure defined by a function S, positive, nondecreasing, right-continuous, bounded on every bounded interval, such that:

$$\forall(s,t),\ 0 \leq s < t,\ m_S(]s,t]) = S(t) - S(s) \tag{15.1}$$

where S is called the *cumulative distribution function* (cdf) of m_S.

The *stress intensity* s is the Stieltjes density of this measure. As usual, we will use the same notation S for S and m_S. For point stresses, s is the density of m_S with respect to the counting measure of occurrence times. Note that the cumulative hazard rate Λ defines a Stieltjes measure called *failure measure.*

The model is based on a mathematical hypothesis: at each time, conditionally to all stresses occurred, the failure measure is proportional to the stress measure.

$$\Lambda_{cond}(t) = \alpha S(t) + B(t) \tag{15.2}$$

where B is the proper failure measure (the proper failures caused by internal conditions) and α is a positive stress sensitivity coefficient. This means that the stress weakens the component. More generally, the stress sensitivity coefficient could be considered as depending both on time and environmental variables.

Let T be the random lifetime of the component subjected to stress. The reliability function is:

$$R_s(t) = P(T > t) = E\left[e^{-\Lambda_{cond}(t)}\right] = E\left[e^{-\alpha S(t) - B(t)}\right] \tag{15.3}$$

The reliability of the component is always less when it is subjected to stress than in an unstressed environment.

In the same manner, the failure rate of the component subjected to stress is:

$$\lambda_s(t) = \frac{dB(t)}{dt} - \frac{d}{dt}\ln E\left[e^{-\alpha S(t)}\right] \tag{15.4}$$

which is the sum of the proper failure rate λ, due to the internal failure conditions and the failure rate due to the external (stress) failure conditions.

A *point stress* is a succession of impulsions, which occur at random times $0 \leq X_1 < X_2 < ...$ with random amplitudes $Z_1, Z_2, ...$. Let N_t be the number of stresses occurred between 0 and t.

- The point stress measure is: $\sum_{i=1}^{\infty} Z_i \, \delta_{X_i}$ so its cdf is: $S(t) = \sum_{i=1}^{N_t} Z_i$.

- The stress intensity is:

$$s(t) = \sum_{i=1}^{N_t} Z_i \, 1_{X_i}(t) = \begin{cases} 0 & \text{if } t \text{ is not a stress occurrence time} \\ Z_i & \text{if } t = X_i \end{cases} \qquad (15.5)$$

We assume that:

- The stress occurrence time and stress amplitude processes are independent.

- The stress occurrence time process $\{N_t\}_{t \geq 0}$ is an homogeneous Poisson process (HPP), with intensity ξ.

- The amplitudes of the stresses are independent and identically distributed (i.i.d) positive random variables, with cdf H and Laplace transform ϕ, like a r.v. Z.

15.3 System of n Identical Components Subjected to an Homogeneous Poisson Stress Process

We are concerned with a system of n identical components subjected to a common homogeneous Poisson stress process. The stress sensitivity coefficient $\alpha > 0$ and the proper failure rate λ are identical for all n components.

We suppose that conditionally to the stress, the random lifetimes $T_1, T_2, .., T_n$ are independent.

For all $t_1, t_2, ..., t_n > 0$, the survival function of the random vector $(T_1, T_2, ..., T_n)$ is obtained as:

$$\begin{aligned} R(t_1, t_2, ..., t_n) &= P(T_1 > t_1, T_2 > t_2, ..., T_n > t_n) \\ &= E\left[e^{-\sum_{i=1}^{n} \Lambda(t_i)} \right] = e^{-\lambda \sum_{i=1}^{n} t_i} \, E\left[e^{-\alpha \sum_{i=1}^{n} S(t_i)} \right] \end{aligned}$$

$$= e^{-\lambda\sum_{i=1}^{n} t_i} E\left[e^{-\alpha \sum_{i=1}^{n}\sum_{j=1}^{N_{t_i}} Z_j} \right]$$

$$= e^{-\lambda\sum_{i=1}^{n} t_i} \prod_{i=1}^{n} E\left[e^{-\alpha(n-i+1) \sum_{j=N^*_{t_{i-1}}+1}^{N^*_{t_i}} Z_j} \right]$$

$$= e^{-\lambda\sum_{i=1}^{n} t_i} \prod_{i=1}^{n} e^{-\xi\,[1-\phi(\alpha(n-i+1))]\,(t^*_i - t^*_{i-1})}$$

So,

$$R(t_1, t_2, ..., t_n) = e^{-\sum_{i=1}^{n}(\theta_{n-i+1} - \theta_{n-i} + \lambda)t^*_i} \tag{15.6}$$

where $\theta_i = \xi\,[1 - \phi(\alpha i)]$ and $t^*_1 < t^*_2 < ... < t^*_n$ are the ordered values of $t_1, t_2, ..., t_n$.

There it is easy to see that the distribution of the random vector $(T_1, T_2, ..., T_n)$ is a symmetrical multivariate exponential distribution of Marshall and Olkin (abbreviated MVE) [Marshall and Olkin (1967)] with parameters $\mu_1, \mu_2, ..., \mu_n$ defined by:

$$\forall i \in \{1, 2, ...n\},\ \sum_{k=1}^{i} \mu_k\, C_{i-1}^{k-1} = \theta_{n-i+1} - \theta_{n-i} + \lambda \tag{15.7}$$

Using the properties of the MVE [Barlow and Proschan (1981)], we deduce that $P(T_i > t) = e^{-(\lambda+\theta_1)t}$. So, the lifetime of each system component is an exponential distribution with parameter $\lambda + \theta_1$.

For $t_1 = t_2 = ... = t_n = t$ and $\theta_n = \xi\,[1 - \phi(\alpha n)]$ the survival function is:

$$P\,(T_1 > t, T_2 > t, ..., T_n > t) = e^{-(n\lambda + \theta_n)t} \tag{15.8}$$

which represents the reliability function for a series configuration of n identical components.

15.4 Some Particular Configurations of the n Identical Component System

Let A_i be the number of paths having i components. For a n identical component system, with a proper failure rate λ, the reliability function in an unstressed environment is [Birnbaum and Esary (1961)]:

$$R_p(t) = \sum_{i=1}^{n} A_i e^{-i\lambda t} \left(1 - e^{-\lambda t}\right)^{n-i} = \sum_{j=1}^{n} b_j \, e^{-j\lambda t} \tag{15.9}$$

where $b_j = \sum_{i=1}^{j} (-1)^{j-i} A_i C_{n-i}^{n-j}$. Note that $R_p(0) = 1$, so $\sum_{j=1}^{n} b_j = 1$.

Proposition 15.4.1 *The reliability function of a n identical components system subjected to a common homogeneous Poisson stress process is:*

$$R_s(t) = \sum_{j=1}^{n} b_j \, e^{-(j\lambda+\theta_j)t} \tag{15.10}$$

PROOF.

$$\begin{aligned} R_s(t) &= E\left[\sum_{i=1}^{n} A_i e^{-i(\alpha S(t)+\lambda t)} \left(1 - e^{-(\alpha S(t)+\lambda t)}\right)^{n-i}\right] \\ &= \sum_{j=1}^{n} \left(\sum_{i=1}^{j} (-1)^{j-i} A_i C_{n-i}^{n-j}\right) e^{-j\lambda t} \, E\left[e^{-j\alpha S(t)}\right] \\ &= \sum_{j=1}^{n} b_j \, e^{-(j\lambda+\theta_j)t}. \end{aligned}$$

With (15.10) we can calculate the reliability function for some particular configurations.

1) Series configuration ($A_i = 0$ for $i \leq n-1$, $A_n = 1$)

$$R_{s,n/n}(t) = e^{-(n\lambda+\theta_n)t}. \tag{15.11}$$

2) Parallel configuration ($A_i = C_n^i$ for $i \leq n$)

$$R_{s,1/n}(t) = \sum_{i=1}^{n} (-1)^{i-1} C_n^i \, e^{-(i\lambda+\theta_i)t}. \tag{15.12}$$

3) k-out-of-n (k/n) configuration ($A_i = C_n^i$ for $i \geq k$)

$$R_{s,k/n}(t) = \sum_{i=k}^{n} (-1)^{i-k} C_n^i C_{i-1}^{k-1} \, e^{-(i\lambda+\theta_i)t}. \tag{15.13}$$

15.5 Architecture and Stress Influence

We are interested in the study of the dependence between the system architecture and the stress influence. Thus, we want to find the system configurations which are more sensitive to the stress. We propose as an indicator of the stress influence on the system reliability:

$$I = 1 - MTTF_s / MTTF_p \tag{15.14}$$

where $MTTF_p$ represents the mean time to failure of the system in an unstressed environment and $MTTF_s$ represents the mean time to failure of the system subjected to stress. This indicator represents the relative reduction rate of $MTTF_p$ due to the stress.

Using (15.11), (15.12), (15.13) we obtain the indicator I of the stress influence for:

1) Series configuration

$$I_{n/n} = 1 - \frac{n\lambda}{n\lambda + \theta_n} \,. \tag{15.15}$$

2) Parallel configuration

$$I_{1/n} = 1 - \frac{\sum_{i=1}^{n} (-1)^{i-1} C_n^i \dfrac{1}{i\lambda + \theta_i}}{\dfrac{1}{\lambda} \sum_{i=1}^{n} \dfrac{1}{i}} \,. \tag{15.16}$$

3) k-out-of-n (k/n) configuration

$$I_{k/n} = 1 - \frac{\sum_{i=k}^{n} (-1)^{i-k} C_n^i C_{i-1}^{k-1} \dfrac{1}{i\lambda + \theta_i}}{\dfrac{1}{\lambda} \sum_{i=k}^{n} \dfrac{1}{i}} \,. \tag{15.17}$$

Proposition 15.5.1 $\left(I_{n/n}\right)_{n \geq 1}$ *is a decreasing sequence.*

PROOF. From (15.9) for $t = \alpha Z/\lambda$, we can write:

$$R_p(\tfrac{\alpha Z}{\lambda}) = \sum_{j=1}^{n}\sum_{i=1}^{j}(-1)^{j-i}A_iC_{n-i}^{n-j}\,e^{-j\alpha Z}.$$

Using the inequality $0 \le E\left[R_p(\tfrac{\alpha Z}{\lambda})\right] \le 1$ and $\theta_j = \xi E\left[1 - e^{-j\alpha Z}\right]$ we obtain:

$$0 \le \sum_{j=1}^{n}\sum_{i=1}^{j}(-1)^{j-i}A_iC_{n-i}^{n-j}\,E\left[e^{-j\alpha Z}\right] \le 1$$

or

$$0 \le \sum_{j=1}^{n}\sum_{i=1}^{j}(-1)^{j-i}A_iC_{n-i}^{n-j}\,\theta_j \le \xi. \tag{15.18}$$

For the particular case of the fail-safe configuration $n-1/n$, the relation (15.18) is:

$$0 \le n\theta_{n-1} - (n-1)\theta_n \le \xi \tag{15.19}$$

then:

$$\left(\frac{n}{n\lambda+\theta_n}\right)_{n\ge1} \text{ is an increasing sequence.}$$

So, $I_{n-1/n-1} - I_{n/n} = \lambda\left(\dfrac{n}{n\lambda+\theta_n} - \dfrac{n-1}{(n-1)\lambda+\theta_{n-1}}\right) > 0.$

This means that we obtain a configuration less sensitive to the stress if we increase the number of the components in a series configuration.

Results

1)

$$I_{n-1/n} - I_{n/n} = \frac{n^2}{2n-1}\left(I_{n-1/n-1} - I_{n/n}\right) > 0. \tag{15.20}$$

2)

$$I_{n/n+1} - I_{n/n} = \frac{n^2}{2n+1}\left(I_{n/n} - I_{n+1/n+1}\right) > 0. \tag{15.21}$$

3)

$$I_{1/n} - I_{n/n} > 0 \text{ for } n \le 5 \text{ at least.} \tag{15.22}$$

For $n \le 5$ the parallel configuration is more sensitive to the stress that a series configuration.

The inequality (15.22) is equivalent to:

$$\sum_{j=1}^{n-1} \left(\sum_{i=1}^{j} (-1)^{i-1} C_n^i \frac{1}{i} \right) \left(I_{j/j} - I_{j+1/j+1} \right) > 0 \tag{15.23}$$

where: $\sum_{i=1}^{j} (-1)^{i-1} C_n^i \frac{1}{i}$ is proven to be positive for $n \leq 5$.

With these results we can conjecture that the reliability and the stress sensitivity increase in the same order. But, for instance it is very difficult to show that:

$$I_{n/n} < I_{n-1/n} < \ldots < I_{2/n} < I_{1/n} \tag{15.24}$$

for all $n \geq 1$.

15.6 Example - System of Two Identical Components Subjected to a Common, Homogeneous Poisson Stress Process

- The reliability function of each component is:
  - unstressed environment: $R_{p,1}(t) = R_{p,2}(t) = e^{-\lambda t}$
  - stressed environment : $R_{s,1}(t) = R_{s,2}(t) = e^{-(\lambda+\theta_1)t}$

- The reliability function of the series configuration is:
  - unstressed environment: $R_{p,2/2}(t) = e^{-2\lambda t}$
  - stressed environment: $R_{s,2/2}(t) = e^{-(2\lambda+\theta_2)t}$

- The reliability function of the parallel configuration is:
  - unstressed environment:
    $R_{p,1/2}(t) = 2e^{-\lambda t} - e^{-2\lambda t}$
  - stressed environment:
    $R_{s,1/2}(t) = 2e^{-(\lambda+\theta_1)t} - e^{-(2\lambda+\theta_2)t}$

- The failure rate of the series configuration is:

- unstressed environment: $\lambda_{p,2/2}(t) = 2\lambda =$ constant
- stressed environment: $\lambda_{s,2/2}(t) = 2\lambda + \theta_2 =$ constant.

- The failure rate of the parallel configuration is:
  - unstressed environment: $\lambda_{p,1/2}(t) = 2\lambda \dfrac{1 - e^{-\lambda t}}{2 - e^{-\lambda t}}$

    with $\lim_{t\to\infty} \lambda_{p,1/2}(t) = \lambda$ and $\lambda_{p,1/2}(0) = 0$
  - stressed environment:

$$\lambda_{s,1/2}(t) = \frac{2(\lambda + \theta_1) - (2\lambda + \theta_2)\, e^{-(\lambda+\theta_2-\theta_1)t}}{2 - e^{-(\lambda+\theta_2-\theta_1)t}}$$

    with $\lim_{t\to\infty} \lambda_{s,1/2}(t) = \lambda + \theta_1$ and $\lambda_{s,1/2}(0) = 2\theta_1 - \theta_2 > 0.$

  For the series configuration the stress contribution to the failure rate is constant. For the parallel configuration the stress contribution to the failure rate starts from $2\theta_1 - \theta_2$, is first increasing, then decreasing and converges rapidly to $\lambda + \theta_1$.

- The relative reduction rate of MTTF due to the stress action is:
  - series configuration: $I_{2/2} = 1 - \dfrac{2\lambda}{2\lambda + \theta_2}$
  - parallel configuration: $I_{1/2} = 1 - \dfrac{2\lambda}{3}\left(\dfrac{2}{\lambda + \theta_1} - \dfrac{1}{2\lambda + \theta_2}\right).$

Example. For $\lambda = 1$, $\alpha = 1$, $\xi = 6$ and Z exponentially distributed with parameter 1 ($\theta_1 = 3$, $\theta_2 = 4$), $I_{2/2} = 2/3$ and $I_{1/2} = 7/9$, which confirms that $I_{2/2} < I_{1/2}$. The parallel configuration is more stress sensitive than the series configuration.

15.7 Conclusions

We have proposed a modelling of the reliability of a complex system under stress environment. This model assumes that the stress process is an homogeneous Poisson process, which is common for all n identical components of the system.

It appears clearly that the stress environment reduces the lifelength of the system. With the results on the indicator I for different particular configurations, we can conjecture that the reliability and the stress sensitivity increase in the same order: so, the more reliable a configuration will be, the more stress sensitive it will be.

For the two identical components system and its configurations, subjected to a common homogeneous Poisson stress process, the failure rate is constant or increasing and converges rapidly to a constant. This example confirms that the most reliable configuration is the most sensitive to the stress.

References

1. Barlow, R. E. and Proschan, F. (1981). *Statistical Theory of Reliability and Life Testing - Probability Models: Multivariate Distributions Based for Dependent Components*, pp. 127–156, Silver Spring.

2. Birnbaum, Z. and Esary, J. (1961). Multi-component system and structures and their reliability, *Technometrics*, **3**, 55–77.

3. Gaudoin, O. and Soler, J. L. (1997). Failure rate behaviour of components subject to random stresses, *Reliability Engineering and System Safety*, **58**, 19–30.

4. Marshall, W. and Olkin, I. (1967). A multivariate exponential distribution, *Journal of Applied Statistical Association*, **62**, 31–44.

16

On the Failure Rate

Gheorghe Oprişan

Politehnica University of Bucharest, Bucharest, Romania

Abstract: Some considerations concerning the concept of failure rate are presented. For a semi-Markov process are given equations of Markov renewal type using the failure rates of passing from a state into another. The existence and uniqueness of the solutions is discussed.

Keywords and phrases: Failure rate, failure processes, transition rate

16.1 Introduction

Let us consider a system for which we measure time in terms of operating time between failures, thus ignoring repair times. If $T_1, T_2, ...$ are the successive failure times and $X_1 = T_1,\ X_2 = T_2 - T_1, ..., X_n = T_n - T_{n-1}, ...$, then the random variables $X_1, X_2, ...$ are not independent, generally. However, in many reliability problems they are supposed to be i.i.d. If this is the case, let us denote by $f(t)$ and $F(t)$ their common density and distribution function, respectively. A random variable X with the distribution function F is called the system's lifetime. We can consider $X_1, X_2, ...,$ as the lifetime X of independent copies of the given system.

For an arbitrary failure process, as that described above, let $N(t)$ be the number of failures in $[0, t]$ which is a point process. If the random variables $X_1, X_2, ...$ are i.i.d., then the point process $\{N(t);\ t \in R_+\}$ is studied in the Renewal theory.

Generally, a system unfolding in time can be described by a stochastic process $\{X(t),\ t \in R_+\}$ with a discrete set of states E.

16.2 Failure Process

As we know, the function

$$r(t) = \lim_{h \to 0} \frac{P(t \leq X < t + h | X > t)}{h} \tag{16.1}$$

is called the failure rate (or force of mortality) of the lifetime distribution and the distribution F is of type IRF or DFR as $r(t)$ does not decrease or it does not increase.

Obviously, we have

$$r(t) = \frac{f(t)}{\bar{F}(t)} = \frac{d}{dt}\left[-\log \bar{F}(t)\right] \tag{16.2}$$

$$f(t) = r(t) \exp\left[-\int_0^t r(x)dx\right] \tag{16.3}$$

where

$$\bar{F}(t) = 1 - F(t) = P(X > t) = \exp\left[-\int_0^t r(x)dx\right] \tag{16.4}$$

is the survival function.

We shall assume that

$$F(0-) = 0 < F(x) < 1 \quad \text{for all} \quad x > 0.$$

The conditional probability

$$P(X - t \leq y | X > t) = \frac{F(t+y) - F(t)}{\bar{F}(t)}, \quad y > 0 \tag{16.5}$$

is called the distribution of the residual lifetime. It is the additional life from t onward of an equipment (with life distribution F) that has survived to time t.

The mean residual life function is defined by

$$s(t) = \frac{1}{\bar{F}(t)} \int_0^\infty F(t + dy) = \frac{1}{\bar{F}(t)} \int_0^\infty \bar{F}(y)dy \tag{16.6}$$

and the expected value of X is $s(0) = \int_0^\infty \bar{F}(t)dt$.

The survival function, the failure rate function, and the mean residual life function are basic tools for the study of reliability characteristics. They are equivalent in the sense that knowing anyone of them, the others two can be determined. The question of an ordering of two life distributions with respect to a reliability measure was discussed by Gupta and Kirmani (1987).

If the simultaneous failures do not occur, Leadbetter (1970) has proved that

$$\frac{d}{dt}[EN(t)] = \lim_{h\to 0} \frac{P[N(t+h) - N(t) \geq 1]}{h} \tag{16.7}$$

provided that the derivative of the function $U(t) = EN(t)$ exists. In this case the common value of the two members of (16.7) is called the failure rate of the process [Thompson (1981)] and we shall denote it by $\lambda(t)$.

Under the assumed conditions, the failure rate of the process $\lambda(t)$ verifies the following equation, which is a consequence of the Renewal equation

$$\lambda(t) = f(t) + \int_0^t \lambda(t-x) f(x) dx. \tag{16.8}$$

A connection between $r(t)$ and $\lambda(t)$ is given by the following relation derived from (16.3) and (16.8)

$$\lambda(t) = r(t) \exp\left[-\int_0^t r(x)dx\right] + \int_0^t \left[\lambda(t-x) r(x) \exp\left(-\int_0^x r(y)dy\right)\right] dx. \tag{16.9}$$

From a well known result of the Renewal theory, we derive:

Proposition 16.2.1 *If the failure rate* $r(t)$ *is bounded on finite intervals, then*

$$\lambda(t) = f(t) + \int_0^t f(t-x) U(dx)$$

is bounded on finite intervals and is the unique solution of (16.8) [or (16.9)] with this property.

The following proposition comes from the Renewal theorem.

Proposition 16.2.2 *If* $\mu = E(X) = \int_0^\infty F(dx) < \infty$ *we have*

$$\lim_{t\to\infty} \lambda(t) = \frac{1}{\mu} \tag{16.10}$$

provided that either the limit exists or λ *is uniformly continuous on* $[0, \infty)$.

We note that, for small h, $\lambda(t)h$ is approximately the probability of failure in the interval $(t, t+h]$. To see the connection between the failure rate of the life time distribution [i.e., $r(t)$] and the failure rate of the process [i.e., $\lambda(t)$] we can consider n independent copies of the system and $N_n(t)$ as the number of copies failed before age t. Then, for each t, under the assumption that the n lifetimes are i.i.d., $N_n(t)$ is binomial. Hence

$$P(N_n(t) = k) = \binom{n}{k} [F(t)]^k [1 - F(t)]^{n-k}. \tag{16.11}$$

From (16.11) and Glivenko-Cantelli theorem we can deduce [Thompson (1981)]

$$P\left[\lim_{n\to\infty} \frac{N_n(t+h)-N_n(t)}{h(n-N_n(t))} = r(t)+0(1)\right] = 1 \qquad (16.12)$$

provided that $f(t)$ is continuous. Here $0(1)$ is a function such that $\lim_{h\to 0} 0(1) = 0$.

So, for short intervals and large samples, $r(t)$ is the number of failures per unit of time and per individuals at risk.

16.3 Semi-Markov Process

Generally, a system evaluates in a set (usually finite) E of states and $T_1, T_2, \ldots$ are the successive jumps from a state to another. The random variables $X_1, X_2, \ldots$ are neither independent nor identically distributed. Such a system can be described by a semi-Markov process $\{Z(t),\ t \in \mathbf{R}_+\}$ [Çinlar (1969)].

Let $Q_{ij}(t)$, $(Q_{ij}(\infty) = p_{ij})$, $t \in \mathbf{R}_+$, $i,j \in E$, be the semi-Markov matrix corresponding to the process $\{Z(t),\ t \in R_+\}$ and we set $J_n = Z(t)$ on the set $\{N(t) = n\}$. In other words $Z(t) = J_{N(t)}$, $t \in \mathbf{R}_+$. The two components processes (J_n, T_n) and (J_n, X_n) with the state space $E \times \mathbf{R}_+$ are called Markov renewal process and $J - X$ process respectively.

For the real stochastic systems the transitions of the Markov renewal process from a state to another is determined by a finite number of random independent factors (e.g. the failures of a component). Each of these factors is characterized by a random time of influence at the end to whom the process changes the state. Usually, this random time depends on the "departure state". Therefore, for each $i \in E$ we consider the family $\{\tau_{ik},\ k \in E\}$ of random independent times of influence of the factors which determine the "jump" from i. We assume that the random variables τ_{ik}, $k \in E$, take values in $\bar{\mathbf{R}}_+$. The time spent in i is $\eta_i = \min_{k\in E} \tau_{ik}$.

More precisely we have the following

Proposition 16.3.1 *For each $i \in E$ there is a family of independent random variables [Korolyuk and Turbin (1976)] $\{\tau_{ik},\ k \in E\}$, $\tau_{ik} : \omega \to \bar{\mathbf{R}}_+$, with the distribution functions*

$$A_{ik}(t) = \begin{cases} 0, & \text{if } t \le 0 \\ 1 - \exp\left[-\int_0^t \frac{Q_{ik}(du)}{1-H_i(u)}\right], & \text{if } t > 0 \end{cases} \qquad (16.13)$$

where

$$H_i(t) = \sum_{j\in E} Q_{ij}(t),\ i \in E.$$

The semi-Markov matrix $Q_{ij}(t)$ *is given by*

$$Q_{ij}(t) = \int_0^t h_{ij}(u)A_{ij}(du),\ i,j \in E \tag{16.14}$$

where

$$h_{ij}(u) = \frac{1 - H_i(u)}{1 - A_{ij}(u)} = E[I_{ij}|\tau_{ij} = u] \tag{16.15}$$

and I_{ij}, $i,j \in E$, *is the characteristic function of the event* $\{\min_{k\in E} \tau_{ik} = \tau_{ij}\}$.

PROOF. Obviously, the function A_{ik} given by (16.13) are mass functions and hence they are the distribution functions of some random variables τ_{ik}, $i,k \in E$, with values in $\bar{\mathbf{R}}_+$ (we have $A_{ik}(\infty) \leq 1$). ■

The survival functions of the random variables

$$\eta_i = \min_{k\in E} \tau_{ik},\ i \in E,$$

are

$$\begin{aligned} P(\eta_i \geq t) &= \prod_{j\in E}(1 - A_{ij}(t)) \\ &= \exp\left[-\sum_{j\in E}\int_0^t \frac{Q_{ij}du}{1 - H_i(u)}\right] = \exp\left[-\int_0^t \frac{H_i(du)}{1 - H_i(u)}\right] = 1 - H_i(t). \end{aligned} \tag{16.16}$$

The random variable η_i gives the time spent by the process in the state $i \in E$.

From (16.13) we obtain the measure equality

$$\frac{A_{ik}(dt)}{1 - A_{ik}(t)} = \frac{Q_{ik}(dt)}{1 - H_i(t)}$$

which proves (16.14).

Now, from the independence of the random variables τ_{ij}, we have

$$\begin{aligned} E[I_{ij}|\tau_{ij} = u] &= P[I_{ij} = 1|\tau_{ij} = u] = P\left[\min_{k\in E} \tau_{ik} = \tau_{ij}|\tau_{ij} = u\right] \\ &= P\left[\min_{k\neq j} \tau_{ik} \geq u|\tau_{ij} = u\right] \\ &= \prod_{k\neq j}(1 - A_{ik}(u)) = \frac{1 - H_i(u)}{1 - A_{ij}(u)} = h_{ij}(u) \end{aligned} \tag{16.17}$$

and (16.15) is proved.

From (16.15) we obtain

$$E[I_{ij}] = \int_0^\infty h_{ij}(u) A_{ij}(du) = Q_{ij}(\infty) = p_{ij}, \quad i,j \in E. \tag{16.18}$$

If we set

$$\wedge_{ij}(t) = \int_0^t \frac{Q_{ij}(du)}{1 - H_i(u)}, \quad i,j \in E,\ t > 0$$

then

$$A_{ik}(t) = \begin{cases} 0, & \text{if} \quad t \leq 0 \\ 1 - e^{-\wedge_{ik}(t)} & \text{if} \quad t > 0. \end{cases}$$

The function $\lambda_{ij}(t) = \wedge'_{ij}(t), \quad i,j \in E$, (provided that the derivative of the function $\wedge_{ij}$ exists) is called the rate of transition from i to j.

From (16.17), we have

$$h_{ij}(u) = \frac{\exp\left[-\sum_{k\in E} \wedge_{ik}(u)\right]}{\exp[-\wedge_{ij}(u)]}$$

which, together with (16.14), gives

$$Q_{ij}(t) = \int_0^t e^{-\wedge_i(u)} \wedge_{ij}(du) \tag{16.19}$$

where

$$\wedge_i(u) = \sum_{k\in E} \wedge_{ik}(u), \quad i \in E.$$

The distribution function of η_i, $i \in E$, can be computed from (16.16)

$$H_i(t) = P(\eta_i \leq t) = \begin{cases} 0, & \text{if} \quad t \leq 0 \\ 1 - e^{-\int_0^t \lambda_i(u)du}, & \text{if} \quad t > 0 \end{cases}$$

where the function $\lambda_i(t) = \wedge'_i(t)$, $i \in E$, is called the rate of leaving the state i.

We consider the following functions useful in reliability:

$$R_{ij}(t) = E[N_j(t)|Z(0) = i] = \sum_{n\in N} Q_{ij}^{*n}(t), \quad i,j \in E$$

where $N_j(t)$ is the number of visits paid to j in the interval $[0,t]$

$$p_{ij}(t) \quad = \quad P(Z(t) = j|Z(0) = i), \quad i,j \in E$$

$G_{ij}(t)$ = the distribution function of the first jump time from i to j.

These function verify some Markov renewal equations which can be written in terms of transition rate as follows:

$$R_{ij}(t) = \sum_{k\in E}\int_0^t R_{kj}(t-s)e^{-\Lambda_i(s)}\lambda_{ik}(s)ds + \delta_{ij}, \quad i,j\in E \tag{16.20}$$

$$p_{ij}(t) = \sum_{k\in E}\int_0^t P_{kj}(t-s)e^{-\Lambda_i(s)}\lambda_{ik}(s)ds + \delta_{ij}e^{-\Lambda_i(t)}, \quad i,j\in E \tag{16.21}$$

$$G_{ij}(t) = \sum_{k\neq j}\int_0^t G_{kj}(t-s)e^{-\Lambda_i(s)}\lambda_{ik}(s)ds + Q_{ij}(t) \tag{16.22}$$

Proposition 16.3.2 *Let $A_i = \{t \in \mathbf{R}_+ | \lambda_i(t) > 0\}$ and $A = \bigcap_{i\in E} A_i$. We assume that the set A has a strictly positive Lebesgue measure. Then $\sup_{i\in E}\sum_{j\in E} R_{ij}(t) < \infty$ and the Equations (16.20)-(16.22) have bounded solutions which are unique.*

PROOF. The assumption made in this proposition implies that $\sup_{i\in E} H_i(t) < 1$ for some $t > 0$ and the conclusion can be deduced from a result of Çinlar (1969). ■

Remark If the set E is finite the assumption made in Proposition 16.3.2 is fulfilled.

Let $g : E \to \mathbf{R}$ be a positive function and $W_t = \int_0^t g(Z_s)ds$ an additive functional of the semi-Markov process. If $\alpha_i(t,w) = P(W_t \leq w | Z(0) = i)$ is the distribution function of W_t, then $1 - \alpha_i(t)$ is called the performability over a utilization interval $[0,t]$. The equation verified by $\alpha_i(t,w)$ [Limnios and Oprişan (1997)] can be expressed in terms of failure rates as follows

$$\alpha_i(t,w) = u_{[0,w]}(g(i)t)e^{-\Lambda_i(t)} + \sum_{k\in E}\int_0^t \alpha_k(w - g(i)s)e^{-\Lambda_i(t)}\lambda_{ik}(s)ds. \tag{16.23}$$

Using the Laplace transform we can derive some conclusion concerning the existence and the uniqueness of the solution of the Equation (16.23).

The major problem related with the utilization of the semi-Markov processes in the Reliability of the complex systems is to find the functions matrix $\{Q_{ij}(t); i,j \in E\}$ by means of the distribution functions of the operating and repairing periods of the component elements. For this one can use the failure or the repair rates of the components to derive the transition rates $\lambda_{ij}(t)$, $i,j \in E$, to pass from a state to another.

References

1. Çinlar, E. (1969). Markov renewal theory, *Advances in Applied Probability*, **1**, 123–187.

2. Gupta, R. C. and Kirmani, S. N. U. A. (1987). On order relations between reliability measures, *Communications in Statistics—Stochastic Models*, **3**, 149–156.

3. Korolyuk, V. S. and Turbin, A. F. (1976). *Semi-Markov Processes and Its Applications*, Kiev: Naukova Dumka (in Russian).

4. Leadbetter, M. R. (1970). On basic results of point process theory, *Proceedings of the 6-th Berkeley Symposium on Mathematics and Statistics on Probability*, pp. 449–462, Berkeley, CA: University of California Press.

5. Limnios, N. and Oprişan, Gh. (1997). A general framework for reliability and performability modelling of semi-Markov process, Contributed papers, *VIII International Symposium on Applied Stochastic Models and Data Analysis*, pp. 261–266, Napoli-Italy, June 11-14, Rocca Curto Editore.

6. Thompson, W. A., Jr. (1981), On the foundation of reliability, *Technometrics*, **23**, 1–13.

17

Asymptotic Results for the Failure Time of Consecutive k-out-of-n Systems

Brahim Ksir
University of Constantine, Constantine, Algeria

Abstract: We consider a consecutive k-out-of-n system. This system fails if and only if there are at least k consecutive failed components. Our goal is to state asymptotic results concerning the failure time Z_n of the system. Z_n is given by: $Z_n = \min_{1 \le j \le n-k+1} \max_{j \le i \le j+k-1} T_i$, where $T_1, ..., T_n$ denote the failure times of components. We establish strong laws of Erdos-Rényi-Shepp type for the statistics Z_n and $k_n \frac{Z_n - a}{\log k_n}$ for $k_n = [c \log n]$ and a suitable constant a. We suppose that the components are independent and with equal failure distributions.

Keywords and phrases: Consecutive k-out-of-n systems, failure time, strong laws

17.1 Introduction

A consecutive k-out-of-n system consists of n components disposed linearly. The system fails if and only if at least k consecutive components are failed. Each component and the system has two states: it is functional or failed. This kind of systems has great importance in application. They have been proposed to model telecommunication systems and oil pipelines. Many papers gave methods to calculate reliability of such systems [see, for example, Derman, Lieberman and Ross (1982), Canfield and McCormick (1992)]. Some limit theorems are proved in [Papastavridis (1987), Chrissaphinou and Papastavridis (1990)]. In our paper we suppose that k grows with n (we note k_n) and we establish strong laws of Erdos-Rényi-Shepp type [Ksir (1989), Shepp (1964), Deheuvels, Devroye and Lynch (1986)] for the failure time of the system when the components are supposed independent and identically distributed.

17.2 Strong Laws for the Failure Time of the System

We suppose that the components are statically independent and with equal failure distributions. So, the failure times $T_1, ..., T_n$ of the components of the system are (positive) independent random variables identically distributed. The failure time of the system is: $Z_n = \min_{1\leq j\leq n-k_n+1} \max_{j\leq i\leq j+k_n-1} T_i$. We are interested with the asymptotic behaviour of the random variables Z_n and $k_n \frac{Z_n-a}{\log k_n}$ for the choice $k_n = [c\log n], c > 0$, ($[x]$ represents the integer part of x), and a constant satisfying an equation given below.

Let us consider $m = E(T_1)$, $M = ess-supT_1$, h the Cramer transformation of T_1 and t^* verifying: $h(a) = at^* - \text{logE}(\exp(t^*T_1))$. We take $a \in]m, M[$ such that: $\exp(-h(a)) = \exp\left(-\frac{1}{c}\right)$. Under these considerations we state the following results:

Theorem 17.2.1 *$\lim_{n\to\infty}\sup Z_n \leq a$ a.s.*

PROOF.

$$\begin{aligned}
\Pr(Z_n > a) &= \Pr\left(\min_{1\leq j\leq n-k_n+1} \max_{j\leq i\leq j+k_n-1} T_i > a\right) \\
&= \Pr\left(\min_{1\leq j\leq n-k_n+1} Y_j > a\right)
\end{aligned}$$

where $Y_j = \max_{j\leq i\leq j+k_n-1} T_i$. But $\min_{1\leq j\leq n-k_n+1} Y_j \leq \frac{1}{n-k_n+1}\sum_{j=1}^{n-k_n+1} Y_j$, so:

$$\begin{aligned}
\Pr(Z_n > a) &\leq \Pr\left(\frac{1}{n-k_n+1}\sum_{j=1}^{n-k_n+1} Y_j > a\right) \\
&\leq \Pr\left(\sum_{j=1}^{n-k_n+1} Y_j > (n-k_n+1)a\right) \\
&\leq \Pr\left(\sum_{j=1}^{n-k_n+1} Y_j - (n-k_n+1)a > 0\right) \\
&\leq E\left(\exp\left(t\left(\sum_{j=1}^{n-k_n+1} Y_j - (n-k_n+1)a\right)\right)\right), \quad \forall t > 0 \\
&\leq E\left(\exp\left(t^*\left(\sum_{j=1}^{n-k_n+1} Y_j - (n-k_n+1)a\right)\right)\right) \quad \text{for } t = t^*
\end{aligned}$$

$$\begin{aligned}
&\leq \exp\left(-ant^* + ak_n t^* - at^*\right) E\left(\exp\left(t^* \sum_{j=1}^{n-k_n+1} Y_j\right)\right) \\
&\leq \exp\left(-ant^* + at^* c \log n - at^*\right) E \exp\left(t^* \sum_{i=1}^{n} T_i\right) \\
&\leq E\left(\exp\left(t^*\left(\sum_{i=1}^{n} T_i - na\right)\right)\right) \exp\left(at^* c \log n - at^*\right) \\
&\leq \exp\left(-nh\left(a\right)\right) n^{act^*} \exp\left(-at^*\right) \\
&\leq n^{act^*} \exp\left(-\frac{n}{c}\right).
\end{aligned}$$

The sequence $\sum_{n\geq 1} n^{act^*} \exp\left(-\frac{n}{c}\right) < +\infty$ from D'Alembert's criterium.

Indeed:

$$\lim_{n\to\infty} \left((n+1)^{act^*} \exp\left(-\frac{n+1}{c}\right)\right)\left(n^{-act^*} \exp\left(\frac{n}{c}\right)\right) = \exp\left(-\frac{1}{c}\right) < 1.$$

Hence

$$\sum_{n\geq 1} \Pr\left(Z_n > a\right) < +\infty.$$

From Borel-Cantelli lemma we deduce: $\lim_{n\to\infty} \sup Z_n \leq a$ a.s. ■

Theorem 17.2.2 $\lim_{n\to\infty} \inf k_n \frac{z_n - a}{\log k_n} \geq \frac{1}{t^*}$ *a.s.*

PROOF. For $j \in N^*$, we define the number $n_j = \inf\{n : [c \log n] = j\}$. For $n_j \leq n \leq n_{j+1}$, we have $Z_{n_j} \leq Z_n \leq Z_{n_{j+1}}$:

$$\begin{aligned}
\Pr\left(j \frac{Z_{n_j} - a}{\log j} < \frac{1}{t^*}\right) &= \Pr\left(Z_{n_j} < a + \frac{\log j}{jt^*}\right) \\
&= \Pr\left(\min_{1\leq l\leq n_j - j+1} \max_{l\leq i\leq l+j-1} T_i < a + \frac{\log j}{jt^*}\right) \\
&\leq \sum_{l=1}^{n_j - j+1} \Pr\left(\max_{l\leq i\leq l+j-1} T_i < a + \frac{\log j}{jt^*}\right) \\
&\leq \sum_{l=1}^{n_j} \Pr\left(\max_{l\leq i\leq j} T_i \circ \theta_l < a + \frac{\log j}{jt^*}\right) \\
&\leq \sum_{l=1}^{n_j} \Pr\left(\max_{l\leq i\leq j} T_i < a + \frac{\log j}{jt^*}\right) \\
&\leq n_j \Pr\left(\frac{1}{j} \sum_{i=1}^{j} T_i < a + \frac{\log j}{jt^*}\right) \\
&\leq n_j \Pr\left(\sum_{i=1}^{j} T_i < j\left(a + \frac{\log j}{jt^*}\right)\right)
\end{aligned}$$

where θ is a translation operator.

By using [Ksir (1989, Corollary 1)], we obtain:

$$\begin{aligned} &\leq n_j \exp\left(-jh\left(a\right)\right) \exp\left(\left(-jt^* \log j\right)/jt^*\right) \\ &\leq n_j \exp\left(-\frac{j}{c}\right) \end{aligned}$$

$[c \log n_j] = j$, for $c' > c$ such that $[c' \log n_j] = j$ we have: $n_j \leq \exp(\frac{j+1}{c'})$. Then,

$$\Pr\left(j\frac{Z_{n_j} - a}{\log j} \leq \frac{1}{t^*}\right) \leq \exp\left(\frac{1}{c'}\right)\left(\exp\left(\frac{1}{c'} - \frac{1}{c}\right)^j\right).$$

Hence,

$$\sum_{j\geq 1} \Pr\left(j\frac{Z_{n_j} - a}{\log j} < \frac{1}{t^*}\right) \leq \exp\left(\frac{1}{c}\right)\sum_{j\geq 1} \exp j\left(\frac{1}{c'} - \frac{1}{c}\right) < \infty.$$

By applying again Borel-Cantelli lemma we obtain:

$$\lim_{n\to\infty} \inf k_n \frac{Z_n - a}{\log k_n} \geq \frac{1}{t^*}\ .$$

Corollary 17.2.1 $\lim_{n\to\infty} Z_n = a \quad a.s.$

PROOF. In Theorem 17.2.1 we have proved that:

$$\lim_{n\to\infty} \sup Z_n \leq a \quad a.s. \tag{17.1}$$

Now, we have to prove the following inequality:

$$\lim_{n\to\infty} \inf Z_n \geq a \quad a.s.$$

By using the same considerations as in Theorem 17.2.2 we must prove that:

$$\sum_{j\geq 1} \Pr\left(Z_{n_j} < a\right) < \infty.$$

But,

$$\Pr\left(Z_{n_j} < a\right) \leq \Pr\left(Z_{n_j} < a + \frac{\log j}{jt^*}\right).$$

The sequence $\sum_{j\geq 1} \Pr(Z_{n_j} < a + \frac{\log j}{jt*})$ converges (from Theorem 17.2.2). So, $\sum_{j\geq 1} \Pr(Z_{n_j} < a) < \infty$. From Borel-Cantelli lemma we deduce:

$$\lim_{n\to\infty} \sup Z_n \geq a \quad a.s. \tag{17.2}$$

By combining (17.1) and (17.2) we conclude. ∎

Corollary 17.2.2 $\lim_{n\to\infty} k_n \frac{Z_n - a}{\log k_n} = \frac{1}{t^*}$ *a.s.* $\lim_{n\to\infty} Z_n = a$ *a.s.*

PROOF. In Theorem 17.2.2 we have established the inequality:

$$\lim_{n\to\infty} \inf k_n \frac{Z_n - a}{\log k_n} \geq \frac{1}{t^*} \quad a.s. \tag{17.3}$$

Now we have to state the following inequality:

$$\lim_{n\to\infty} \sup k_n \frac{Z_n - a}{\log k_n} \geq \frac{1}{t^*} \quad a.s.$$

We can write:

$$\begin{aligned} \Pr\left(k_n \frac{Z_n - a}{\log k_n} > \frac{1}{t^*}\right) &= \Pr\left(Z_n - a > \frac{\log k_n}{k_n t^*}\right) \\ &= \Pr\left(Z_n > a + \frac{\log k_n}{k_n t^*}\right) \\ &\leq \Pr(Z_n > a). \end{aligned}$$

From Theorem 17.2.1 we have:

$$\sum_{n\geq 1} \Pr(Z_n > a) < \infty.$$

So:

$$\sum_{n\geq 1} \Pr\left(k_n \frac{Z_n - a}{\log k_n} \frac{1}{t^*}\right) < \infty.$$

From Borel-Cantelli lemma we obtain:

$$\lim_{n\to\infty} \sup k_n \frac{Z_n - a}{\log k_n} \geq \frac{1}{t^*} \quad a.s. \tag{17.4}$$

We conclude by combining inequalities (17.3) and (17.4). ■

References

1. Derman, C., Lieberman, G. J. and Ross, S. M. (1982). On the consecutive k-out-of-n: F system, *IEEE Transactions on Reliability*, **31**, 57–63.

2. Canfield, E. R. and McCormick, W. P. (1992). Asymptotic reliability of consecutive k-out-of-n systems, *Journal of Applied Probability*, **29**, 142–155.

3. Papastavridis, S. G. (1987). A limit theorem for the reliability of a consecutive k-out-of-n system, *Advances in Applied Probability*, **19**, 746–748.

4. Chrissaphinou, O. and Papastavridis, S. (1990). Limit distribution for a consecutive k-out-of-n system, *Advances in Applied Probability*, **22**, 491–493.

5. Ksir, B. (1989). Shepp statistic for Markov chains. Application to a long-run average cost criterion, *Journal of Applied Probability*, **27**, 767–775.

6. Shepp, L. A. (1964). A limit theorem concerning moving averages, *Annals of Mathematical Statistics*, **35**, 424–428.

7. Deheuvels, P., Devroye, L. and Lynch, J. (1986). Exact convergence rate in the limit theorems of Erdos-Rényi and Shepp type, *Annals of Probability*, **14**, 209–223.

PART III

Special Techniques and Applications

18

Two-Stage Start-Up Demonstration Testing

N. Balakrishnan and P. S. Chan

McMaster University, Hamilton, Ontario, Canada
Chinese University of Hong Kong, Shatin, Hong Kong

Abstract: Start-up demonstration tests and various extensions and generalizations of them (in order to accommodate dependence between the trials, to allow for corrective action to be taken once the equipment fails for the first time, etc.) have been discussed quite extensively in the literature. In this paper, we propose a start-up demonstration test to be performed in two stages which would facilitate an early rejection of a potentially bad equipment and would also enable the experimenter to place a more stringent requirement for acceptance upon observing a certain number of failures. Specifically, the decision procedure proposed is as follows. Perform start-up demonstration tests on the equipment under study consecutively and decide to:

1. Accept the equipment (in the first stage) if a run of c_1 successes occurs before d_1 failures.

2. Accept the equipment if no run of c_1 successes occurs before d_1 failures, but a run of c_2 successes is observed before the next d_2 failures.

3. Reject the equipment if no run of c_1 successes occurs before d_1 failures and also no run of c_2 successes occurs before the next d_2 failures.

We then derive the probability generating function of the waiting time for the termination of the start-up demonstration testing, and the mean of this waiting time. We also establish some recurrence relations satisfied by the probability mass function which will facilitate easy recursive computation of probabilities. We also discuss the distributions of some related random variables such as the numbers of successes and failures.

Keywords and phrases: Start-up demonstration testing, Bernoulli trials, probability generating function, runs, waiting time, mean waiting time, two-stage testing

18.1 Introduction

A start-up demonstration test, as first discussed by Hahn and Gage (1985), involves successive attempted start-ups of an equipment with each attempt resulting in either a success or a failure and accepts the equipment if a pre-specified number (c) of consecutive successfull start-ups occur on or before a pre-specified number of attempts. While Hahn and Gage (1985) explained probability calculations for such start-up demonstration tests, Viveros and Balakrishnan (1993) discussed some inferential methods (for the unknown probability of a successfull start-up, p) based on data obtained from start-up demonstration tests. Viveros and Balakrishnan (1993) and Balakrishnan, Balasubramanian and Viveros (1995) studied the start-up demonstration testing problem after allowing the outcomes of successive attempted start-ups to be dependent in a Markovian fashion [instead of being independent as in the original formulation of Hahn and Gage (1985)]. Balakrishnan, Balasubramanian and Viveros (1995) also considered the start-up demonstration testing under corrective action model which allows for a corrective action to be taken by the experimenter immediately after observing the first failure (i.e., after the first failed attempt). An extension of the start-up demonstration testing in which the outcomes of successive attempted start-ups have a higher-order Markov dependence has been discussed recently by Aki, Balakrishnan and Mohanty (1996). In a similar vein, Balakrishnan, Mohanty and Aki (1997) have discussed the start-up demonstration testing with corrective actions under the higher-order Markov dependence model. For a synthesis of all these developments and work on related waiting time problems, we refer the interested readers to the book by Balakrishnan and Koutras (1998).

Now, let us consider the following scenario. A customer wishes to purchase a power generation equipment. He/she will decide to buy the equipment if a certain number of consecutive successfull start-ups (c_1) are achieved before a certain number (d_1) of failures occur. If this does not happen (meaning that d_1 failures occur before a run of c_1 consecutive successfull start-ups), then the customer may wish to place a more stringent requirement on the equipment and demand c_2 consecutive successfull start-ups to occur before the next d_2 failures in order to decide to buy that equipment. This is a typical example of a two-stage start-up demonstration test.

Formally, let X be the number of trials required to terminate the experiment based on the following criteria:

1. Accept the equipment if a run of c_1 successes occurs before d_1 failures.

2. Accept if no run of c_1 successes occurs before d_1 failures, but a run of c_2 successes occurs before the next d_2 failures.

3. Reject if no run of c_1 successes occurs before d_1 failures and also no run of c_2 successes occurs before the next d_2 failures.

In this paper, we study the distribution of the waiting time variable X through its probability generating function. We then derive some recurrence relations that will facilitate the computation of the probabilities associated with the random variable X and also some related random variables such as the numbers of successes and failures in the start-up demonstration testing experiment.

18.2 Probability Generating Function

Assume that the successive start-ups are independent events; further, let

p = Probability of a success in a single start-up

q = Probability of a failure in a single start-up, i.e., $q = 1 - p$

S = number of successes observed at the time of termination of the experiment

F = number of failures observed at the time of termination of the experiment

X = the total number of trials required to terminate the experiment.

We observe that there are three possible cases that lead to the termination of the start-up experiment.

1. *Acceptance of the item in the first stage*: The corresponding sequence is

$$\underbrace{S\cdots S}_{r_1}\underbrace{F}_{1^{\text{st}}}\underbrace{S\cdots S}_{r_2}\underbrace{F}_{2^{\text{nd}}}\cdots\underbrace{S\cdots S}_{r_k}\underbrace{F}_{k^{\text{th}}}\underbrace{S\cdots S}_{c_1}, \tag{18.1}$$

with $0 \le r_i \le c_1 - 1$, $i = 1, 2, \ldots, k$, $\sum_{i=1}^{k} r_i + k + c_1 = r$, and $0 \le k \le d_1 - 1$.

Its contribution to the joint probability generating function of (S, F) is

$$\begin{aligned}
\pi_1(t,u) &= \sum_{k=0}^{d_1-1} \left\{(1 + pt + \cdots + p^{c_1-1}t^{c_1-1})qu\right\}^k p^{c_1}t^{c_1} \\
&= \sum_{k=0}^{d_1-1} \left(\frac{1 - p^{c_1}t^{c_1}}{1 - pt}qu\right)^k p^{c_1}t^{c_1} \\
&= \frac{1 - \left(\frac{1-p^{c_1}t^{c_1}}{1-pt}qu\right)^{d_1}}{1 - \left(\frac{1-p^{c_1}t^{c_1}}{1-pt}qu\right)} p^{c_1}t^{c_1} \\
&= \frac{(1 - pt)^{d_1} - (1 - p^{c_1}t^{c_1})^{d_1}(qu)^{d_1}}{(1 - pt)^{d_1-1}(1 - pt - qu + qp^{c_1}t^{c_1}u)} p^{c_1}t^{c_1}\,.
\end{aligned} \tag{18.2}$$

2. *Acceptance of the item in the second stage*: The corresponding sequence is given by

$$\underbrace{S\cdots S}_{\leq c_1-1}\underbrace{F}_{1^{\text{st}}}\underbrace{S\cdots S}_{\leq c_1-1}\underbrace{F}_{2^{\text{nd}}}\cdots\underbrace{S\cdots S}_{\leq c_1-1}\underbrace{F}_{d_1^{\text{th}}}\underbrace{S\cdots S}_{\leq c_2-1}\underbrace{F}_{1^{\text{st}}}\underbrace{S\cdots S}_{\leq c_2-1}\underbrace{F}_{2^{\text{nd}}}\cdots\underbrace{F}_{r^{\text{th}}}\underbrace{S\cdots S}_{c_2}, \tag{18.3}$$

with $0 \leq r \leq d_2 - 1$. Its contribution to the probability generating function of (S, F) is

$$\begin{aligned}
\pi_2(t,u) &= \{(1+pt+\cdots+p^{c_1-1}t^{c_1-1})qu\}^{d_1} \\
&\quad \times \sum_{r=0}^{d_2-1}\{(1+pt+\cdots+p^{c_2-1}t^{c_2-1})qu\}^r p^{c_2}t^{c_2} \\
&= \left\{\left(\frac{1-p^{c_1}t^{c_1}}{1-pt}\right)qu\right\}^{d_1}\sum_{r=0}^{d_2-1}\left\{\left(\frac{1-p^{c_2}t^{c_2}}{1-pt}\right)qu\right\}^r p^{c_2}t^{c_2} \\
&= \left\{\left(\frac{1-p^{c_1}t^{c_1}}{1-pt}\right)qu\right\}^{d_1}\left\{\frac{1-\left(\frac{1-p^{c_2}t^{c_2}}{1-pt}\right)^{d_2}(qu)^{d_2}}{1-\left(\frac{1-p^{c_2}t^{c_2}}{1-pt}\right)qu}\right\}p^{c_2}t^{c_2} \\
&= \frac{(1-p^{c_1}t^{c_1})^{d_1}(qu)^{d_1}\{(1-pt)^{d_2}-(1-p^{c_2}t^{c_2})^{d_2}(qu)^{d_2}\}p^{c_2}t^{c_2}}{(1-pt)^{d_1+d_2-1}\{1-pt-qu+qp^{c_2}t^{c_2}u\}}.
\end{aligned} \tag{18.4}$$

3. *Rejection of the item*: The corresponding sequence is given by

$$\underbrace{S\cdots S}_{\leq c_1-1}\underbrace{F}_{1^{\text{st}}}\underbrace{S\cdots S}_{\leq c_1-1}\underbrace{F}_{2^{\text{nd}}}\cdots\underbrace{S\cdots S}_{\leq c_1-1}\underbrace{F}_{d_1^{\text{th}}}\underbrace{S\cdots S}_{\leq c_2-1}\underbrace{F}_{1^{\text{st}}}\underbrace{S\cdots S}_{\leq c_2-1}\underbrace{F}_{2^{\text{nd}}}\cdots\underbrace{S\cdots S}_{\leq c_2-1}\underbrace{F}_{d_2^{\text{th}}}. \tag{18.5}$$

Its contribution to the joint probability generating function of (S, F) is

$$\begin{aligned}
\pi_3(t,u) &= \{(1+pt+\cdots+p^{c_1-1}t^{c_1-1})qu\}^{d_1} \\
&\quad \times \{(1+pt+\cdots+p^{c_2-1}t^{c_2-1})qu\}^{d_2} \\
&= \left(\frac{1-p^{c_1}t^{c_1}}{1-pt}\right)^{d_1}\left(\frac{1-p^{c_2}t^{c_2}}{1-pt}\right)^{d_2}q^{d_1+d_2}u^{d_1+d_2} \\
&= \frac{(1-p^{c_1}t^{c_1})^{d_1}(1-p^{c_2}t^{c_2})^{d_2}}{(1-pt)^{d_1+d_2}}q^{d_1+d_2}u^{d_1+d_2}.
\end{aligned} \tag{18.6}$$

Therefore, the joint probability generating function of (S, F) is obtained as

$$\begin{aligned}
\pi(t,u) &= \pi_1(t,u)+\pi_2(t,u)+\pi_3(t,u) \\
&= \frac{(1-pt)^{d_1}-(1-p^{c_1}t^{c_1})^{d_1}(qu)^{d_1}}{(1-pt)^{d_1-1}(1-pt-qu+qp^{c_1}t^{c_1}u)}p^{c_1}t^{c_1} \\
&\quad +\frac{(1-p^{c_1}t^{c_1})^{d_1}(qu)^{d_1}\{(1-pt)^{d_2}-(1-p^{c_2}t^{c_2})^{d_2}(qu)^{d_2}\}p^{c_2}t^{c_2}}{(1-pt)^{d_1+d_2-1}\{1-pt-qu+qp^{c_2}t^{c_2}u\}}
\end{aligned}$$

$$+\frac{(1-p^{c_1}t^{c_1})^{d_1}(1-p^{c_2}t^{c_2})^{d_2}}{(1-pt)^{d_1+d_2}}q^{d_1+d_2}u^{d_1+d_2}. \tag{18.7}$$

From (18.7), the probability generating function of F is obtained (by setting $t=1$) to be

$$\begin{aligned}\pi_F(u) &= \frac{\left\{q^{d_1}-(1-p^{c_1})^{d_1}q^{d_1}u^{d_1}\right\}p^{c_1}}{q^{d_1-1}(1-p-qu+qp^{c_1}u)} \\ &\quad +\frac{(1-p^{c_1})^{d_1}(qu)^{d_1}\left\{q^{d_2}-(1-p^{c_2})^{d_2}(qu)^{d_2}\right\}p^{c_2}}{q^{d_1+d_2-1}\{1-p-qu+qp^{c_2}u\}} \\ &\quad +(1-p^{c_1})^{d_1}(1-p^{c_2})^{d_2}u^{d_1+d_2} \\ &= (1-p^{c_1})^{d_1}(1-p^{c_2})^{d_2}u^{d_1+d_2}+\frac{p^{c_1}}{1-u(1-p^{c_1})}-\frac{(1-p^{c_1})^{d_1}p^{c_1}u^{d_1}}{1-u(1-p^{c_1})} \\ &\quad +\frac{(1-p^{c_1})^{d_1}p^{c_2}u^{d_1}}{1-u(1-p^{c_2})}-\frac{(1-p^{c_1})^{d_1}(1-p^{c_2})^{d_2}p^{c_2}u^{d_1+d_2}}{1-u(1-p^{c_2})} \\ &= \frac{p^{c_1}}{1-u(1-p^{c_1})}+(1-p^{c_1})^{d_1}u^{d_1}\left\{\frac{p^{c_2}}{1-u(1-p^{c_2})}-\frac{p^{c_1}}{1-u(1-p^{c_1})}\right\} \\ &\quad +(1-p^{c_1})^{d_1}(1-p^{c_2})^{d_2}u^{d_1+d_2}\left\{1-\frac{p^{c_2}}{1-u(1-p^{c_2})}\right\}.\end{aligned} \tag{18.8}$$

Similarly, from (18.7), the probability generating function of S is obtained (by setting $u=1$) to be

$$\begin{aligned}\pi_S(t) &= \frac{(1-p^{c_1}t^{c_1})^{d_1}(1-p^{c_2}t^{c_2})^{d_2}q^{d_1+d_2}}{(1-pt)^{d_1+d_2}} \\ &\quad +\frac{\{(1-pt)^{d_1}-(1-p^{c_1}t^{c_1})^{d_1}q^{d_1}\}p^{c_1-1}t^{c_1}}{(1-pt)^{d_1-1}\{1-t+qp^{c_1-1}t^{c_1}\}} \\ &\quad +\frac{(1-p^{c_1}t^{c_1})^{d_1}p^{c_2-1}t^{c_2}q^{d_1}\{(1-pt)^{d_2}-(1-p^{c_2}t^{c_2})^{d_2}q^{d_2}\}}{(1-pt)^{d_1+d_2-1}\{1-t+qp^{c_2-1}t^{c_2}\}}.\end{aligned} \tag{18.9}$$

Finally, from (18.7), the probability generating function of X is obtained (by setting $u=t$) to be

$$\begin{aligned}\pi_X(t) &= \frac{(1-p^{c_1}t^{c_1})^{d_1}(1-p^{c_2}t^{c_2})^{d_2}q^{d_1+d_2}t^{d_1+d_2}}{(1-pt)^{d_1+d_2}} \\ &\quad +\frac{\{(1-pt)^{d_1}-(1-p^{c_1}t^{c_1})^{d_1}q^{d_1}t^{d_1}\}p^{c_1}t^{c_1}}{(1-pt)^{d_1-1}\{1-t+qp^{c_1}t^{c_1+1}\}} \\ &\quad +\frac{(1-p^{c_1}t^{c_1})^{d_1}q^{d_1}p^{c_2}t^{d_1+c_2}\{(1-pt)^{d_2}-(1-p^{c_2}t^{c_2})^{d_2}q^{d_2}t^{d_2}\}}{(1-pt)^{d_1+d_2-1}\{1-t+qp^{c_2}t^{c_2+1}\}}.\end{aligned} \tag{18.10}$$

18.3 Probabilities and Recurrence Relations

From (18.8), we can explicitly write down the probability mass function of F as

$$\Pr(F=f)=\begin{cases} p^{c_1}(1-p^{c_1})^f, & f=0,1,\ldots,d_1-1,\\ p^{c_2}(1-p^{c_1})^{d_1}(1-p^{c_2})^{f-d_1}, & f=d_1,d_1+1,\ldots,d_1+d_2-1\\ (1-p^{c_1})^{d_1}(1-p^{c_2})^{d_2}, & f=d_1+d_2\\ 0, & \text{otherwise.}\end{cases} \tag{18.11}$$

The moments of F can also be obtained by differentiating (18.8) with respect to u and then setting $u=1$. For example, we obtain the mean of F as

$$\begin{aligned} E(F) &= \pi_F'(1)\\ &= \frac{(1-p^{c_1})}{p^{c_1}}+\frac{(1-p^{c_1})^{d_1}(1-p^{c_2})}{p^{c_2}}-\frac{(1-p^{c_1})^{d_1}(1-p^{c_1})}{p^{c_1}}\\ &\quad -(1-p^{c_1})^{d_1}(1-p^{c_2})^{d_2}\frac{(1-p^{c_2})}{p^{c_2}}\\ &= \frac{(1-p^{c_1})}{p^{c_1}}\{1-(1-p^{c_1})^{d_1}\}+(1-p^{c_1})^{d_1}\frac{(1-p^{c_2})}{p^{c_2}}\{1-(1-p^{c_2})^{d_2}\}\\ &= \sum_{i=1}^{d_1}(1-p^{c_1})^i+(1-p^{c_1})^{d_1}\sum_{i=1}^{d_2}(1-p^{c_2})^i. \end{aligned} \tag{18.12}$$

It is very difficult to write down explicitly the probability mass function of S from (18.9). However, the moments of S can be obtained easily by differentiating (18.9) with respect to t and then setting $t=1$. For example, we obtain the mean of S as

$$\begin{aligned} E(S) &= \pi_S'(1)\\ &= (1-p^{c_1})^{d_1}(1-p^{c_2})^{d_2}\left(\frac{p}{q}-\frac{1}{qp^{c_2-1}}\right)\\ &\quad +\frac{1}{qp^{c_1-1}}\left\{1-\left(1-p^{c_1}\right)^{d_1}\right\}+\frac{1}{qp^{c_2-1}}\left(1-p^{c_1}\right)^{d_1}-\frac{p}{q}. \end{aligned} \tag{18.13}$$

From (18.12) and (18.13), we readily obtain

$$\begin{aligned} E(F+S) &= E(X)\\ &= (1-p^{c_1})^{d_1}(1-p^{c_2})^{d_2}\left\{\frac{p}{q}-\frac{1}{qp^{c_2-1}}-\frac{1}{p^{c_2}}+1\right\}+\frac{1}{p^{c_1}}-1\\ &\quad +\frac{(1-p^{c_1})^{d_1}}{p^{c_2}}-(1-p^{c_1})^{d_1}-\frac{(1-p^{c_1})^{d_1}}{p^{c_1}}+(1-p^{c_1})^{d_1} \end{aligned}$$

$$
\begin{aligned}
& +\frac{1}{qp^{c_1-1}}\{1-(1-p^{c_1})^{d_1}\}+\frac{1}{qp^{c_2-1}}(1-p^{c_1})^{d_1}-\frac{p}{q} \\
= & \frac{1-(1-p^{c_1})^{d_1}}{qp^{c_1-1}}+\frac{1-(1-p^{c_1})^{d_1}}{qp^{c_2-1}}-\frac{1}{q} \\
& +(1-p^{c_1})^{d_1}(1-p^{c_2})^{d_2}\frac{1}{q}\left(1-\frac{1}{p^{c_2}}\right) \\
= & \frac{1}{q}\left\{\left(\frac{1}{p^{c_1}}-1\right)\left[1-(1-p^{c_1})^{d_1}\right]\right. \\
& \left.+\left(\frac{1}{p^{c_2}}-1\right)\left(1-p^{c_1}\right)^{d_1}\left[1-(1-p^{c_2})^{d_2}\right]\right\}.
\end{aligned}
\tag{18.14}
$$

The probability mass function of X does not take on a simple explicit form from its probability generating function given in (18.10). However, a recurrence relation may be derived from (18.10) for the probabilities of X which will facilitate an easy recursive computation of them. First, let us write

$$(1+t+t^2+\cdots+t^n)^m=\sum_{i=0}^{mn}A_m^{(n)}(i)t^i.$$

Then, from (18.10) we can write

$$
\begin{aligned}
\pi_X(t) = & \; q^{d_1+d_2}t^{d_1+d_2}\sum_{i=0}^{d_1(c_1-1)}A_{d_1}^{(c_1-1)}(i)p^it^i\sum_{j=0}^{d_2(c_2-1)}A_{d_2}^{(c_2-1)}(j)p^jt^j \\
& +\frac{p^{c_1}t^{c_1}(1-pt)}{1-t+qp^{c_1}t^{c_1+1}}\left\{1-q^{d_1}t^{d_1}\sum_{i=0}^{d_1(c_1-1)}A_{d_1}^{(c_1-1)}(i)p^it^i\right\} \\
& +\frac{q^{d_1}p^{c_2}t^{d_1+c_2}(1-pt)}{1-t+qp^{c_2}t^{c_2+1}}\sum_{i=0}^{d_1(c_1-1)}A_{d_1}^{(c_1-1)}(i)p^it^i \\
& \times\left\{1-q^{d_2}t^{d_2}\sum_{j=0}^{d_2(c_2-1)}A_{d_2}^{(c_2-1)}(j)p^jt^j\right\}.
\end{aligned}
\tag{18.15}
$$

Mutliplying both sides of (18.15) by $(1-t+qp^{c_1}t^{c_1+1})(1-t+qp^{c_2}t^{c_2+1})$, we have

$$
\begin{aligned}
& (1-t+qp^{c_1}t^{c_1+1})(1-t+qp^{c_2}t^{c_2+1})\pi_X(t) \\
& = (1-t+qp^{c_1}t^{c_1+1})(1-t+qp^{c_2}t^{c_2+1})q^{d_1+d_2}t^{d_1+d_2} \\
& \quad \times\sum_{i=0}^{d_1(c_1-1)}\sum_{j=0}^{d_2(c_2-1)}A_{d_1}^{(c_1-1)}(i)A_{d_2}^{(c_2-1)}(j)p^{i+j}t^{i+j} \\
& \quad +(1-t+qp^{c_2}t^{c_2+1})p^{c_1}t^{c_1}(1-pt)
\end{aligned}
$$

$$
\begin{aligned}
&+(1-t+qp^{c_1}t^{c_1+1})p^{c_2}q^{d_1}t^{c_2+d_1}(1-pt)\sum_{i=0}^{d_1(c_1-1)} A_{d_1}^{(c_1-1)}(i)p^i t^i \\
&-(1-t+qp^{c_1}t^{c_1+1})p^{c_2}q^{d_1+d_2}t^{c_2+d_1+d_2}(1-pt) \\
&\quad\times \sum_{i=0}^{d_1(c_1-1)}\sum_{j=0}^{d_2(c_2-1)} A_{d_1}^{(c_1-1)}(i)A_{d_2}^{(c_2-1)}(j)p^{i+j}t^{i+j}.
\end{aligned} \tag{18.16}
$$

The above equation can be readily rewritten as

$$
\begin{aligned}
&\pi_X(t)(1-2t+t^2+qp^{c_1}t^{c_1+1}-qp^{c_1}t^{c_1+2}+qp^{c_2}t^{c_2+1}-qp^{c_2}t^{c_2+2}+q^2p^{c_1+c_2}t^{c_1+c_2+2}) \\
&= (1-2t+t^2+qp^{c_1}t^{c_1+1}-qp^{c_1}t^{c_1+2}+qp^{c_2}t^{c_2+1}-qp^{c_2}t^{c_2+2}+q^2p^{c_1+c_2}t^{c_1+c_2+2}) \\
&\quad\times q^{d_1+d_2}t^{d_1+d_2}\sum_{i=0}^{d_1(c_1-1)}\sum_{j=0}^{d_2(c_2-1)} A_{d_1}^{(c_1-1)}(i)A_{d_2}^{(c_2-1)}(j)p^{i+j}t^{i+j} \\
&\quad+\{1-(1+p)t+pt^2+qp^{c_2}t^{c_2+1}-qp^{c_2+1}t^{c_2+2}\}p^{c_1}t^{c_1} \\
&\quad-\{1-(1+p)t+pt^2+qp^{c_2}t^{c_2+1}-qp^{c_2+1}t^{c_2+2}\}p^{c_1}q^{d_1}t^{c_1+d_1} \\
&\quad\times \sum_{i=0}^{d_1(c_1-1)} A_{d_1}^{(c_1-1)}(i)p^i t^i \\
&\quad+\{1-(1+p)t+pt^2+qp^{c_1}t^{c_1+1}-qp^{c_1+1}t^{c_1+2}\}p^{c_2}q^{d_1}t^{c_2+d_1} \\
&\quad\times \sum_{i=0}^{d_1(c_1-1)} A_{d_1}^{(c_1-1)}(i)p^i t^i \\
&\quad-\{1-(1+p)t+pt^2+qp^{c_1}t^{c_1+1}-qp^{c_1+1}t^{c_1+2}\}p^{c_2}q^{d_1+d_2}t^{c_2+d_1+d_2} \\
&\quad\times \sum_{i=0}^{d_1(c_1-1)}\sum_{j=0}^{d_2(c_2-1)} A_{d_1}^{(c_1-1)}(i)A_{d_2}^{(c_2-1)}(j)p^{i+j}t^{i+j} \\
&= p^{c_1}t^{c_1}-(1+p)p^{c_1}t^{c_1+1}+p^{c_1+1}t^{c_1+2}+qp^{c_1+c_2}t^{c_1+c_2+1}-qp^{c_1+c_2+1}t^{c_1+c_2+2} \\
&\quad+\{-p^{c_1}q^{d_1}t^{c_1+d_1}+(1+p)p^{c_1}q^{d_1}t^{c_1+d_1+1}-p^{c_1+1}q^{d_1}t^{c_1+d_1+2} \\
&\qquad+p^{c_2}q^{d_1}t^{c_2+d_1}-(1+p)p^{c_2}q^{d_1}t^{c_2+d_1+1}+p^{c_2+1}q^{d_1}t^{c_2+d_1+2}\} \\
&\quad\times \sum_{i=0}^{d_1(c_1-1)} A_{d_1}^{(c_1-1)}(i)p^i t^i \\
&\quad+\{q^{d_1+d_2}t^{d_1+d_2}-2q^{d_1+d_2}t^{d_1+d_2+1}+q^{d_1+d_2}t^{d_1+d_2+2}+p^{c_1}q^{d_1+d_2+1}t^{d_1+d_2+c_1+1} \\
&\quad-p^{c_1}q^{d_1+d_2+1}t^{d_1+d_2+c_1+2}-p^{c_2}q^{d_1+d_2}t^{d_1+d_2+c_2}+2p^{c_2}q^{d_1+d_2}t^{d_1+d_2+c_2+1} \\
&\quad-p^{c_2}q^{d_1+d_2}t^{d_1+d_2+c_2+2}-p^{c_1+c_2}q^{d_1+d_2+1}t^{d_1+d_2+c_1+c_2+1} \\
&\quad+p^{c_1+c_2}q^{d_1+d_2+1}t^{d_1+d_2+c_1+c_2+2}\} \\
&\quad\times \sum_{i=0}^{d_1(c_1-1)}\sum_{j=0}^{d_2(c_2-1)} A_{d_1}^{(c_1-1)}(i)A_{d_2}^{(c_2-1)}(j)p^{i+j}t^{i+j}.
\end{aligned} \tag{18.17}
$$

Now, let us define the quantities

$$R_1(x) = \begin{cases} p^{c_1} & \text{if } x = c_1, \\ -(1+p)p^{c_1} & \text{if } x = c_1+1, \\ p^{c_1+1} & \text{if } x = c_1+2, \\ qp^{c_1+c_2} & \text{if } x = c_1+c_2+1, \\ -qp^{c_1+c_2+1} & \text{if } x = c_1+c_2+2, \\ 0 & \text{otherwise;} \end{cases} \tag{18.18}$$

$$R_2(x) = \begin{cases} -p^{c_1}q^{d_1}A_{d_1}^{(c_1-1)}(0) \\ \qquad \text{if } x = c_1+d_1, \\ -p^{c_1+1}q^{d_1}A_{d_1}^{(c_1-1)}(1) + (1+p)p^{c_1}q^{d_1}A_{d_1}^{(c_1-1)}(0) \\ \qquad \text{if } x = c_1+d_1+1, \\ -p^{x-d_1}q^{d_1}A_{d_1}^{(c_1-1)}(x-c_1-d_1) \\ \quad +(1+p)p^{x-d_1-1}q^{d_1}A_{d_1}^{(c_1-1)}(x-c_1-d_1-1) \\ \quad -p^{x-d_1-1}q^{d_1}A_{d_1}^{(c_1-1)}(x-c_1-d_1-2) \\ \qquad \text{if } c_1+d_1+2 \le x \le c_1(d_1+1)+2 \\ 0 \\ \qquad \text{otherwise;} \end{cases} \tag{18.19}$$

$$R_3(x) = \begin{cases} p^{c_2}q^{d_1}A_{d_1}^{(c_1-1)}(0) \\ \qquad \text{if } x = c_2+d_1, \\ p^{c_2+1}q^{d_1}A_{d_1}^{(c_1-1)}(1) - (1+p)p^{c_2}q^{d_1}A_{d_1}^{(c_1-1)}(0) \\ \qquad \text{if } x = c_2+d_1+1, \\ p^{x-d_1}q^{d_1}A_{d_1}^{(c_1-1)}(x-c_2-d_1) \\ \quad -(1+p)p^{x-d_1-1}q^{d_1}A_{d_1}^{(c_1-1)}(x-c_2-d_1-1) \\ \quad +p^{x-d_1-1}q^{d_1}A_{d_1}^{(c_1-1)}(x-c_2-d_1-2) \\ \qquad \text{if } c_2+d_1+2 \le x \le c_2+c_1d_1+2 \\ 0 \\ \qquad \text{otherwise;} \end{cases} \tag{18.20}$$

$$S_1(x) = \begin{cases} q^{d_1+d_2}A_{d_1}^{(c_1-1)}(0)A_{d_2}^{(c_2-1)}(0) \\ \qquad \text{if } x = d_1+d_2 \\ pq^{d_1+d_2}\{A_{d_1}^{(c_1-1)}(0)A_{d_2}^{(c_2-1)}(1) + A_{d_1}^{(c_1-1)}(1)A_{d_2}^{(c_2-1)}(0)\} \\ \quad -2q^{d_1+d_2}A_{d_1}^{(c_1-1)}(0)A_{d_2}^{(c_2-1)}(0) \\ \qquad \text{if } x = d_1+d_2+1 \\ p^{x-d_1-d_2}q^{d_1+d_2}\sum_{i=0}^{x-d_1-d_2} A_{d_1}^{(c_1-1)}(i)A_{d_2}^{(c_2-1)}(x-d_1-d_2-i) \\ \quad -2p^{x-d_1-d_2-1}q^{d_1+d_2}\sum_{i=0}^{x-d_1-d_2-1} A_{d_1}^{(c_1-1)}(i)A_{d_2}^{(c_2-1)}(x-d_1-d_2-1-i) \\ \quad +p^{x-d_1-d_2-2}q^{d_1+d_2}\sum_{i=0}^{x-d_1-d_2-2} A_{d_1}^{(c_1-1)}(i)A_{d_2}^{(c_2-1)}(x-d_1-d_2-2-i) \\ \qquad \text{if } d_1+d_2+2 \le x \le c_1d_1+c_2d_2+2 \\ 0 \\ \qquad \text{otherwise;} \end{cases} \tag{18.21}$$

$$S_2(x) = \begin{cases} p^{c_1} q^{d_1+d_2+1} A_{d_1}^{(c_1-1)}(0) A_{d_2}^{(c_2-1)}(0) \\ \qquad \text{if } x = d_1+d_2+c_1+1 \\ p^{x-d_1-d_2-1} q^{d_1+d_2+1} \sum_{i=0}^{x-d_1-d_2-c_1-1} A_{d_1}^{(c_1-1)}(i) \\ \quad \times A_{d_2}^{(c_2-1)}(x-d_1-d_2-c_1-1-i) \\ \quad -p^{x-d_1-d_2-1} q^{d_1+d_2+1} \sum_{i=0}^{x-d_1-d_2-c_1-2} A_{d_1}^{(c_1-1)}(i) \\ \quad \times A_{d_2}^{(c_2-1)}(x-d_1-d_2-c_1-2-i) \\ \qquad \text{if } d_1+d_2+c_1+2 \le x \le c_1(d_1+1)+c_2d_2+2 \\ 0 \\ \qquad \text{otherwise;} \end{cases} \tag{18.22}$$

$$S_3(x) = \begin{cases} -p^{c_2} q^{d_1+d_2} A_{d_1}^{(c_1-1)}(0) A_{d_2}^{(c_2-1)}(0) \\ \qquad \text{if } x = d_1+d_2+c_2 \\ -p^{c_2} q^{d_1+d_2} \{A_{d_1}^{(c_1-1)}(0) A_{d_2}^{(c_2-1)}(1) + A_{d_1}^{(c_1-1)}(1) A_{d_2}^{(c_2-1)}(0)\} \\ \quad +2p^{c_2} q^{d_1+d_2} A_{d_1}^{(c_1-1)}(0) A_{d_2}^{(c_2-1)}(0) \\ \qquad \text{if } x = d_1+d_2+c_2+1 \\ -p^{x-d_1-d_2} q^{d_1+d_2} \sum_{i=0}^{x-d_1-d_2-c_2} A_{d_1}^{(c_1-1)}(i) A_{d_2}^{(c_2-1)}(x-d_1-d_2-c_2-i) \\ \quad +2p^{x-d_1-d_2-1} q^{d_1+d_2} \sum_{i=0}^{x-d_1-d_2-c_2-1} A_{d_1}^{(c_1-1)}(i) \\ \quad \times A_{d_2}^{(c_2-1)}(x-d_1-d_2-c_2-1-i) \\ \quad -p^{x-d_1-d_2-2} q^{d_1+d_2} \sum_{i=0}^{x-d_1-d_2-c_2-2} A_{d_1}^{(c_1-1)}(i) \\ \quad \times A_{d_2}^{(c_2-1)}(x-d_1-d_2-c_2-2-i) \\ \qquad \text{if} d_1+d_2+c_2+2 \le x \le c_1d_1+c_2(d_2+1)+2 \\ 0 \\ \qquad \text{otherwise;} \end{cases} \tag{18.23}$$

$$S_4(x) = \begin{cases} -p^{c_1+c_2} q^{d_1+d_2+1} A_{d_1}^{(c_1-1)}(0) A_{d_2}^{(c_2-1)}(0) \\ \qquad \text{if } x = d_1+d_2+c_1+c_2+1 \\ -p^{x-d_1-d_2-1} q^{d_1+d_2+1} \\ \quad \times \sum_{i=0}^{x-d_1-d_2-c_1-c_2-1} A_{d_1}^{(c_1-1)}(i) A_{d_2}^{(c_2-1)}(x-d_1-d_2-c_1-c_2-1-i) \\ \quad +p^{x-d_1-d_2-2} q^{d_1+d_2+1} \\ \quad \times \sum_{i=0}^{x-d_1-d_2-c_1-c_2-2} A_{d_1}^{(c_1-1)}(i) A_{d_2}^{(c_2-1)}(x-d_1-d_2-c_1-c_2-2-i) \\ \qquad \text{if } d_1+d_2+c_1+c_2+2 \le x \le c_1(d_1+1)+c_2(d_2+1)+2 \\ 0 \\ \qquad \text{otherwise.} \end{cases} \tag{18.24}$$

Then, upon comparing the coefficients of t^x on both sides of (18.17), we obtain the recurrence relation for the probabilities of X as

$$\begin{aligned} & P(x) - 2P(x-1) + P(x-2) + qp^{c_1}P(x-c_1-1) - qp^{c_1}P(x-c_1-2) \\ & \quad +qp^{c_2}P(x-c_2-1) - qp^{c_2}P(x-c_2-2) + q^2p^{c_1+c_2}P(x-c_1-c_2-2) \\ & = \sum_{i=0}^{3} R_i(x) + \sum_{i=1}^{4} S_i(x) \\ & \qquad \text{for } x = M, M+1, M+2, \ldots, M^*, \end{aligned} \tag{18.25}$$

where $M = \min(c_1, d_1 + c_2, d_1 + d_2)$, $M^* = c_1 d_1 + c_2 d_2$, and $R_i(x), i = 1, 2, 3$ and $S_i(x), i = 1, 2, 3, 4$ are as defined in Eqs (18.18) – (18.24).

We may note that the relation in (18.25) will be trivially satisfied for values of $x = M^* + 1, \ldots, M^* + c_1 + c_2 + 2$.

From the probability generating function of X in (18.10), we obtain the mean of X to be

$$
\begin{aligned}
E(X) &= \pi'_X(1) \\
&= -d_1(1-p^{c_1})^{d_1-1} c_1 p^{c_1} (1-p^{c_2})^{d_2} - d_2(1-p^{c_1})^{d_1}(1-p^{c_2})^{d_2-1} c_2 p^{c_2} \\
&\quad +(d_1+d_2)(1-p^{c_1})^{d_1}(1-p^{c_2})^{d_2} \\
&\quad +(d_1+d_2)(1-p^{c_1})^{d_1}(1-p^{c_2})^{d_2} p/q \\
&\quad +\frac{p^{c_1}}{q^{d_1} p^{c_1}} \Bigg[c_1 q^{d_1} \left\{ 1-p^{c_1})^{d_1} \right\} - d_1 p q^{d_1-1} \\
&\qquad + d_1(1-p^{c_1})^{d_1-1} c_1 p^{c_1} q^{d_1} - d_1(1-p^{c_1})^{d_1} q^{d_1} \Bigg] \\
&\quad +\frac{(d_1-1)p}{q^{d_1} q p^{c_1}} p^{c_1} q^{d_1} \left\{ 1-(1-p^{c_1})^{d_1} \right\} \\
&\qquad + \frac{1-(c_1+1) q p^{c_1}}{q^{d_1-1} q^2 (p^{c_1})^2} p^{c_1} q^{d_1} \left\{ 1-(1-p^{c_1})^{d_1} \right\} \\
&\quad +\frac{q^{d_1} p^{c_2}}{q^{d_1+d+2} p^{c_2}} \Bigg[-d_1(1-p^{c_1})^{d_1-1} c_1 p^{c_1} q^{d_2} \left\{ 1-(1-p^{c_2})^{d_2} \right\} \\
&\qquad + (d_1+c_2)(1-p^{c_1})^{d_1} q^{d_2} \left\{ 1-(1-p^{c_2})^{d_2} \right\} \\
&\qquad - d_2(1-p^{c_1})^{d_1} q^{d_2-1} p/q \\
&\qquad + d_2(1-p^{c_1})^{d_1}(1-p^{c_2})^{d_2-1} c_2 p^{c_2} q^{d_2} \\
&\qquad - d_2(1-p^{c_1})^{d_1}(1-p^{c_2})^{d_2} q^{d_2} \Bigg] \\
&\quad +\frac{(d_1+d_2-1)p}{q^{d_1+d_2} q p^{c_2}} (1-p^{c_1})^{d_1} q^{d_1} p^{c_2} q^{d_2} \{ 1-(1-p^{c_2})^{d_2} \} \\
&\quad +\frac{1-(c_2+1) q p^{c_2}}{q^{d_1+d_2-1} q^2 (p^{c_2})^2} (1-p^{c_1})^{d_1} q^{d_1} p^{c_2} q^{d_2} \{ 1-(1-p^{c_2})^{d_2} \} \\
&= \frac{1}{q} \Bigg[\left(\frac{1}{p^{c_1}} - 1 \right) \left\{ 1-(1-p^{c_1})^{d_1} \right\} \\
&\qquad + \left(\frac{1}{p^{c_2}} - 1 \right) (1-p^{c_1})^{d_1} \left\{ 1-(1-p^{c_2})^{d_2} \right\} \Bigg]. \qquad (18.26)
\end{aligned}
$$

This is exactly the same expression as derived earlier in Eq. (18.14).

We may observe the following two special cases of interest.

1. No stopping due to failures and c consecutive successes, i.e., when $d_1 \to \infty$ and $c_1 = c$, (18.26) reduces to

$$E(X) = \frac{1}{q}\left(\frac{1}{p^c - 1}\right) = \frac{1 - p^c}{qp^c}.$$

This is the result corresponding to the classical start-up demonstration testing [Viveros and Balakrishnan (1993)].

2. Stop as soon as c consecutive successes or as soon as d failures, i.e., when $c_2 \to \infty$ and $d_2 \to 0$, (18.26) reduces to

$$E(X) = \frac{1}{q}\left(\frac{1}{p^c - 1}\right)\{1 - (1 - p^c)^d\}.$$

This is a known result for the sooner waiting time problem [see, for example, Ebneshahrashoob and Sobel (1990) and Balasubramanian, Viveros and Balakrishnan (1993)].

Acknowledgements. The first author thanks the Natural Sciences and Engineering Research Council of Canada for funding this research. The authors also express their thanks to Mrs. Debbie Iscoe for the excellent typing of the manuscript.

References

1. Aki, S., Balakrishnan, N. and Mohanty, S. G. (1996). Sooner and later waiting time problems for success and failure runs in higher order Markov dependent trials, *Annals of the Institute of Statistical Mathematics*, **48**, 773–787.

2. Balakrishnan, N., Balasubramanian, K. and Viveros, R. (1995). Start-up demonstration tests under correlation and corrective action, *Naval Research Logistics*, **42**, 1271–1276.

3. Balakrishnan, N. and Koutras, M. V. (1998). *Runs and Patterns with Applications*, John Wiley & Sons, New York (to appear).

4. Balakrishnan, N., Mohanty, S. G. and Aki, S. (1997). Start-up demonstration tests under Markov dependence model with corrective actions, *Annals of the Institute of Statistical Mathematics*, **49**, 155–169.

5. Balasubramanian, K., Viveros, R. and Balakrishnan, N. (1993). Sooner and later waiting time problems for Markovian Bernoulli trials, *Statistics & Probability Letters*, **18**, 153–161.

6. Ebneshahrashoob, M. and Sobel, M. (1990). Sooner and later waiting time problems for Bernoulli trials: frequency and run quotas, *Statistics & Probability Letters*, **9**, 5–11.

7. Hahn, G. J. and Gage, J. B. (1986). Evaluation of a start-up demonstration test, *Journal of Quality Technology*, **15**, 103–105.

8. Koutras, M. V. (1996). On a waiting time distribution in a sequence of Bernoulli trials, *Annals of the Institute of Statistical Mathematics*, **48**, 789–806.

9. Viveros, R. and Balakrishnan, N. (1993). Statistical inference from start-up demonstration test data, *Journal of Quality Technology*, **22**, 119–130.

19

Optimal Prophylaxis Policy for Systems With Partly Observable Parameters

B. P. Harlamov

Russian Academy of Sciences, Saint-Petersburg, Russia

Abstract: An optimal controlling problem for a partly observable one-server system with simple prophylaxis and a partly observable two-server system with prophylaxis and cold reserve are solved. These systems are proposed to have observable hazard rates which are realizations of some random processes. The optimal times of prophylactic switches are proved to be the first exit times from some regions depending on distributions of these random hazard rate functions. The precise forms of these regions are obtained under assumption that the hazard rates are distributed as continuous semi-Markov processes.

Keywords and phrases: Prophylaxis, cold reserve, random hazard rate function, stopping time, first exit time, optimal rule, semi-Markov process, inverse process with positive independent increments

19.1 Introduction

In the reliability theory a system is called completely observable if its failure time is predictable on the base of information about all its observable parameters. An example of such a predictable time is the first exit time of a continuous process from an open set when the process is observable till this time [Dellacherie (1972)]. In most cases for real systems the situation is typical when the failure time is not predictable but depends on values of observable parameters.

In this paper we consider such partly observable systems where one can estimate reliability of the system more precise using properties of observable parameters. Let ξ be such an observable parameter of the system. The parameter ξ varies from one regeneration cycle to another. We regard it random, independent, with some distribution $Q(d\xi)$ which is the same for each cycle. In this case we can use a conditional distribution function $P(\zeta < t|\xi)$ $(t \geq 0)$ where ζ is a failure time. Namely we assume observable hazard rates in every cycle of operating which are realizations of some random process. We consider two kinds of the systems: the first one is a partly observable one-server system with simple prophylaxis and the second one is a partly observable two-server system with prophylaxis and cold reserve. In both cases optimal times of prophylactic switches are proved to be the first exit times from some regions depending on distributions of these random hazard rate functions.

The precise forms of these regions are evaluated under assumption that hazard rates depend on continuous semi-Markov processes. An inverse strictly increasing process with independent increments [Itô and McKean (1965)] is a simple non-trivial example of a continuous semi-Markov process [Harlamov (1983)]. This process can be a good mathematical model for abrasion [Vinogradov et al. (1990), Gerzbach and Kordonski (1966)] and for other processes which can affect reliability.

19.2 One-Server System

19.2.1 Mathematical model

In this section we consider a server system with one channel. Its action consists of alternating work and repair periods. The repair periods are of two forms: a repair after a failure and a prophylactic repair. An average duration of the first kind of repair is more than of the second. The prophylactic repair begins at the instant defined by an operator of the system. A rule of this prophylactic switches is an object of optimization. A coefficient of readiness (a stationary probability for the system to be in a work state at the time $t = 0$) is a criterion of optimization.

Let the process $(X(t))_{t \geq 0}$ describe the system action. It is generated by the sequence of random evolutions $(\xi_i, T_i)_{i=1}^{\infty}$ where ξ_i is a partial random process $(\xi_i(t))_{t \geq 0,}$; T_i is a positive random value defining an interval on which this partial process is considered:

$$X(t) = \xi_i(t - \sum_{k=1}^{i-1} T_k), \quad \text{where} \quad \sum_{k=1}^{i-1} T_k \leq t < \sum_{k=1}^{i} T_k.$$

For given system this partial process can be of the following three types:

- degenerate process taking a constant value $\xi_i(t) = S_1$ on the whole interval $[0, T_i)$; one interprets the state S_1 as a repair after failure; T_i is a duration of the repair; for every cycle with this evolution T_i has the same distribution function $F_1(t)$ $(t \geq 0)$;

- degenerate process taking a constant value $\xi_i(t) = S_2$ on the whole interval $[0, T_i)$; one interprets the state S_2 as a prophylaxis repair; T_i is a duration of the repair; for every cycle with this evolution T_i has the same distribution function $F_2(t)$ $(t \geq 0)$;

- non-degenerate process with continuous from the right trajectories having limits from the left $\xi_i : [0, \infty) \to R^d$ $(d \geq 1)$ where $S_1, S_2 \notin R^d$; one interprets $\xi_i(t)$ as a value of a parameter of the system defining its reliability, for example, it is a hazard rate in the i-th cycle of acting; T_i is a minimum among a failure time ζ_i and a time τ_i of switching the system for prophylaxis in the i-th cycle of acting: $T_i = \zeta_i \wedge \tau_i$.

We assume every partial proccss of thc third type to have the same distribution $Q(d\xi)$. A distribution of the failure time ζ_i is completely defined by a realization of the process ξ_i in such a way that the hazard rate $r_i(t)$ is a function of ξ_i: $r_i(t) = \alpha(\xi_i(t))$, where α is some positive function (may by with infinite values) on the region of all realizations of the process ξ_i. We have

$$P(\zeta_i < t \,|\xi_i) = 1 - \exp\left(- \int_0^t \alpha(\xi_i(s))\, ds \right)$$

is a distribution function of ζ_i for the partial random process of the third type. It is important to note that the partial process is not a function of ξ_i but its distribution depends on ξ_i. So in the model to be considered the failure time is twice random. It formalizes dependence of the failure time on inner and outer reasons.

The time τ_i has to be choose on every cycle of the third type. It depends on a realization of the process ξ_i like a so called Markov time: the event $\tau_i \geq t$ depends on the process ξ_i only till the point t.

To complete the model one has to define a rule of transitions between types of cycles. It is the following:

- after the first type the third type follows;

- after the second type the third type follows;

- if $T_i < \tau_i$ then after the third type the first type follows;

- if $T_i = \tau_i$ then after the third type the second type follows;

19.2.2 Coefficient of readiness

To avoid difficulties with existence of a limit distribution we assume all partial distributions are non-lattice. In the frame of our model we solve the problem of optimal prophylactic switching with the optimality criterion

$$1 - \gamma = 1 - \lim_{t\to\infty} (P(X(t) = S_1) + P(X(t) = S_2))$$

From the theory of stepped semi-Markov processes (see [Korolyuk and Turbin (1976)]) one can easily find that $\gamma = U/(U+V)$ where[1]

$$
\begin{aligned}
U &= Q(m_1 - (m_1 - m_2)\exp\left(-\int_0^{\tau(\xi)} \alpha(\xi(s))ds\right), \\
V &= Q(\int_0^{\tau(\xi)} \exp\left(-\int_0^t \alpha(\xi(s))ds)dt\right), \\
m_1 &= E(T_1|X(0) = S_1) = \int_0^\infty (1 - F_1(t))dt, \\
m_2 &= E(T_1|X(0) = S_2) = \int_0^\infty (1 - F_2(t))dt.
\end{aligned}
$$

19.3 Two-Server System

19.3.1 Mathematical model

A mathematical model of the system is a pair of dependent processes $X(t) = (X_1(t), X_2(t))$ $(t \geq 0)$ where like in one-server case every process is defined by its sequence of random evolutions $(\xi_{1i}, T_{1i})_{i=1}^\infty$ and $(\xi_{2i}, T_{2i})_{i=1}^\infty$. Here ξ_{ki} is a partial random process $(\xi_{ki}(t))_{t\geq 0,}$; T_{ki} is a positive random value defining an interval the partial process ξ_{ki} to be considered on:

$$X_k(t) = \xi_{ki}\left(t - \sum_{j=1}^{i-1} T_{kj}\right), \quad \sum_{j=1}^{i-1} T_{kj} \leq t < \sum_{j=1}^{i} T_{kj} \quad (k = 1, 2).$$

For each $k \in \{1, 2\}$ the partial random process can be of the following four types:

- degenerate process taking constant value $\xi_{ki}(t) = S_1$ on the whole interval $[0, T_{ki})$; one interprets the state S_1 as a repair after failure; the value T_{ki} is a duration of the repair; for every cycle with the partial process of this type T_{ki} has the same distribution function $F_1^{(k)}(t)$ $(t \geq 0)$;
- degenerate process taking constant value $\xi_{ki}(t) = S_2$ on the whole interval $[0, T_{ki})$; one interprets the state S_2 as a prophylactic repair; the value T_{ki} is a duration of the repair; for every cycle with the partial process of this type T_{ki} has the same distribution function $F_2^{(k)}(t)$ $(t \geq 0)$;

[1]Here and in what follows we denote $Q(f) = \int_{\mathcal{D}} f(\xi)Q(d\xi)$.

- non-degenerate process with continuous from the right trajectories having limits from the left $\xi_{ki} : [0,\infty) \to R^d$ $(d \geq 1)$ where $S_1, S_2 \notin R^d$; one interprets $\xi_{ki}(t)$ as a value of a parameter of the system defining its reliability, for example, it is a hazard rate in the i-th cycle of acting; T_{ki} is a minimum among a failure time ζ_{ki} and a time τ_{ki} of switching the system for prophylaxis in the i-th cycle of acting: $T_{ki} = \zeta_{ki} \wedge \tau_{ki}$; an enlarged state corresponding to the set of all partial processes of the third type is denoted S_3 (work state); we assume each partial process of the third type has the same distribution $Q_k(d\xi)$ $(k = 1, 2)$; a distribution of the failure time ζ_{ki} is completely defined by a realization of the random process ξ_{ki} in such a way that the hazard rate $r_{ki}(t)$ depends on the realization $r_{ki}(t) = \alpha_{ki}(\xi_{ki}(t))$ where α_{ki} is some positive function (may be, with infinite values) on the set of all realizations of the random process ξ_{ki}; we have

$$P(\zeta_{ki} < t\,|\xi_{ki}) = 1 - \exp\left(- \int_0^t \alpha_{ki}(\xi_{ki}(s))\, ds \right)$$

  is a distribution function of ζ_{ki} for the partial process of the third type; the time τ_{ki} is to be chosen in every cycle of the process X_k with the partial process of the third type; it depends on a realization ξ_{ki} like some Markov time;

- degenerate process taking constant value $\xi_{ki}(t) = S_4$ on the whole interval $[0, T_{ki})$; one interprets the state S_4 as waiting period for reserved channel; the value T_{ki} is a duration of this period; it depends on the sequence $(\xi_{ni}, T_{ni})_{i=1}^{\infty}$ $(n \neq k)$, begins at the end of the partial process of the first or second type if at the same time the process X_n is of the third type and continues till the time when the type of X_n changes.

A directed graph of transitions between states S_i $(i \in \{1, 2, 3, 4\})$ of one channel is represented by the matrix:

$\to$	1	2	3	4
1	0	0	1	1
2	0	0	1	1
3	1	1	0	0
4	0	0	1	0

For full description of the model we consider all possible double states of our two-server system:

S_{11} — both channels are in repair after failure;
S_{12} — the first channel is in repair after failure, the second one is in prophylactic repair;
S_{21} — the second channel is in repair after failure, the first one is in prophylactic repair;
S_{13} — the first channel is in repair after failure, the second one is working;
S_{31} — the second channel is in repair after failure, the first one is working;
S_{23} — the first channel is in prophylactic repair, the second one is working;
S_{32} — the second channel is in prophylactic repair, the first one is working;
S_{43} — the first channel is in reserve, the second one is working;
S_{34} — the second channel is in reserve, the first one is working;

Note that there exists a transition through the state S_{43} only in chains

$$S_{13} \to \left\{ \begin{array}{l} S_{11} \\ S_{32} \\ S_{43} \quad \to \quad \left\{ \begin{array}{l} S_{32} \\ S_{31} \end{array} \right. \end{array} \right.$$

$$S_{23} \to \left\{ \begin{array}{l} S_{21} \\ S_{32} \\ S_{43} \quad \to \quad \left\{ \begin{array}{l} S_{32} \\ S_{31} \end{array} \right. \end{array} \right.$$

A transition through S_{34} exists only in two other corresponding chains. Hence we can add the state S_{43} to the previous state S_{13} for to obtain some new state $\bar{S}_{13}$. Correspondingly the pair (S_{23}, S_{43}) turns into $\bar{S}_{23}$, the pair (S_{31}, S_{34}) turns into $\bar{S}_{31}$, the pair (S_{32}, S_{34}) turns into $\bar{S}_{32}$. So we have a directed graph of transitions for 7 states of the system:

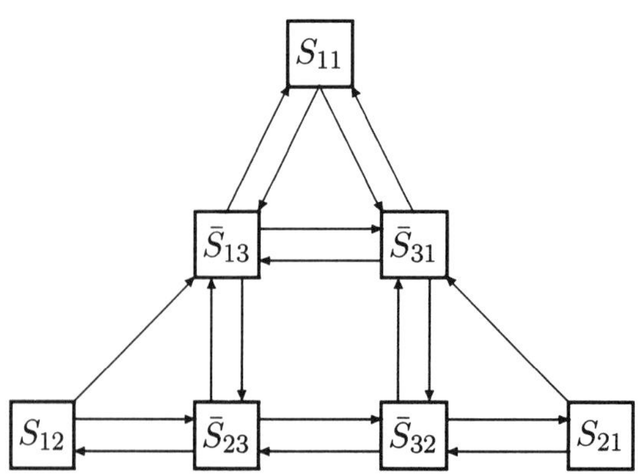

Figure 19.1: Transition graph for two-server system

19.3.2 Coefficient of readiness

In frames of our model we choose an optimal rule of prophylactic switching with the optimality criterion

$$1 - \gamma = \lim_{t \to \infty} P(X(t) \notin \{S_{11}, S_{12}, S_{21}\})$$

(coefficient of readiness).

In this paper we assume that duration of repair after failure has an exponential distribution. So does prophylactic repair. In this case α_{k1} and α_{k2} are constant. It permits us to embed our two-component process in some stepped semi-Markov process and to receive rather simple formulae for a stationary distribution. Besides we assume that our system is symmetric. It means all coefficients do not depend on the index $k = 1, 2$ which denotes a number of a channel.

After simple but awkward evaluations we obtain the following formula:

$$\gamma = \frac{p_A m_A + p_D m_D}{p_A m_A + p_D m_D + p_B m_B + p_C m_C},$$

where

$$\begin{aligned} p_A &= C_1 \left(1 - e(\tau) - J_2(\tau) + \frac{\alpha_2}{\alpha_1 + \alpha_2} J_2\right) J_1, \\ p_B &= C_1 \left(1 - e(\tau) - J_2(\tau) + \frac{\alpha_2}{\alpha_1 + \alpha_2} J_2\right), \\ p_C &= C_1(e(\tau) - J_1 + J_1(\tau)), \\ p_D &= C_1(e(\tau) - J_1 + J_1(\tau)) J_2, \end{aligned}$$

where

$$\begin{aligned} J_i(\tau) &= Q\left(\int_0^\tau \alpha_3(\xi(t)) \exp\left(-\int_0^t (\alpha_i + \alpha_3(\xi(s)))ds\right) dt\right) \quad (i = 1, 2), \\ J_i &= J_i(\infty), \\ e(\tau) &= Q\left(\exp\left(-\int_0^\tau \alpha_3(\xi(t))dt\right)\right), \end{aligned}$$

C_1 is a norming factor;

$$\begin{aligned} m_A &= \frac{1}{2\alpha_1}, \\ m_B &= Q\left(\int_0^\tau \exp\left(-\int_0^t \alpha_3(\xi(s))ds\right) + \int_\tau^\infty \exp\left(-\int_0^t (\alpha_1 + \alpha_3(\xi(s))\right) ds\right), \end{aligned}$$

$$m_C = Q\left(\int_0^\tau \exp\left(-\int_0^t \alpha_3(\xi(s))ds\right) + \int_\tau^\infty \exp\left(-\int_0^t (\alpha_2+\alpha_3(\xi(s))\right) ds\right),$$

$$m_D = \frac{1}{\alpha_1+\alpha_2}\,.$$

19.4 Optimization

19.4.1 Functional equation

A necessary condition for a smooth functional $A(\tau)$ to have minimum at the point τ_0 is the equality $A'(\tau_0) = 0$ where

$$A'(\tau) = \lim_{h\to 0}\frac{1}{h}(A(\tau+h\eta) - A(\tau))$$

and η is any measurable positive function on the set of all trajectories. So we admit $A(\tau)$ being defined for any measurable function τ. In our case functionals A representing coefficients of readiness have real sense only for τ of special forms namely for Markov times. To find necessary conditions in this case we define them formally on the set of all measurable functions τ. Then we obtain necessary condition for some τ to be minimum point in this wide set. And then we show that this τ is a Markov time. So we obtain necessary conditions for a Markov time to be minimum point of the functional A on the set of all measurable function τ and particularly on the set of all Markov times τ.

In the case of the one-server system the equation $A'(\tau) = 0$ gives

$$Q(\eta(\xi)\ \exp\left(-\int_0^{\tau(\xi)} \alpha(\xi(s))ds)(U(\tau) - \alpha(\xi(\tau))(m_1-m_2)V(\tau))\right) = 0,$$

where $A(\tau) = U(\tau)/(U(\tau)+V(\tau))$. Because η is an arbitrary measurable positive function we conclude that Q-almost sure

$$\alpha(\xi(\tau)) = \frac{U(\tau)}{(m_1-m_2)V(\tau)}\,. \tag{19.1}$$

In the case of the two-server system the equation $A'(\tau) = 0$ gives Q-almost sure

$$\alpha(\xi(\tau)) = \frac{p_B J_1/(2\alpha_1) + p_C J_2/(\alpha_1+\alpha_2)}{C_1(m_C J_1/(2\alpha_1) - m_B J_2/(\alpha_1+\alpha_2))}\,, \tag{19.2}$$

where $\alpha = \alpha_3$. Note that the left part of the Equation (19.1) depends formally on ξ but its right part is a ratio of some expectations (does not depend on ξ).

Denote this functional $Z(\tau)$. It makes the time τ random (it does depend on ξ itself). For a continuous random process such a random time is, for example, σ_b the first exit time from the region $\{x \in R^d : \alpha(x) < b\}$ for some b which is to be found from an equation of the form $b = Z(\sigma_b)$. The same is true for the Equation (19.2). For an increasing random process $\alpha(\xi(t))$ $(t \geq 0)$ this random time will be unique if any. And it does exist if Q-almost sure $\alpha(\xi(t)) \to \infty$ $(t \to \infty)$.

To solve equation $b = Z(\sigma_b)$ we must know distribution Q. We exclude degenerate case. For the most known random processes solving this process is a very difficult problem. Fortunately there exists a class of random processes for which it possible. It is a class of continuous semi-Markov processes. A distribution of such a process is defined by distributions of its first exit points from some open sets. So it is adapted to a problem like above and permits to evaluate functionals of the form

$$Q\left(\exp(-\int_0^{\sigma_b} \alpha(\xi(s))ds)\right), \quad Q\left(\int_0^{\sigma_b} \exp\left(-\int_0^t \alpha(\xi(s))ds\right) dt\right)$$

and so on.

19.4.2 Continuous semi-Markov process

A random process $\xi(t)$ $(t \geq 0)$ is called semi-Markov if it possesses Markov property with respect to any first exit time from an open set [see Harlamov (1983, 1985) and others]. Such a process is not obliged to be Markov (with respect to any non-random times). Although any strictly Markov process is, at the same time, semi-Markov one too [Dynkin (1963)]. In this paper we deal with a subclass of one-dimensional monotone semi-Markov processes. A non-decreasing process of this subclass is an inverse process with independent positive increments. Varying of diagnostic parameter under the rule of such a process seems to be reasonable. For example, in an abrasion process the demands of this model imply independence of durations which take non-overlapped portions of material to be erased. It is well verified with statistical data for abrasion of autocar tires or electro-machine contact brushes.

19.4.3 Evaluation of functionals

Consider an inverse process with independent positive increments $\xi(t)$ where $\xi(0) = x$ $(x \in R)$. A Laplace image of a semi-Markov transition function of this process [Itô and McKean (1965), Harlamov (1983)] for the first hitting time of the level b $(b > x)$ has the following form (Levi-Khinchin expansion):

$$\begin{aligned} E_x e^{-\lambda\sigma_b} &= \exp\Bigl(-\lambda(g(\sigma_b)-g(\sigma_x)) \\ &\quad - \int_{0+}^{\infty}(1-e^{-\lambda u})n(du\times[\sigma_x,\sigma_b)) + \sum_{x\le x_i<b}\log E_{x_i}(e^{-\lambda\tau_0})\Bigr), \end{aligned}$$

where σ_c is the first hitting time of the level c,

$$g(\sigma_c) = \int_0^c \rho(x)\,dx$$

is a determinate component,

$$n(B\times[\sigma_c,\sigma_d)) = \int_c^d \nu(B\,|x)dx$$

is a Poissonian component (in the case when these components are absolutely continuous with respect to Lebesgue measure on the line), τ_0 is the first exit time from the initial state, (x_i) is a sequence of fixed levels, on which the trajectory ξ has intervals of constancy. Each of these components may be absent. The case with the first component only means deterministic character of the process. The third component is absent in a space-homogeneous process. In what follows we consider the case without the third component.

From the Levi-Khinchin expansion one can easily derive that

$$Q\exp\left(-\int_0^{\sigma_b}\alpha(\xi(s))ds\right) = \exp\left(-\int_0^b \alpha(x)m(\alpha,x)dx\right),$$

where

$$m(\alpha,x) = \rho(x) + \int_{0+}^{\infty}\frac{1-e^{\alpha(x)u}}{\alpha(x)}\,\nu(du\,|x),$$

and also

$$\begin{aligned} &Q\left(\int_0^{\sigma_b}\alpha(\xi(t))\exp\left(-\int_0^t \beta(\xi(s))ds\right)dt\right) \\ &= \int_0^b \alpha(x)\exp\left(-\int_0^x \beta(s)m(\beta,s)ds\right)m(\beta,x)dx. \end{aligned}$$

As a result we obtain analytical equations for optimal levels. The Equation (19.1) turns into the equation

$$b = \frac{m_1 - (m_1 - m_2)\exp\left(-\int_0^b \alpha(x) m(\alpha, x)dx\right)}{(m_1 - m_2)\int_0^b \exp\left(-\int_0^x \alpha(s) m(\alpha, s)ds\right) m(\alpha, x)dx}. \tag{19.3}$$

The Equation (19.2) we put on in the form

$$\begin{aligned}
b \cdot \Bigg(B_0 \int_0^b &\exp\left(-\int_0^x \alpha(s) m(\alpha, s)ds\right) m(\alpha, x)dx \\
&+ B_1 \int_b^\infty \exp\left(-\int_0^x (\alpha_2 + \alpha(s)) m(\alpha_2 + \alpha, s)ds\right) m(\alpha_2 + \alpha, s)dx\Bigg) \\
&- B_2 \int_b^\infty \exp\left(-\int_0^x (\alpha_1 + \alpha(s)) m(\alpha_1 + \alpha, s)ds\right) m(\alpha_1 + \alpha, s)dx \\
= \; & A - B_0 \exp\left(-\int_0^b \alpha(s) m(\alpha, s)ds\right) \\
&- B_1 \int_0^b \alpha(x) \exp\left(-\int_0^x (\alpha_2 + \alpha(s)) m(\alpha_2 + \alpha, s)ds\right) m(\alpha_2 + \alpha, x)dx \\
&+ B_2 \int_0^b \alpha(x) \exp\left(-\int_0^x (\alpha_1 + \alpha(s)) m(\alpha_1 + \alpha, s)ds\right) m(\alpha_1 + \alpha, x)dx,
\end{aligned} \tag{19.4}$$

where

$$\begin{aligned}
A &= \frac{1}{2\alpha_1} J_1 + \frac{\alpha_2 - 2\alpha_1}{2\alpha_1(\alpha_1 + \alpha_2)} J_1 J_2, \\
B_0 &= B_1 - B_2, \\
B_1 &= \frac{1}{2\alpha_1} J_1, \\
B_2 &= \frac{1}{\alpha_1 + \alpha_2} J_2, \\
J_i &= \int_0^\infty \alpha(x) \exp\left(-\int_0^x (\alpha_i + \alpha(s)) m(\alpha_i + \alpha, s)ds\right) m(\alpha_i + \alpha, x)dx \\
&\qquad (i = 1, 2).
\end{aligned}$$

If parameters of the process are known these equations can be solved at least numerically with the help of a computer. When $m_1 > m_2$ ([Equation (19.3)] and $\alpha_1 < \alpha_2$ [Equation (19.4)] a sufficient condition for existence and uniqueness of solutions of these equations is Q-almost sure increasing of the corresponding function $\alpha(\xi(t))$ to infinity as $t \to \infty$. These equations can be simplified for a space-homogeneous case, when $\rho(x)$, $\nu(du\,|x)$ do not depend on x. We show it by two examples.

19.4.4 Process of maximal values

An important example of an inverse processes with independent positive increments is a process of maximal values of any one-dimensional continuous semi-Markov process. For a continuous one-dimensional process $\xi(t)$ with an initial point $x = \xi(0)$ its process of maximal values is a process $\theta(t) = \max(\xi(s) : 0 \le s \le t)$. In this case if ξ is a semi-Markov process then the same is true for the process θ. Besides it is monotone. Hence it is an inverse processes with independent positive increments. A simple example of such a process is a transformed Wiener process. For the process of maximal values of a standard Wiener process the parameter ρ is equal to zero and a functional parameter ν does not depend on a space variable and is of the form $\nu(du) = \frac{du}{\sqrt{2\pi u^3}}$. From identity

$$\int_{0+}^{\infty} (1 - e^{-\alpha u}) \frac{1}{\sqrt{2\pi u^3}}\, du = \sqrt{2\alpha}$$

[see Borodin and Salminen (1996)] one can derive $m(\alpha, x) = \sqrt{\frac{2}{\alpha(x)}}$. For a homogeneous Brownian motion with a parameter of local variance D and local drift μ we have an one-dimensional density

$$p_t(x) = \frac{1}{\sqrt{2\pi D t}} \exp\left(-\frac{(x-\mu t)^2}{2Dt}\right),$$

and also [see Borodin and Salminen (1996)]

$$\begin{aligned} E_0 e^{-\lambda \sigma_b} &= \exp\left(-b\left(\sqrt{2\frac{\lambda}{D} + \mu^2} - \mu\right)\right), \\ m(\alpha, x) &= \frac{1}{\alpha(x)}\left(\sqrt{2\frac{\alpha(x)}{D} + \mu^2} - \mu\right). \end{aligned}$$

19.4.5 Inversed Gamma-process

Gamma process is a process with independent positive increments whose distributions are gamma-distributions with corresponding parameters. An inverse gamma-process is not a process of maximal values but it can be used for description of the hazard rate process. The gamma-distribution is used for description of a random failure time in the work [Gerzbach and Kordonski (1966)] for a constant level of abrasion. For the inverse gamma-process

$$Q(\sigma_x \in dt) = \frac{1}{\Gamma(x)}\, h^x t^{x-1} e^{-ht} dt \;(x > 0),$$

where $h > 0$ is a parameter of the process. And also

$$Q(\exp(-\lambda\sigma_x)) = \frac{h^x}{(h+\lambda)^x}\,.$$

From this formula one can obtain parameters of the Levi-Khinchin expansion

$$\rho = 0, \quad \nu(du) = \frac{e^{-hu}}{u}\, du.$$

And also

$$m(\alpha, x) = \frac{1}{\alpha(x)} \log\left(1 + \frac{\alpha(x)}{h}\right).$$

The main advantage of the inverse gamma-process comparatively with other models of changing of a diagnostic parameter is its simplicity, sufficient flexibility due to its two parameters, and also its well known properties which one can find in books and tables with values of the gamma-distributions [see Pagurova (1963) and others].

References

1. Borodin, A. N. and Salminen, P. (1996). *Handbook of Brownian Motion—Facts and Formulae*, Berlin: Birkhäuser Verlag.

2. Dellacherie, C. (1972). *Capacités et processus stochastiques*, New York: Springer-Verlag.

3. Dynkin, E. B. (1963). *Markov Processes*, Moskow: Fizmatgiz (in Russian).

4. Gerzbach, I. B. and Kordonski, H. B. (1966). *Models of Failure*, Moskow: Nauka (in Russian).

5. Harlamov, B. P. (1983). Transition functions of continuous semi-Markov process on the line, *Zapiski nauch. semin. LOMI*, **130**, 190–205 (in Russian).

6. Harlamov, B. P. (1985). Distribution of time move with respect to sequence of states in semi-Markov process, *Zapiski nauch. semin. LOMI*, **142**, 167–173 (in Russian).

7. Itô, K. and McKean Jr., H. P. (1965). *Diffusion Processes and Their Sample Paths*, New York: Springer-Verlag.

8. Korolyuk, V. S. and Turbin, A. F. (1976) *Semi-Markov Processes and Their Applications*, Kiev: Naukova Dumka (in Russian).

9. Pagurova, V. N. (1963). *Tables of Non-complete Gamma-function*, Moskow: Nauka (in Russian).

10. Vinogradov, V. N., Sorokin, G. M. and Kolokol'nikov, M. G. (1990). *Abrasion*, Moskow: Nauka (in Russian).

20

Exact Methods to Compute Network Reliability

Corinne Lucet and Jean-François Manouvrier

Université de Technologie de Compiègne, Compiègne, France

Abstract: In this chapter we present and compare some exact methods to resolve network reliability problems. These problems concern all kinds of networks, such as computer, communication or power networks. When components of the network are subject to random failures, the network may or may not continue functioning after the failures of some components. The probability that the network will function is its reliability. Networks are modeled by a graph $G = (V, E)$ composed of elements that fail independently one another with known probabilities. The K-terminal reliability problem has been studied extensively. It consists in evaluating the probability that a given subset of vertices, denoted K, is connected. This problem is NP-hard. We propose here to present the main methods, developed since the 1970s. We first consider the enumeration methods using elementary states, paths or cuts. Then we explain the factoring method performed by reductions. These allow to handle series-parallel graphs to be processed in linear time. Finally, we present the decomposition method implemented as a table-based reduction algorithm allowing to resolve the reliability problem in linear time for graphs with bounded pathwidth.

Keywords and phrases: Reliability, network, enumeration, reduction, factoring, decomposition, pathwidth

20.1 Introduction

There are a lot of works in literature devoted to the problem of network reliability. For general networks, the K-terminal reliability problem is NP-hard [Ball (1980)]. Network reliability has been approached and resolved with different methods in the literature. The methods that provide an exact reliability are called exact methods, in opposition to simulation methods that provide an approximate result. The aim of this chapter is to present and compare some exact methods to resolve network reliability problems.

The methods of long standing are the enumeration of states and the enumeration of paths or cuts. Their implementations are light but their complexities grow exponentially with the size of the network. Nevertheless, there exist some works about the probability computation improvement [Abraham (1979), Jain and Gopal (1988), Locks (1978) and Rai, Veeraraghavan and Trivedi (1995)]. More recent methods such as factoring, reduction and decomposition, take the network topology into consideration. These are generally more efficient than enumeration. For instance, reductions provide a result in linear time for special graph families such as series-parallel graphs. But for general graphs, reduction and factoring must be combined. The complexity of such methods is directly dependent on the graph topology. So research works about reductions tend to define new reductions in order to handle larger graph families [Gadani (1981), Satyanarayana and Chang (1983), Theologou and Carlier (1991) and Wood (1985)]. The decomposition presented here belongs to a special algorithm class, called table-based reduction algorithms. It is based on the dynamic programming. We have shown that the network reliability problems can be efficiently solved by such a method for graphs with bounded pathwidth [Carlier and Lucet (1996) and Lucet (1993)].

This chapter is organized as follows. First, in Section 20.2, we describe the problems and give some definitions. Then in Section 20.3, we present the enumeration methods which are state enumeration and path or cut enumeration. Next in Section 20.4, we explain the factoring method performed by the reduction method, which is linear for series-parallel graphs. Then in Section 20.5, we present the decomposition method, that allows to compute reliability in linear time for graphs with bounded pathwidth. Finally, in Section 20.6, we compare the efficiency of these methods.

20.2 Definitions and Notation

Notation

$G = (V, E)$	a graph with V, the set of vertices in G, and $E \subset V \times V$, the set of arcs (or edges) in G.
K	a subset of vertices of G, $K \subset V$. A vertex that belongs to K is called a K-vertex.
p_e and q_e	p_e : the reliability of the component e ($e \in V$ or $e \in E$). q_e=1-p_e.

x_e	a Boolean variable for the state of the component $e \in G$: e functions ($x_e = 1$) or not ($x_e = 0$).
$\mathbf{R}(G)$	reliability of G.
$\mathcal{G}_i$	a state of G.
$G(\mathcal{G}_i)$	partial graph of G associated with $\mathcal{G}_i$.
$\mathbf{E}(X)$	Probability for the Boolean expression X to be true. X will be a Bernouilli variable, so $\mathbf{E}(X) = \Pr(X = 1)$.
$\overline{x}$, $+$, $.$	the Boolean operators "not", "or" and "and".
$\lvert K\rvert$, $\lvert V\rvert$, $\lvert E\rvert$	denote the cardinality of the sets K, V, E.

A stochastic graph

Our network model is a graph $G = (V, E)$. This graph is directed if E is a set of ordered pairs, i.e. a set of arcs, or undirected if E is a set of unordered pairs, i.e. a set of edges. The graph is said to be stochastic because each element (vertices, arcs or edges) can fail, statistically independently, with known probability. The failure probability of the element e is denoted q_e ($q_e \in [0,1]$) and its reliability is $p_e = 1 - q_e$.

Subgraph and partial graph

A subgraph of a given graph $G = (V, E)$ is a graph $G' = (V', E')$ such that $V' \subset V$ and $E' = (V' \times V') \cap E$. A partial graph of a given graph $G = (V, E)$ is a graph $G'' = (V, E'')$ such that $E'' \subset E$.

Graph state, associated probability and associated partial graph

Each element of the stochastic graph is subject to failure. As there are two states for an element (the element functions or fails), there are $2^{|V|+|E|}$ possible states for the network. One state $\mathcal{G}_i$ of the stochastic graph $G = (V, E)$ is denoted $< x_1, x_2, ..., x_{|V|+|E|} >$ where the Boolean x_e stands for the state of element e, i.e., $x_e = 0$ when element e fails and $x_e = 1$ The associated probability of $\mathcal{G}_i$ is:

$$\Pr(\mathcal{G}_i) = \prod_{e \in G} [x_e \cdot p_e + (1 - x_e) \cdot q_e]$$

With each state $\mathcal{G}_i$ of $G = (V, E)$ is associated a partial graph $G(\mathcal{G}_i) = (V'', E'')$, such that:

$u \in V'' \Leftrightarrow (u \in V \text{ and } x_u = 1)$,
and $(u, v) \in E'' \Leftrightarrow ((u, v) \in E,\ u \in V, v \in V, \text{ and } x_{(u,v)} = x_u = x_v = 1)$.

In Section 20.5, we will also consider states and partial graphs of some subgraphs H and L of G, a stochastic graph with perfect vertices. $\mathcal{H}_i = \langle x'_1, x'_2, ..., x'_{|E'|}\rangle$ denotes one state of $H = (V', E')$ and $H(\mathcal{H}_i)$ its associated partial graph.

Reliability

The reliability of a network $G = (V, E)$ is the probability that G supports a given operation. We distinguish three kinds of operation and hence three kinds of reliability: 2-terminal reliability (also called terminal-pair reliability), K-terminal reliability and all-terminal reliability. When the operation requires that each pair of vertices is able to communicate via at least one operational path, this is all-terminal reliability. When the operation requires only a few vertices, a subset K of V, to communicate each other, this is K-terminal reliability. 2-terminal reliability is the probability that two given vertices, called the source and the sink, can communicate. Note that 2-terminal and all-terminal reliability are particular cases of K-terminal reliability. With a directed stochastic graph, the reliability can consider a single source, and the other vertices of K are the sinks. In the literature, the usual stochastic graphs are undirected, often with perfect vertices, i.e. each vertex has a reliability that is equal to one.

20.3 Enumeration

Among the enumeration methods, we distinguish between the enumeration of states and the enumeration of paths or cuts.

20.3.1 State enumeration

A very basic method to compute reliability consists in enumerating all the possible states of the stochastic graph and keeping those that allow the network to function or to keep all the failure states, if they are less numerous.

$$\mathbf{R}(G) = \sum_{G(\mathcal{G}_i) \text{ functions}} \Pr(\mathcal{G}_i) = 1 - \sum_{G(\mathcal{G}_i) \text{ fails}} \Pr(\mathcal{G}_i)$$

Because of the exponential number of possible states, this formula needs to be improved. Instead of enumerating all the states of the network, a method involves classifying them in a rooted binary tree. The nodes of this binary tree are partial graphs associated with a set of states, the root stands for the graph with every element that functions, and the two branches below a node correspond to a state of a given element: whether it functions or not. So the height of this binary tree is $|V| + |E|$ when vertices and edges (or arcs) can fail, and the leaves of the tree are the states of the graph. With this method we look for the failure states of the graph. If a node stands for a failure network, all the states below this node are failure states, and as we know the probability of this set of failure states, we use it to reduce the number of terms in the sum of the previous given reliability formula. Note that a similar method can be used by looking for the operating states; in that case, the root of the binary tree is the network with all element that fail.

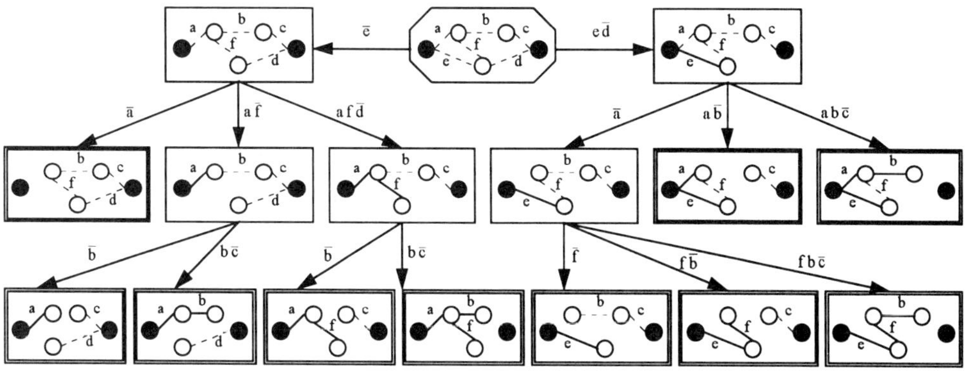

$$\mathbf{R}(G) = 1 - (\, q_e q_a + q_e p_a q_f q_b + q_e p_a q_f p_b q_c + q_e p_a p_f q_d q_b + q_e p_a p_f q_d p_b q_c + p_e q_d q_a q_f + p_e q_d q_a p_f q_b + p_e q_d q_a p_f p_b q_c + p_e q_d p_a q_b + p_e q_d p_a p_b q_c)$$

Figure 20.1: State factorization with the path method. Search for the invalid state sets. 2-terminal reliability

So this method looks for the failure states by considering the failure of each element one by one in an arbitrary order. In order to find these states faster, we can choose an efficient order: for 2-terminal reliability, the classical technique is to find at each step a path from the source to the sink, and to enumerate the sets of states that prevent this path from functioning [Dotson and Gobien (1979) and Hasanuddin and Jamil (1987)]. (Figure 20.1)

20.3.2 Path enumeration – Cut enumeration

These methods are composed of two steps. First the enumeration of the paths or the enumeration of the cuts provide a Boolean expression, required for the second step, which is the computation of this Boolean expression probability. There are two ways to do this second step: the inclusion-exclusion method and the technique of the sum of disjoint products.

Path enumeration

Another method to compute the network reliability is the enumeration of the minimal paths that provide a working network. The reliability is the probability for the network to have at least a functioning minimal path. These minimal paths are different for each reliability problem. For 2-terminal reliability, a minimal path is a path from the source to the sink that does not contain another such path. For K-terminal reliability, the minimal paths are Steiner trees and, for all-terminal reliability, they are spanning trees.

For 2-terminal reliability, there are several ways to obtain the minimal paths between two vertices [31]. We present here those which use the connection matrix of size $|V| \times |V|$ (or adjacency matrix), denoted A, with the element $a_{ij} = x_{ij}$ if $(i,j) \in E$, where x_{ij} is a Boolean variable stand for the arc (or edge) (i,j), else $a_{ij} = 0$. The matrix $B = A^k$ is composed of elements b_{ij}. b_{ij} is a Boolean expression whose terms represent all the paths from vertex i to vertex j composed of k arcs (or edges). So we can compute all the matrices A^k with $k \leq |V|$ to obtain all minimal paths [Misra and Rao (1970)]. A more efficient method is in [Rai and Aggarwal (1978)].

For K-terminal reliability with $|K| > 2$, we cite [Jain and Gopal (1988) and Piekarski (1965)] for the search of Steiner trees.

After the enumeration step, a sum is obtained, that is a Boolean expression denoted $\Phi(G)$, whose terms are all minimal paths, and each term is the product of the Boolean variable associated with the elements that composed the path: $\Phi(G) = \sum_i P_i$ with $P_i = \prod_k x_{ik}$. An example of $\Phi(G)$ is on Figure 20.2.

This Boolean expression will be used to compute the reliability in the two final sections of 20.3.2: $\mathbf{R}(G) = \mathbf{E}(\Phi(G))$.

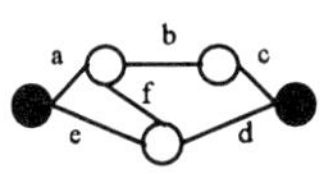

The four paths are {a,b,c}, {a,f,d}, {e,d} and {e,f,b,c}.
$\Phi(G) = x_a x_b x_c + x_a x_f x_d + x_e x_d + x_e x_f x_b x_c$
The six cuts are {a,e}, {a,d,f}, {b,e,f}, {b,d}, {c,d} and {c,e,f}.
$\Phi_c(G) = \bar{x}_a \bar{x}_e + \bar{x}_a \bar{x}_d \bar{x}_f + \bar{x}_b \bar{x}_e \bar{x}_f + \bar{x}_b \bar{x}_d + \bar{x}_c \bar{x}_d + \bar{x}_c \bar{x}_e \bar{x}_f$

Figure 20.2: Path (and cut) enumeration. 2-terminal reliability

Cut enumeration

A cut is a set of edges whose failures produce network failure. A minimal cut is a cut that does not include another cut. Another Boolean expression can be obtained with the enumeration of the minimal cuts. This Boolean expression, $\Phi_c(G)$, is the complement of $\Phi(G)$. So, in the Boolean algebra, $\Phi_c(G) = \overline{\Phi(G)}$.

An example is on Figure 20.2. This inversion principle is discussed in [Locks (1978)].

Once $\Phi_c(G)$ is known, the reliability is computed by: $\mathbf{R}(G) = 1 - \mathbf{E}(\Phi_c(G))$.

Cut enumeration is essentially used for 2-terminal reliability. A modified connection matrix can be used as in the previous section [Martelli (1976) and Shier and Whited (1986)]. Other means of enumerating all minimal cuts are in [Hasanuddin (1988), Jasmon and Kai (1985), Tsukiyama, Shirakawa and Ozaki (1980) and Yan, Taha and Landers (1994)].

Inclusion-exclusion formula

One method of transforming a Boolean expression such as $\Phi(G)$ or $\Phi_c(G)$ into a probability expression is to use Poincaré's theorem, also called the inclusion-exclusion method [Buzacott and Chang (1984), Buzacott (1987) and Locks

(1985)]. Let us consider an example with two minimal paths: $\Phi(G) = P_1 + P_2$ so $\mathbf{E}(\Phi(G)) = \mathbf{E}(P_1+P_2) = \mathbf{E}(P_1)+\mathbf{E}(P_2)-\mathbf{E}(P_1 \cdot P_2)$. The inclusion-exclusion is a generalization of this principle.

Here is Poincaré's formula for m minimal paths:

$$\begin{aligned}\mathbf{E}(\Phi(G)) &= \sum_{1 \leq i \leq m} \mathbf{E}(P_i) - \sum_{1 \leq i_1 < i_2 \leq m} \mathbf{E}(P_{i_1} \cdot P_{i_2}) + \cdots \\ &+ (-1)^{m+1} \cdot \mathbf{E}(P_1 \cdot P_2 \cdot P_3 \cdots P_m)\end{aligned}$$

This formula is less efficient than the following one in next section, and it is essentially used for an approximation of the reliability by keeping only the first terms of the sum.

Sum of disjoint products

Another method of evaluating $\mathbf{E}(\Phi(G))$ (or $\mathbf{E}(\Phi_c(G))$) consists in developing the Boolean expression so that each term is an event that does not include another event of the sum, i.e., all the terms are disjoint. This is done by using the following Boolean formula: $\Phi(G) = \sum_{i=1}^{m}(P_i) = P_1 + \sum_{i=2}^{m}[P_i \cdot \prod_{j=1}^{i-1}(\overline{P}_j)]$. Once all the terms of the Boolean sum are disjoint, the probability can be easily deduced: $\mathbf{E}(\Phi(G)) = \mathbf{E}(P_1) + \sum_{i=2}^{m} \mathbf{E}[P_i \cdot \prod_{j=1}^{i-1}(\overline{P}_j)]$.

The disjunction must be applied for the elementary variables of the sum. For the development of $\Phi(G)$, the following formula is required: if $P_i = \prod_{j=1}^{p} x_{ij}$ then $\overline{P}_i = \overline{x}_{i1} + \sum_{j=2}^{p}[\overline{x}_{ij} \cdot \prod_{k=1}^{j-1}(x_{ik})]$.

Different ways to compute this sum of disjoint products can simplify the calculation [Abraham (1979), Beichelt, Spross (1987), Locks (1987) and Rai, Veeraraghavan and Trivedi (1995)]. The most popular one is the Abraham method.

20.4 Reduction With Factoring

A reduction is a topological method that can be applied to stochastic undirected graphs [Satyanarayana and Chang (1983), Theologou (1990) and Wood (1985)]. We first suppose that the vertices are perfect. In fact, we will consider graphs with imperfect vertices later. The reduction principle consists in reducing the size of the network G to get a new graph G', such that $\mathbf{R}(G) = \Omega \cdot \mathbf{R}(G')$. For this, a part of the network, with a specific topology, is replaced by another one with new edges whose failure probabilities depend on the previous edge probabilities. The reductions have to be applied until the obtained graph G' cannot be reduced. There are three kinds of reductions, series-parallel reductions (Table 20.1), polygon-to-chain reductions [Choi and Jun (1995)] (Table 20.2) and delta-star reductions [Gadani (1981)] (star-delta reductions are the

Table 20.1: The three series-parallel reductions

● = a K-vertex ○ = any vertex ○ = a vertex not in K

Original form of the network	Substituted elements provided by the reduction	Reduction formulas Formulas for the new probabilities and reliability	Reduction type
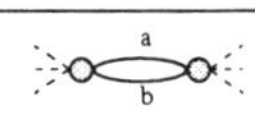	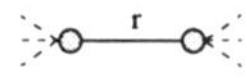	$p_r = 1 - q_a q_b$ $\Omega = 1$	Parallel reduction
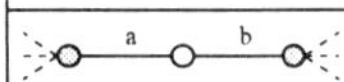	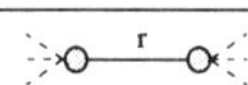	$p_r = p_a p_b$ $\Omega = 1$	Series reduction
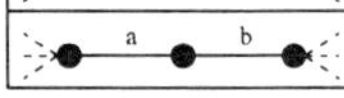	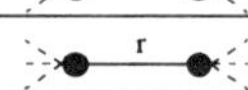	$p_r = p_a p_b / (1 - q_a q_b)$ $\Omega = 1 - q_a q_b$ $\lvert K\rvert \leftarrow \lvert K\rvert - 1$	Series reduction

Table 20.2: The seven polygon-to-chain reductions

Original form of the network	Substituted elements provided by reduction	Reduction formulas	Formulas for the new probabilities and reliability
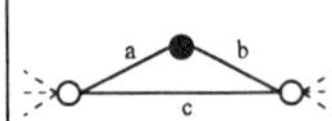		$\alpha = q_a p_b q_c$, $\beta = p_a q_b q_c$ $\gamma = p_a p_b p_c \left(1 + \frac{q_a}{p_a} + \frac{q_b}{p_b} + \frac{q_c}{p_c}\right)$	$p_r = \frac{\delta}{\alpha + \delta}$ $p_s = \frac{\delta}{\beta + \delta}$ $\Omega = \frac{(\alpha+\delta)(\beta+\delta)}{\delta}$
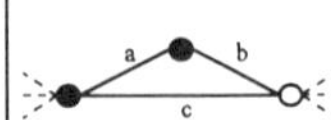	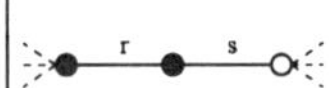	$\alpha = q_a p_b q_c$, $\beta = p_a q_b q_c$ $\gamma = p_a p_b p_c \left(1 + \frac{q_a}{p_a} + \frac{q_b}{p_b} + \frac{q_c}{p_c}\right)$	
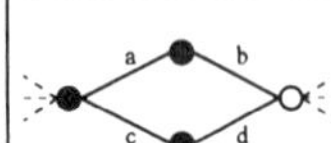		$\alpha = p_a q_b q_c p_d + q_a p_b p_c q_d + q_a p_b q_c p_d$, $\beta = p_a q_b p_c q_d$ $\gamma = p_a p_b p_c p_d \left(1 + \frac{q_a}{p_a} + \frac{q_b}{p_b} + \frac{q_c}{p_c} + \frac{q_d}{p_d}\right)$, $\lvert K\rvert \leftarrow \lvert K\rvert - 1$	
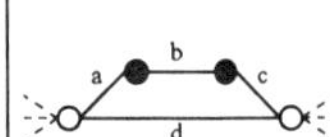	if $\lvert K\rvert = 2$ then 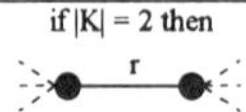 else 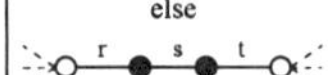	$\alpha = q_a p_b p_c q_d$, $\beta = p_a q_b p_c q_d$, $\delta = p_a p_b q_c q_d$ $\gamma = p_a p_b p_c p_d \left(1 + \frac{q_a}{p_a} + \frac{q_b}{p_b} + \frac{q_c}{p_c} + \frac{q_d}{p_d}\right)$	if $\lvert K\rvert=2$ then $\Omega = p_b + p_a q_b p_c$ $p_r = (p_b + p_a q_b p_c p_d) / \Omega$ else $p_r = \gamma / (\alpha + \gamma)$ $p_s = \gamma / (\beta + \gamma)$, $p_t = \gamma / (\delta + \gamma)$ $\Omega = (\alpha + \gamma)(\beta + \gamma)(\delta + \gamma) / \gamma$

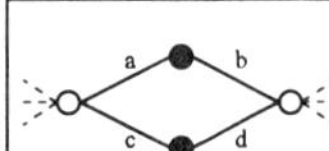	if $\lvert K\rvert = 2$ then 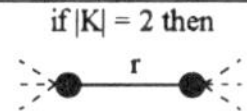
	else 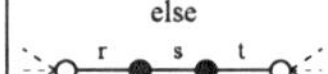
		$\alpha = q_a p_b q_c p_d$
		$\beta = p_a q_b q_c p_d + q_a p_b p_c q_d$
		$\delta = p_a q_b p_c q_d$
		$\gamma = p_a p_b p_c p_d \left(1 + \frac{q_a}{p_a} + \frac{q_b}{p_b} + \frac{q_c}{p_c} + \frac{q_d}{p_d}\right)$
			if $\lvert K\rvert = 2$ then
			$\Omega = \alpha + \beta + \delta + \gamma$
			$p_r = (\alpha + \delta + \gamma) / \Omega$
			else
			$p_r = \frac{\gamma}{\alpha + \gamma}$
			$p_s = \frac{\gamma}{\beta + \gamma}$
			$p_t = \frac{\gamma}{\delta + \gamma}$
			$\Omega = \frac{(\alpha+\gamma)(\beta+\gamma)(\delta+\gamma)}{\gamma}$

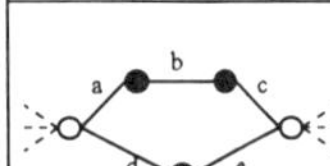	if $\lvert K\rvert = 3$ then 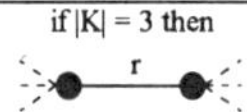 else 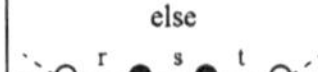	$\alpha = q_a p_b p_c q_d q_e$ $\beta = p_a q_b p_c (p_d q_e + q_d p_e) + p_b (q_a p_c p_d q_e + p_a q_c q_d p_e)$ $\delta = p_a p_b q_c p_d q_e$ $\gamma = p_a p_b p_c p_d p_e \left(1 + \frac{q_a}{p_a} + \frac{q_b}{p_b} + \frac{q_c}{p_c} + \frac{q_d}{p_d} + \frac{q_e}{p_e}\right)$ $\lvert K\rvert \leftarrow \lvert K\rvert - 1$	
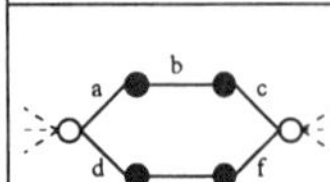	if $\lvert K\rvert = 4$ then 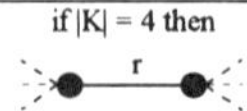 else 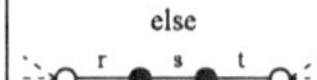	$\alpha = q_a p_b p_c q_d p_e p_f$, $\delta = p_a p_b q_c p_d p_e q_f$ $\beta = p_a q_b p_c (p_d q_e p_f + q_d p_e p_f + p_d p_e q_f)$ $+ p_a p_b q_c p_f (p_d q_e + q_d p_e) + p_a p_b q_c p_d (p_e q_f + q_e p_f)$ $\gamma = p_a p_b p_c p_d p_e p_f \left(1 + \frac{q_a}{p_a} + \frac{q_b}{p_b} + \frac{q_c}{p_c} + \frac{q_d}{p_d} + \frac{q_e}{p_e} + \frac{q_f}{p_f}\right)$ $\lvert K\rvert \leftarrow \lvert K\rvert - 2$	

Table 20.3: Delta-star reductions

Original form of the network	New elements provided by the reduction	Reduction formulas Formulas for the new probabilities and reliability
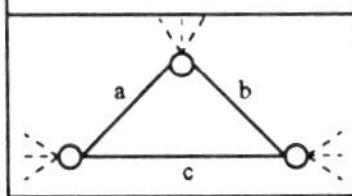	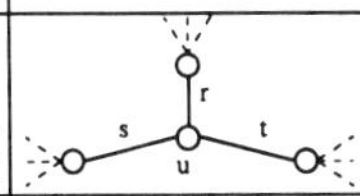	$\alpha = p_a p_b + p_a p_c + p_b p_c - 2\, p_a p_b p_c$ $\beta_1 = q_a q_b p_c \;,\; \beta_2 = q_a p_b q_c \;,\; \beta_3 = p_a q_b q_c$
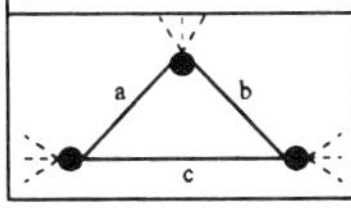	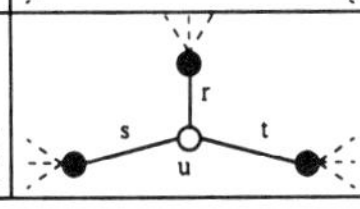	$p_r = \dfrac{\alpha}{\alpha + \beta_1} \;,\; p_s = \dfrac{\alpha}{\alpha + \beta_2} \;,\; p_t = \dfrac{\alpha}{\alpha + \beta_3} \;,\; p_u = \dfrac{(\alpha + \beta_1)(\alpha + \beta_2)(\alpha + \beta_3)}{\alpha^2}$ $\Omega = 1$

reverse reductions) (Table 20.3). These reduction methods are very effective for certain classes of graphs, like series-parallel graphs, which are reduced to a tree with series-parallel reductions. In this way, the reliability can be computed in linear time.

As it is not possible to reduce all graphs totally with the reduction method, it is often combined with the factoring method [Satyanarayana and Chang (1983), Theologou (1990), and Wood (1985)]. The factoring formula divides the reliability problem into two sub-problems:

$$\mathbf{R}(G) = p_e \cdot \mathbf{R}(G \;/\; \text{ edge } e \text{ functions}) + (1 - p_e) \cdot \mathbf{R}(G \;/\; \text{ edge } e \text{ fails})$$

The factoring formula must be applied if and only if no reduction on the graph is possible. Then, some reductions could be applied to the induced sub-problems.

Let us now consider stochastic graphs with imperfect vertices. All reductions can be adapted in case of imperfect vertices [Lai (1994)]. The factoring formula is still valid if we consider that e is composed of an edge and its two incident vertices, but the failure probabilities of these vertices are modified for the two graphs of the sub-problem [Theologou and Carlier (1991)].

20.5 Decomposition

The decomposition method can solve the reliability problem for some classes of stochastic undirected graphs in linear time [Carlier and Lucet (1996)]. First we will consider all-terminal reliability problem with perfect vertices in Sections 20.5.1, 20.5.2 and 20.5.3. The other problems will be studied in Section 20.5.4.

20.5.1 The principle

The decomposition principle for reliability problems was introduced by Rosenthal in 1977 [Rosenthal (1977)]. It generalizes the basic decomposition principle of a graph G into two subgraphs H and L separated by an articulation vertex

(Figure 20.3a), which allows the all-terminal reliability to be computed by the formula: $\mathbf{R}(G) = \mathbf{R}(H) \cdot \mathbf{R}(L)$.

$$\mathrm{H} \cap \mathrm{L} = \mathrm{F}\ ,\ \mathrm{H} \cup \mathrm{L} = \mathrm{G}$$

a> an articulation point: |F|=1 b> a general boundary set: |F|=3

Figure 20.3: Decomposition principle

The decomposition method consists in considering two subgraphs H and L, separated by F, the vertex boundary set, that is a separator, i.e., its removal leaves G disconnected (Figure 20.3b). Both subgraphs, H and L, have their set of states, respectively $\{\mathcal{H}_i\}$ and $\{\mathcal{L}_j\}$, so that the reliability can be written as the formula:

$$\mathbf{R}(G) = \sum_{\mathcal{H}_i, \mathcal{L}_j /\ H(\mathcal{H}_i) \cup L(\mathcal{L}_j) \text{ is connected}} \Pr(\mathcal{H}_i) \cdot \Pr(\mathcal{L}_j)$$

To use this formula, it is necessary to combine two by two all the states of H and L. To reduce the corresponding number of combinations, first we can eliminate the failure states $\mathcal{H}_i$, i.e., the states such that there exists a vertex in $H(\mathcal{H}_i)$ disconnected from the boundary set F, so G cannot be connected. Secondly, we remark that the functioning of G depends only on how the boundary vertices are connected, via H and via L, so we gather together all equivalent operating states in a same class: two states are equivalent if they provide the same boundary vertex connection (Figure 20.4). We denote $C_{H,k}$ the k-th class of H. For all-terminal reliability, these classes are the partitions of the boundary set F, where blocks stand for the connected components. For instance, the different classes of H for a boundary set of three vertices u, v and w are the following:

- $C_{H,1} = [uvw]$: u, v and w are connected together via H.
- $C_{H,2} = [uv][w]$: u and v are both connected via H, and w is isolated from the other two.
- $C_{H,3} = [uw][v]$: u and w are both connected via H, and v is isolated from the other two.
- $C_{H,4} = [u][vw]$: v and w are both connected via H, and u is isolated from the other two.
- $C_{H,5} = [u][v][w]$: u, v and w are disconnected via H.

$$\Pr(C_{H,k}) = \sum_{\mathcal{H}_i /\ \mathcal{H}_i \textit{ belongs to } C_{H,x}} \Pr(\mathcal{H}_i)$$

As each class has an associated probability, the previous reliability formula is factored:

$$\mathbf{R}(G) = \sum_{C_{H,x},\ C_{L,y}/C_{H,x} \text{ and } C_{L,y} \text{ are compatible}} \Pr(C_{H,x}) \cdot \Pr(C_{L,y})$$

Two classes $C_{H,x}$ and $C_{L,y}$ are compatible if the connectivity of the boundary set given by $C_{H,x}$ and the connectivity of the boundary set given by $C_{L,y}$ provides the connectivity of the whole graph G. The number of classes is function of $|F|$, as is the efficiency of this method.

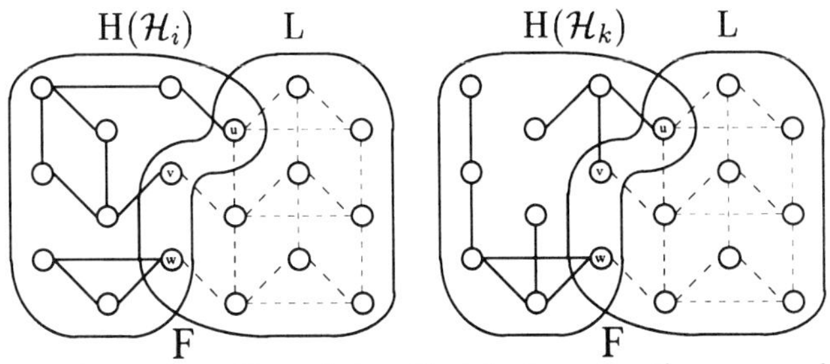

$\mathcal{H}_i$ and $\mathcal{H}_k$ are two operating states that belong to the same class: [uv][w]

Figure 20.4: Equivalent states for all-terminal reliability

20.5.2 Algorithm implementation

We have seen above the decomposition principle which consists in considering two subgraphs H and L of G, and in combining their classes to compute the reliability. Nevertheless, the subgraphs H and L must be solved, i.e., all their class probabilities must be known. Another approach consists in considering only the subgraph H, such that it grows to encompass G. With this end in view, we will describe in this section an algorithm [Lucet (1993)], belonging to the table-based reduction algorithm family [Arnborg (1985)]. At each step of the algorithm, we consider a resolved subgraph whose classes are known and associated probabilities stored in a table, and we enlarge this resolved subgraph by vertex insertions until we have resolved the whole graph.

Vertices are numbered from 1 to $|V|$ by the bijection $\pi : \{1, 2, ..., |V|\} \rightarrow V$. We denote $H^{(k)}$ the subgraph induced by the subset $\{\pi(1), \pi(2), ..., \pi(k)\}$. There are $|V|$ subgraphs $H^{(1)}, H^{(2)}, ..., H^{(|V|)}$ and $|V|$ corresponding boundary sets $F^{(1)}, F^{(2)}, ..., F^{(|V|)}$. For each iteration k, there will be classes and associated probabilities to compute. At the k-th iteration, the vertex $\pi(k)$ and its incident edges are added to $H^{(k-1)}$ in order to build $H^{(k)}$ and $F^{(k)}$. To the subgraph $H^{(|V|)}$, there corresponds only one class denoted $C_{H(|V|)),1}$, whose associated probability is the network reliability. The method consists in computing iteratively the class probabilities of $H^{(1)}, H^{(2)}, ..., H^{(|V|)}$. Hence, $\mathbf{R}(G) = \Pr(C_{H(|V|)),1})$ is reached in $|V| - 1$ iterations.

Algorithm:

Number the vertices of V.
$H^{(1)} \leftarrow (\{\pi(1)\},\{\})$.
$F^{(1)} \leftarrow \pi(1)$.
$\Pr(C_{H^{(1)},1}) \leftarrow 1$.
For $k \leftarrow 2$ to $|V|$, do
    Add π(k) and the α edges $(\pi(k), \pi(h))$ to $H^{(k-1)}$
        such that $h < k$ and $(\pi(k), \pi(h)) \in E$ to obtain $H^{(k)}$.
    Compute $F^{(k)}$.
    For all classes $C_{H^{(k)},y}$ of $H^{(k)}$, do $\Pr(C_{H^{(k)},y}) \leftarrow 0$.
    For all classes $C_{H^{(k-1)},x}$ of $H^{(k-1)}$ with $\Pr(C_{H^{(k-1)},x}) \neq 0$, do
        For the 2^α possible states of connection s_m between
            $\pi(k)$ and $F^{(k-1)}$, do
            Let $\Pr(s_m)$ be the associated probability of s_m.
            Compute the new class $C_{H^{(k)},y}$ of $H^{(k)}$ produced
                by $C_{H^{(k-1)},x}$ and s_m.
            If $C_{H^{(k)},y}$ is a valid class, then
                $\Pr(C_{H^{(k)},y}) \leftarrow \Pr(C_{H^{(k)},y}) + \Pr(C_{H^{(k-1)},x}).\Pr(s_m)$.
$\mathbf{R}(G) \leftarrow \Pr(C_{H^{(|V|)},1})$.

The main difficulty in implementing the algorithm is managing the classes in memory, i.e. their representations and their accesses of probability data. For more, see Carlier and Lucet (1996).

20.5.3 Complexity

The algorithm is efficient only if the sizes of the boundary sets $F^{(k)}$ encountered during the algorithm are not too large. The largest size encountered is called the vertex separation number of the linear ordering, which is equal to the pathwidth of the corresponding path decomposition [Kinnersley (1992)] and will here be denoted F_{max}. The complexity of this algorithm is exponential in relation to the number of classes, and this number of classes (denoted $NC(F_{max})$) grows exponentially with F_{max}. The complexity is upper bounded by $|V| \cdot NC(F_{max}) \cdot 2^{F_{max}} \cdot (F_{max})^2$. With such a complexity, our algorithm is linear for a bounded F_{max}. For more efficiency, a similar algorithm could be implemented using a tree decomposition instead of a path decomposition.

So we are confronted with the problem of finding an optimum linear ordering that gives a minimal F_{max}. These problems, the pathwidth and treewidth problems (given a graph, find a tree decomposition or a path decomposition with the smallest width) are NP-hard [Arnborg, Corneil and Proskurowski (1987)]. Nevertheless, linear solutions exist for graphs with bounded treewidth [Bodlaender (1996)].

The decomposition algorithm can solve the network reliability problem, which is NP-hard, in linear time for a graph with a bounded vertex separation number. It has been proved that some classes of graph problems can be solved in polynomial (or linear time) with such dynamic algorithms using a tree decomposition or a path decomposition with a bounded width [Bodlaender (1988)]. The principle of these algorithms is to use the graph tree or path topology in order to expand a resolved subgraph, until the whole graph is resolved, and to store in memory all partial solutions standing for the resolved elements, i.e., the information required to compute the final solution.

20.5.4 Adaptation for other reliability problems [Lucet (1993)]

This decomposition method can easily be adapted for the general K-terminal reliability problem by considering appropriate classes: These classes are always partitions of F, but a Boolean for their blocks stands for the presence of a K-vertex in the connected component. As an example, for $F = \{u, v\}$, the classes are:

- $[uv]$, $[uv]^k$, $[u][v]$, $[u]^k[v]$, $[u][v]^k$, $[u]^k[v]^k$ (k means that the block contains some K-vertices)

If we consider the reliability with imperfect vertices, then the classes are modified so that a boundary vertex can fail. For $F = \{u, v\}$, the classes are, for the K-terminal reliability problem with imperfect vertices:

- $[uv]$, $[uv]^k$, $[u][v]$, $[u]^k[v]$, $[u][v]^k$, $[u]^k\,[v]^k$, $[u]$, $[u]^k$, $[v]$, $[v]^k$, $[]$

20.6 Conclusion

The complexity of the enumeration methods is $O(|V| \cdot 2^{|V|+|E|})$. Such methods can only handle small size networks. Path or cut enumeration are more commonly employed for 2-terminal reliability. But these enumeration methods cannot be efficient for large networks, i.e., graphs with more than 20 vertices.

A measure of the complexity of a factoring algorithm with reduction is the number of leaves in the computational tree generated by the algorithm. For the factoring using series-parallel reductions, this number depends on the domination of G, denoted $D(G_K)$ [Satyanarayana and Chang (1983)]. For K-terminal reliability on a complete graph, $D(G_K) = (|K| - 1) \cdot (|V| - 2)!$. Satyanarayana and Wood have shown that the factoring algorithm using reductions is more effective than the classical path or cut enumeration methods in [Satyanarayana and Chang (1983)] and [Wood (1985)]. This is confirmed by the experimental works of Theologou and Carlier (1990). But, its running time remains prohibitive for large networks, i.e., the considered networks did not have more than 30 nodes.

The decomposition method allows the reliability for a bounded pathwidth graph to be computed in linear time. Carlier and Lucet have worked out and tested this method for the K-terminal reliability problem with imperfect edges and vertices [Lucet (1993)]. Their results showed that it is more efficient than factoring using reductions and that it can handle real size networks, i.e., the single constraint to apply the decomposition method is the pathwidth, and not the graph size. For example, it computes the reliability of graphs with 140 nodes and a pathwidth of 6 in 6 seconds.

References

1. Abraham, J. A. (1979). An improved algorithm for network reliability, *IEEE Transactions on Reliability*, **28**, 58–61.

2. Arnborg, S. (1985). Efficient algorithms for combinatorial problems on graphs with bounded decomposability - A survey, *BIT*, **25**, 2–23.

3. Arnborg, S., Corneil, D. G. and Proskurowski, A. (1987). Complexity of finding embeddings in a k-tree, *SIAM Journal of Algebraic and Discrete Methods*, **8**, 277–284.

4. Ball, M. O. (1980). Complexity of network reliability computations, *Networks*, **10**, 153–165.

5. Beichelt, F. and Spross, L. (1987). An improved Abraham-Method for generating disjoint sums, *IEEE Transactions on Reliability*, **36**, 70–74.

6. Bodlaender, H. L. (1988). Dynamic programming on graphs with bounded treewidth, *Proceedings of the 15th ICALP88*, **317**, pp. 105–118, Berlin: Springer Verlag.

7. Bodlaender, H. L. (1996). A linear time algorithm for finding tree-decompositions of small treewidth, *SIAM Journal on Scientific Computing*, **25**, 1305–1317.

8. Buzacott, J. A. and Chang, J. S. K. (1984). Cut-set intersections and node partitions, *IEEE Transactions on Reliability*, **33**, 385–389.

9. Buzacott, J. A. (1987). Node partition formula for directed graph reliability, *Networks*, **17**, 227–240.

10. Carlier, J. and Lucet, C. (1996). A decomposition algorithm for network reliability evaluation, *Discrete Applied Mathematics*, **65**, 141–156.

11. Choi, M. S. and Jun, C. H. (1995). Some variant of polygon-to-chain reductions in evaluating reliability of undirected network, *Microelectron. Reliab.*, **35**, 1–11.

12. Dotson, W. P. and Gobien, J. O. (1979). A new analysis technique for probabilistic graphs, *IEEE Transactions on Circuits and Systems*, **26**, 855–865.

13. Gadani, J. P. (1981). System effectiveness evaluation using star and delta transformations, *IEEE Transactions on Reliability*, **30**, 43–47.

14. Hasanuddin Ahmad, S. and Jamil, A. T. M. (1987). A modified technique for computing network reliability, *IEEE Transactions on Reliability*, **36**, 554–556.

15. Hasanuddin Ahmad, S. (1988). Simple enumeration of minimal cutsets of acyclic directed graph, *IEEE Transactions on Reliability*, **37**, 484–487.

16. Jain, S. P. Gopal, K. (1988). An efficient algorithm for computing global reliability of a network, *IEEE Transactions on Reliability*, **37**, 487–492.

17. Jasmon, G. B. and Kai, O. S. (1985). A new technique in minimal path and cutset evaluation, *IEEE Transactions on Reliability*, **34**, 136–143.

18. Kinnersley, N. G. (1992). The vertex separation number of a graph equals its path width, *Inform. Proc. Letters*, **42**, 345–350.

19. Lai, M. K. F. (1994). Polygon-to-chain reductions work for networks with imperfect vertices, *Microelectron. Reliab.*, **34**, 267–274.

20. Locks, M. O. (1978). Inverting and minimalizing path sets and cut sets, *IEEE Transactions on Reliability*, **27**, 107–109.

21. Locks, M. O. (1985). Recent developments in computing of system-reliability, *IEEE Transactions on Reliability*, **34**, 425–435.

22. Locks, M. O. (1987). A minimizing algorithm for sum of disjoint products, *IEEE Transactions on Reliability*, **36**, 445–453.

23. Lucet, C. (1993). Méthode de décomposition pour l'évaluation de la fiabilité des réseaux, *Ph.D. Thesis*, Université de Technologie de Compiègne, Compiègne, France.

24. Martelli, A. (1976). A Gaussian elimination algorithm for the enumeration of cut sets in a graph, *Journal of the Assoc. Comp. Machinery*, **23**, 58–73.

25. Misra, K. B. and Rao, T. S. M. (1970). Reliability analysis of redundant networks using flow graphs, *IEEE Transactions on Reliability*, **19**, 19–24.

26. Piekarski, M. (1965). Listing of all possible trees of a linear graph, *IEEE Transactions on Circuit Theory*, **12**, 124–125.

27. Rai, S. and Aggarwal, K. K. (1978). An efficient method for reliability evaluation of a general network, *IEEE Transactions on Reliability*, **27**, 206–211.

28. Rai, S., Veeraraghavan, M. and Trivedi, K. S. (1995). A survey of efficient reliability computation using disjoint products approach, *Networks*, **25**, 147–163.

29. Rosenthal, A. (1977). Computing the reliability of complex networks, *SIAM Journal of Applied Mathematics*, **32**, 384–393.

30. Satyanarayana, A. and Chang, M. (1983). Network reliability and the factoring theorem, *Networks*, **13**, 107–120.

31. Shen, Y. (1995). A new simple algorithm for enumerating all minimal paths and cuts of a graph, *Microelectron. Reliab.*, **35**, 973–976.

32. Shier, D. R. and Whited, D. E. (1986). Iterative algorithms for generating minimal cutsets in directed graphs, *Networks*, **16**, 133–147.

33. Theologou, O. (1990). Contribution a l'évaluation de la fiabilité des réseaux, *Ph.D. Thesis*, Université de Technologie de Compiègne, Compiègne, France.

34. Theologou, O. and Carlier, J. (1991). Factoring and reductions for networks with imperfect vertices, *IEEE Transactions on Reliability*, **40**, 210–217.

35. Tsukiyama, S., Shirakawa, I. and Ozaki, H. (1980). An algorithm to enumerate all cutsets of a graph in linear time per cutset, *Journal of the Assoc. Comp. Machinery*, **27**, 619–632.

36. Wood, R. K. (1985). A factoring algorithm using polygon-to-chain reductions for computing K-terminal network reliability, *Networks*, **15**, 173–190.

37. Yan, L., Taha, H. A. and Landers, T. L. (1994). A recursive approach for enumerating minimal cutsets in a network, *IEEE Transactions on Reliability*, **43**, 383–388.

21

On Matroid Base Families and the Reliability Computation of Totally Amenable Systems

A. Behr and L. Camarinopoulos

RISA, Berlin, Germany
Aristotle University of Thessaloniki, Thessaloniki, Greece

Abstract: In an often-cited paper, Huseby (1989) has shown the relevance of studying matroid theory with respect to the reliability computation of coherent binary systems. In using domination theory he conjectured that under certain circumstances the reliability domination of systems, which are called totally amenable and which exhibit some matroid properties, represents a measure for the complexity of their reliability computation. In this contribution we analyze these circumstances by considering a classical representation of a matroid in terms of its base family. Finally, we establish an algorithm for the reliability computation of these systems and we assess the computational complexity of it.

Keywords and phrases: Coherent binary systems, matroids, total amenability, domination theory

21.1 Preliminaries

The theoretic framework in which the following treatment of matroid base families is located is the theory of coherent binary systems which provides well-known tools for managing the reliability computation of technical systems [Barlow and Iyer (1985), Barlow and Proschan (1975)]. A *binary system* is an ordered pair $(N, \mathbf{P})$ with $N = \{1, ..., n\}$ as the system's *component set* and with a minimal set $\mathbf{P} = \{P_1, ..., P_m\}$, $m \geq 1$, of subsets of N representing the *minimal path set* of the system. We say that $(N, \mathbf{P})$ is *coherent* iff $N = \bigcup_{i=1}^{m} P_i$ [Barlow and Iyer (1985)]. The elements of N may either fail or operate according to known probabilities p_x, $x \in N$. In this contribution they are assumed to be s-independent. Likewise, a binary system may either fail or operate. A binary system is *operating* if and only if the components of at least one minimal path are operating. Otherwise the system is *failed*. The *reliability* $R(N, \mathbf{P})$ of a binary system is the probability that the system is operating.

Reliability computation of coherent binary systems is known to be an NP-hard problem [Garey and Johnson (1979), Rosenthal (1975)], i.e. for an arbitrary system $(N, \mathbf{P})$ there is no algorithm that calculates $R(N, \mathbf{P})$ in a time which is bounded by a polynomial of N or $\mathbf{P}$. The practical applicability of general methods as e.g. the inclusion-exclusion method, the sum of disjoint products method and the reduction principle [Barlow and Proschan (1975)] is therefore often restricted to small, even not medium-sized technical systems. On the other hand, every more sophisticated approach that allows us to analyze larger systems relies on one or several of these general methods. In the following we choose the *reduction principle* as our basic method for computing $R(N, \mathbf{P})$, i.e., we use the fact that:

$$R(N, \mathbf{P}) = p_x \cdot R(N - \{x\}, \mathbf{P}_{+\{x\}}) + (1 - p_x) \cdot R(N - \{x\}, \mathbf{P}_{-\{x\}}) \qquad (21.1)$$

for every $x \in N$. Herein, $(N - \{x\}, \mathbf{P}_{+\{x\}})$ and $(N - \{x\}, \mathbf{P}_{-\{x\}})$ denote binary systems which are obtained from $(N, \mathbf{P})$ by assuming the component x once to operate perfectly and once to be failed. $(N - \{x\}, \mathbf{P}_{+\{x\}})$ and $(N - \{x\}, \mathbf{P}_{-\{x\}})$ are called *minor systems* of $(N, \mathbf{P})$ [Barlow and Iyer (1985), Huseby (1989)]. If $\mathbf{P}(x) = \{P \in \mathbf{P} \mid x \in P\}$, then $\mathbf{P}_{-\{x\}}$ is alternatively described as $\mathbf{P}_{-\{x\}} = \mathbf{P} - \mathbf{P}(x)$ whereas $\mathbf{P}_{+\{x\}}$ is the minimized set $\{P - \{x\} \mid P \in \mathbf{P}\}$. The use of equation (21.1) leads to a reduction of the computational effort involved in determining $R(N, \mathbf{P})$, denoted by $\widetilde{O}(R(N, \mathbf{P}))$, to

$$\widetilde{O}(R(N, \mathbf{P})) = \widetilde{O}(R(N - \{x\}, \mathbf{P}_{+\{x\}})) + \widetilde{O}(R(N - \{x\}, \mathbf{P}_{-\{x\}})) + O(g(N, \mathbf{P})) \qquad (21.2)$$

with $O(g(N, \mathbf{P}))$ - using O-notation [Garey and Johnson (1979)] - denoting the effort needed to perform the pivotization of component x. By applying the reduction principle to $(N, \mathbf{P})$ and, then, iteratively to all minor systems which are generated by applying the reduction principle, one can decompose $(N, \mathbf{P})$ into a set of minor systems all consisting of at most one component whose reliabilities are trivially determined. It is easy to see that such a procedure for computing $R(N, \mathbf{P})$ requires maximally $2^{|N|} - 1$ pivotizations or, equivalently, applications of the reduction principle. Hence, the computational complexity of a corresponding algorithm is bounded by

$$O(2^{|N|}) \cdot O(g^*(N, \mathbf{P})), \qquad (21.3)$$

where $O(g^*(N, \mathbf{P}))$ denotes the effort needed to perform the pivotization of an arbitrary component in $(N, \mathbf{P})$.

A widely used strategy to reduce the complexity of computing $R(N, \mathbf{P})$ is to modularize the system $(N, \mathbf{P})$. We say that a coherent binary system $(A, \mathbf{A})$

is a *modular subsystem* of $(N, \mathbf{P})$ if and only if the minimal path set $\mathbf{P}$ can be partitioned in the way that

$$\mathbf{P} = [\mathbf{P}(A)_{+A} \times \mathbf{A}] \cup \mathbf{P}_{-A} \tag{21.4}$$

with $\mathbf{P}(A) = \{P \in \mathbf{P} \mid P \cap A \neq \oslash\}$ for some fixed $A \subset N$ and $\mathbf{A} = \{P \cap A \mid P \in \mathbf{P}(A)\}$. The reader may verify the equivalence of this definition to other presentations of the module concept in literature [Barlow and Proschan (1975), Chatterjee (1974)]. It is a well-known fact that the components of a modular subsystem can be replaced by a single, new component x_A, $x_A \notin N$, in the minimal paths of $(N, \mathbf{P})$ representing the modular subsystem. Moreover, the reliability of this component is given by $p_{x_A} = R(A, \mathbf{A})$ and may be computed prior to and separately from $R(N, \mathbf{P})$ which in turn can be expressed by $R(N, \mathbf{P}) = R((N - A) \cup \{x_A\}, [\mathbf{P}(A)_{+A} \times \mathbf{A}] \cup \mathbf{P}_{-A})$. This modified system which is generated by modularizing $(N, \mathbf{P})$ is called the *modularized system* and is henceforth abbreviated as $(N^*, \mathbf{P}^*)$. As usually done, we distinguish two types of coherent binary systems whose reliabilities are immediately obtained due to standard formulas: These are the *parallel system* and the *series system* [Barlow and Proschan (1975)], for which $R(N, \mathbf{P}) = 1 - \prod_{x \in N} \cdot (1 - p_x)$ and $R(N, \mathbf{P}) = \prod_{x \in N} p_x$, respectively. All other systems are called *complex*. In light of the modularization, an *s-p-reduction* [Barlow and Iyer (1985), Huseby (1989)] of a given system $(N, \mathbf{P})$ is the modularization of system with respect to all modular parallel and series systems contained in $(N, \mathbf{P})$ and in the - by this way - generated modularized systems. After performing an s-p-reduction one ends up with a modularized system $(N^*, \mathbf{P}^*)$ which is called *s-p-reduced.*

21.2 Algorithmic Complexity of Reliability Computation and Domination Theory

Combining the reduction principle and s-p-reduction the following algorithm for computing the reliability of an arbitrary coherent binary system can be established:

procedure compute_reliability$(N, \mathbf{P}, \mathbf{p})$;
/* $\mathbf{p}$ denotes the vector of component reliabilities */
begin
    perform s-p-reduction$(N, \mathbf{P}, \mathbf{p})$;
    /* an s-p-reduced system $(N^*, \mathbf{P}^*)$ with
        component reliabilities $\mathbf{p}^*$ is generated */
    $x =$ choose_pivot_element$(N^*, \mathbf{P}^*)$;
    $\mathbf{P}^*_{+\{x\}} =$ generate_minor+$(N^*, \mathbf{P}^*, x)$;

$\mathbf{P}^*_{-\{x\}} = \text{generate\_minor-}(N^*, \mathbf{P}^*, x)$;
if $\mathbf{P}^*_{+\{x\}} = \{\}$ **then** $R(N^* - \{x\}, \mathbf{P}^*_{+\{x\}}) = 1$
else $R(N^* - \{x\}, \mathbf{P}^*_{+\{x\}}) = \text{compute\_reliability}(N^* - \{x\}, \mathbf{P}^*_{+\{x\}}, \mathbf{p}^*)$;
if $\mathbf{P}^*_{-\{x\}} = \{\}$ **then** $R(N^* - \{x\}, \mathbf{P}^*_{-\{x\}}) = 0$
else $R(N^* - \{x\}, \mathbf{P}^*_{-\{x\}}) = \text{compute\_reliability}(N^* - \{x\}, \mathbf{P}^*_{-\{x\}}, \mathbf{p}^*)$;
/* now apply the reduction principle */
$R(N, \mathbf{P}) = p_x \cdot R(N^* - \{x\}, \mathbf{P}^*_{+\{x\}}) + (1 - p_x) \cdot R(N^* - \{x\}, \mathbf{P}^*_{-\{x\}})$;
return $R(N, \mathbf{P})$;
end compute_reliability

Algorithm 21.1: An algorithm for reliability computation of an arbitrary coherent binary system $(N, \mathbf{P})$ using s-p-reduction and pivotization

As mentioned above, the computational complexity of this algorithm is dominated by the number of pivotizations required. Obviously, this number depends on the choice of the pivoting element at each recursion stage of the algorithm. Satyanarayana and Chang (1983) and, in the sequel, Barlow and Iyer (1985) and Huseby (1989) have shown that for certain types of coherent binary systems the minimum number of pivotizations needed equals to the absolute value of a particular combinatorial invariant of the system, which is the *(reliability) domination*: It is defined [Huseby (1989)] as

$$D(N, \mathbf{P}) = \left| \sum_{S \subseteq N} (-1)^{|N-S|} \cdot \Phi(S, \mathbf{P}) \right| \tag{21.5}$$

with $\Phi(S, \mathbf{P})$, $S \subseteq N$, as the structure function of the system $(S, \mathbf{P})$ [3]. Now, if for a given coherent binary system $(N, \mathbf{P})$ and for all minor systems of $(N, \mathbf{P})$ with $|N| > 1$ it holds that

$$D(N, \mathbf{P}) = D(N - \{x\}, \mathbf{P}_{+\{x\}}) + D(N - \{x\}, \mathbf{P}_{-\{x\}}) \tag{21.6}$$

for every $x \in N$, then the system $(N, \mathbf{P})$ is called *totally amenable.* In this case one can prove the following proposition:

Proposition 21.2.1 (for a proof see Barlow and Iyer (1985)) *The minimum number of pivotizations needed to compute the reliability of a totally amenable coherent binary system $(N, \mathbf{P})$ by means of algorithm 1 is $D(N, \mathbf{P}) - 1$.*

Barlow and Iyer (1985) as well as Huseby (1989) have shown furthermore that proposition 1 holds since for $(N, \mathbf{P})$ and for every minor system of $(N, \mathbf{P})$ with $|N| > 1$ there exists a component $x \in N$ such that $(N - \{x\}, \mathbf{P}_{+\{x\}})$ and $(N - \{x\}, \mathbf{P}_{-\{x\}})$ are both coherent. By choosing only components that satisfy this condition as pivot elements in algorithm 1 the total number of pivotizations needed for a reliability computation is exactly $D(N, \mathbf{P}) - 1$.

Although yet not formally proven, it is common sense that for most realistic systems $D(N, \mathbf{P}) \ll 2^{|N|}$ [Barlow and Iyer (1985), Huseby (1989)] such that an algorithm for reliability computation whose computational complexity is dominated by $D(N, \mathbf{P})$ should be much faster than all other variants of algorithm 1. However, such an algorithm is not trivially given: In addition to the $D(N, \mathbf{P}) - 1$ pivotizations one has analyze the efforts which are involved in

(a.) performing an s-p-reduction,
(b.) choosing a pivot element that leads to coherent minor systems and
(c.) generating the minor systems w.r.t. to the chosen pivot element

(21.7)

at each recursion stage of Algorithm 21.1. It is far from clear that for an arbitrary totally amenable system these tasks can be performed faster than in $O(D(N, \mathbf{P}))$ time (or at least in the same order of magnitude) which would mean that the involved efforts can be neglected. Thus, up to yet a computational superiority of Algorithm 21.1 in comparison with the use of standard methods (inclusion-exclusion-method etc.) has only been proven for certain undirected network systems [Barlow and Iyer (1985), Satyanarayana and Chang (1983)] which represent a very special subclass of the totally amenable systems. For other system classes no similar results are known. In the next two paragraphs we develop a variant of Algorithm 21.1 for a new and much broader class of totally amenable systems, the so-called matroid base family systems. By this, we help to widen the poor state-of-the-art in this field.

21.3 Matroid Base Families

Huseby (1989) found a close relationship between the domination invariant and the β-invariant for matroids. He specified a procedure for extending a large set of coherent binary systems, which he called regular, to matroids. These regular systems include certain undirected network systems and the k-out-of-n systems. As a point of interest, he proved that those systems where such an extension is possible are totally amenable which means that for these systems proposition 1 holds. A main objection to Huseby's procedure in extending coherent binary systems to matroids is that it seems a bit far from practical application. There are several mathematical operations to be performed on the regular systems for which efficient algorithms are not known. A much more useful (perhaps theoretically weaker) approach regarding this aspect is to analyze coherent binary systems that exhibit matroidal properties directly. One of the commonly used definitions of a matroid is e.g. the one given in terms of a matroid base family [White (1986)]:

A family $\mathbf{B} = \{B_1, ..., B_m\}$, $m \geq 1$, of subsets of an element set N is a *matroid base family*, if and only if for every $B_i, B_j \in \mathbf{B}$ ($1 \leq i \leq m$, $1 \leq j \leq m$) with $x \in B_i$ there exists an element $y \in B_j$ such that

$$(B_i - \{x\}) \cup \{y\} \in \mathbf{B}. \tag{21.8}$$

Clearly, $(N, \mathbf{B})$ forms a binary system. We will call it a *matroid base family system*. Matroid base families and their corresponding systems have several interesting properties which make them particularly attractive to reliability theoreticians:

Proposition 21.3.1 (for a proof see White (1986)) *Let* $\mathbf{B}$ *be a matroid base family on* N. *Then,*

(a.) all members of $\mathbf{B}$ *have the same cardinality and*
(b.) for every $x \in N$, (21.9)
$\mathbf{B}_{+\{x\}}$ *and* $\mathbf{B}_{-\{x\}}$ *are also matroid base families.*

By this, $(N, \mathbf{B})$ belongs to the class of reliability systems with equally long minimal paths, as for instance, the intensively discussed k-out-of-n:G systems (which are matroids k-out-of-n:G systems - base family systems) and consecutive k-out-of-n:G systems [Papastavridis and Koutras (1993)]. In practice, technical systems in which every minimal path contains the same number of components are widespread, in particular, if highly redundant technical systems are considered (see, e.g., the control rod system of a nuclear reactor or a fault redundant processor array). For brevity, we say that $(N, \mathbf{B})$ has a minimal path cardinality k iff $|B| = k$ for all $B \in \mathbf{B}$.

Further, matroid base family systems are totally amenable:

Proposition 21.3.2 *In analogy to the definition of* $D(N, \mathbf{P})$ *in eq. (21.5), let*

$$d(N, \mathbf{P}) = \sum_{S \subseteq N} (-1)^{|N-S|} \cdot \Phi(S, \mathbf{P}) \tag{21.10}$$

be the 'signed domination' of the coherent binary system $(N, \mathbf{P})$ *[Huseby (1989)]. Then, for a matroid base family system* $(N, \mathbf{B})$ *with the minimal path cardinality* k, $k \geq 1$, *and* $|N| = n$ *it holds that*

$$d(N, \mathbf{B}) = (-1)^{n-k} \cdot D(N, \mathbf{B}). \tag{21.11}$$

PROOF. One can prove this proposition by induction on k. Since for $k = 1$, $(N, \mathbf{B})$ is a parallel system and, hence, with eq. (21.10) it follows that $d(N, \mathbf{B}) = (-1)^{n-1} = (-1)^{n-1} \cdot 1 = (-1)^{n-k} \cdot D(N, \mathbf{B})$, eq. (21.11) holds trivially. For $1 < k \leq n$ use the fact that [Agrawal and Barlow (1984)]

$$d(N, \mathbf{B}) = d(N - \{x\}, \mathbf{B}_{+\{x\}}) - d(N - \{x\}, \mathbf{B}_{-\{x\}}) \tag{21.12}$$

for every $x \in N$ such that the minimal path cardinality of the minor system $(N - \{x\}, \mathbf{B}_{+\{x\}})$ takes the value $k - 1$ and allows for an application of the induction hypothesis. For the minor system $(N - \{x\}, \mathbf{B}_{-\{x\}})$ and for all its minors use eq. (21.12) iteratively as long as they have a minimal path cardinality k. Then apply the induction hypothesis to the generated minors. ■

Lemma 21.3.1 *A matroid base family system $(N, \mathbf{B})$ with $|N| > 1$ is totally amenable.*

PROOF. Insert twice eq. (21.11) into eq. (21.12) such that eq. (21.6) is obtained. Further, consider property (21.9b). ■

21.4 On the Complexity of Computing the Reliability of Matroid Base Family Systems

In what follows, we develop bounds for the computational efforts involved in performing the tasks (21.7a)–(21.7c) on matroid base family systems. By this, we get an assessment of the computational complexity of Algorithm 21.1 with respect to matroid base systems. We begin with task (21.7c):

Lemma 21.4.1 *For a given matroid base family system $(N, \mathbf{B})$ the generation of the minor systems $(N - \{x\}, \mathbf{B}_{+\{x\}})$ and $(N - \{x\}, \mathbf{B}_{-\{x\}})$, $x \in N$, can be performed both in $O(|\mathbf{B}|)$ time.*

PROOF. For the minor $\mathbf{B}_{-\{x\}}$, in order to remove from $\mathbf{B}$ all minimal paths which contain the component x one has to check once all elements of $\mathbf{B}$. For the minor $\mathbf{B}_{+\{x\}}$, one has to select all minimal paths in $\mathbf{B}$ containing x and to remove x from the selected paths. This task can be likewise accomplished by going once over the elements of $\mathbf{B}$. ■

Now we consider task (21.7b). Here, an important result is that there are only two possible types of modules in a matroid base family system:

Proposition 21.4.1 *Let $(A, \mathbf{A})$ be a modular subsystem of a matroid base family system $(N, \mathbf{B})$, $A \subset N$, with $|N| > 1$. Then $(A, \mathbf{A})$ is either (i.) a parallel system or (ii.) a system which itself is connected in series to the remaining system, i.e.,*

$$\mathbf{B}_{+A} \times \mathbf{A} = \mathbf{B}. \tag{21.13}$$

PROOF. In case of a minimal path cardinality $k = 1$ the system $(N, \mathbf{B})$ is a parallel system and, similarly, $(A, \mathbf{A})$ is a parallel system (case (i.)). For $k > 1$, it is easy to construct a system $(N, \mathbf{B})$, in which $(A, \mathbf{A})$ is a parallel system (case (i.)) or, alternatively, a system satisfying condition (21.13) (case (ii.)). If it is a parallel system, then it is obviously $|B \cap A| \leq 1$ for every $B \in \mathbf{B}$. Otherwise, if it is a system as in case (ii.), then it is $|B \cap A| = k_a \geq 1$ for every $B \in \mathbf{B}$ with k_a denoting the minimal path cardinality of $(A, \mathbf{A})$. Note that the minimal paths of $(A, \mathbf{A})$ must have all the same cardinality. Otherwise the minimal paths of the system $(N, \mathbf{B})$ would not have all the same cardinality.

Now, we show by contradiction that $(N, \mathbf{B})$ cannot contain a modular subsystem $(A, \mathbf{A})$ with a minimal path cardinality $k_a > 1$ unless condition (21.13) is satisfied. For that, assume $(A, \mathbf{A})$ to be modular subsystem with $k_a > 1$ and condition (21.13) to be ignored, i.e. a modular partitioning of $\mathbf{B}$ renders $\mathbf{B} = (\mathbf{B}(A)_{+A} \times \mathbf{A}) \cup \mathbf{B}_{-A}$ with $\mathbf{B}_{-A} \neq \oslash$ (cf. eq. (21.4)). This means that there is at least one minimal path $B^* \in \mathbf{B}$ having an empty intersection with A. However, since $(N, \mathbf{B})$ is a matroid base family, according to condition (21.8) there is for every $B \in \mathbf{B}(A)$ with $x \in B \cap A$ a component $y \in B^*$ such that $(B - \{x\}) \cup \{y\} = B^{**} \in \mathbf{B}$. Evidently, $|B^{**} \cap A| = |B \cap A| - 1$, thus contradicting the fact that all minimal paths of $(A, \mathbf{A})$ have the same cardinality. As a consequence, B^* cannot exist and it must hold that $\mathbf{B}_{-A} = \oslash$. ∎

From Proposition 21.4.1 we conclude that an s-p-reduction has to affect two types of modular systems, namely (i.) modular parallel systems and (ii.) modular systems which are in series to the remaining system. The computational effort needed to identify these systems is analyzed in the following:

Lemma 21.4.2 *An s-p-reduction of a matroid base family system $(N, \mathbf{B})$ can be performed in $O(|\mathbf{B}|) \cdot O((|N|)^2)$ time.*

PROOF. For the identification of the largest possible and only modular series system in $(N, \mathbf{B})$ one has to form the intersection of all minimal paths of $(N, \mathbf{B})$ by going once over the elements of $\mathbf{B}$. The involved computational effort is hence $O(|\mathbf{B}|)$. For the modular parallel systems, we use the fact that two components $x, y \in N$ belong to the same modular parallel system in $(N, \mathbf{B})$ if and only if there is no element of $\mathbf{B}$ containing both x and y. Thus, we begin by selecting two arbitrary components $x, y \in N$, $x \neq y$, and check whether there is a minimal path of $(N, \mathbf{B})$ which contains both components. If so, then they are not contained in modular parallel system. If not, then they are components of the same modular parallel system. Next, we choose a third component $z \in N$ ($x \neq z$, $y \neq z$) and check whether z is contained together with x or together with y in some minimal path of $(N, \mathbf{B})$. Depending on this test z may be contained in a modular parallel system together with x, with y, with x and y or it may not be together with x and y in a modular parallel system. Continuing this procedure with all remaining components of the system $(N, \mathbf{B})$, the possible modular parallel systems of $(N, \mathbf{B})$ are identified. In the worst case all possible pairs of components have to be tested against all elements of $\mathbf{B}$ which causes a computational effort of $O(|\mathbf{B}|) \cdot O((|N|)^2)$. ∎

Due to the definition of the concept of s-p-reduction an s-p-reduced matroid base family system may contain at most one component that is connected in series to the remaining system. In the following lemma, that copes with task (21.7a), we call this component a *series component*:

Lemma 21.4.3 *In an s-p-reduced matroid base family system* $(N, \mathbf{B})$ *with* $|N| > 1$, *every component* $x \in N$ *that is not a series component is such that* $(N - \{x\}, \mathbf{B}_{+\{x\}})$ *and* $(N - \{x\}, \mathbf{B}_{-\{x\}})$ *are both coherent binary systems.*

PROOF. Let $x \in N$ be a component which is not a series component. Since for every $y \in N$, $x \neq y$, there exists a minimal path of $(N, \mathbf{B})$ containing both x and y (otherwise x and y would be contained in an modular parallel system (cf. Lemma 3)), it follows that $N - \{x\} = \bigcup_{B \in \mathbf{B}_{+\{x\}}}$ and, hence, that $(N - \{x\}, \mathbf{B}_{+\{x\}})$ is coherent. For $(N - \{x\}, \mathbf{B}_{-\{x\}})$ being not coherent, there must be a component $y \in N$, $x \neq y$, which is contained only in the elements of $\mathbf{B}(\{x\})$. Let $B \in \mathbf{B}(\{x, y\})$. Further, let $B^* \in \mathbf{B}_{-\{x\}}$be a minimal path which, consequently, contains neither x nor y. However, in accordance with condition (21.8) there exists a component $z \in B^*$ such that $(B - \{x\}) \cup \{z\} = B^{**} \in \mathbf{B}$. Thus, B^{**} contains y but not x which contradicts the initial assumption that y is contained only in the elements of $\mathbf{B}(\{x\})$. By this, if this assumption does not hold then $(N - \{x\}, \mathbf{B}_{-\{x\}})$ is obviously a coherent system, too. ■

Note that for sake of this lemma task (21.7a) can be trivially performed, since from the s-p-reduction we know whether there is a series component and, if so, which component it is. The computational effort needed for this task is therefore adequately estimated by a constant c, $c \geq 1$.

We come now to the main result of this manuscript:

Theorem 21.4.1 *The reliability of a matroid base family system* $(N, \mathbf{B})$ *with s independent components N can be computed in* $O(D(N, \mathbf{B})) \cdot O(|\mathbf{B}|) \cdot O((|N|)^2)$ *time.*

PROOF. By considering the results of Lemma 21.3.1 to Lemma 21.4.3 together with the fact that there are exactly $D(N, \mathbf{P}) - 1$ pivotizations needed to compute the reliability of a totally amenable system by Algorithm 21.1 it is

$$\begin{aligned} \tilde{O}(R(N, \mathbf{B})) &= O(D(N, \mathbf{B})) \cdot O(g^*(N, \mathbf{B})) \\ &= O(D(N, \mathbf{B})) \cdot \\ &\quad [O(|\mathbf{B}|) \cdot O((|N|)^2) + 2 \cdot O(|\mathbf{B}|) + c] \\ &= O(D(N, \mathbf{B})) \cdot O(|\mathbf{B}|) \cdot O((|N|)^2). \end{aligned} \tag{21.14}$$

■

21.5 Conclusions

The theorem is of considerable relevance to the reliability computation of totally amenable systems and, moreover, to the reliability computation of coherent binary systems in general. By showing that for matroid base family systems the tasks (21.7a)–(21.7c) can be performed in $O(|\mathbf{B}|) \cdot O((|N|)^2)$ time, we have developed an efficient procedure (in terms of minimal paths and components) for all the work in Algorithm 21.1 that have to be done in addition to pivotization. It stands the question whether $O(D(N, \mathbf{B})) > O(|\mathbf{B}|) \cdot O((|N|)^2)$ or not. If so, then $D(N, \mathbf{B})$ is indeed a measure for the computational complexity of the reliability computation of $D(N, \mathbf{B})$. If not, then $R(N, \mathbf{B})$ can be computed efficiently in using Algorithm 21.1. With the presented results at hand, it should be possible to find an answer to this question. In addition to this, one should remember that all matroids have a representation in terms of their base family and that there is no known totally amenable system which at least cannot be transformed into a matroid by the procedure of Huseby (1989). In principle, the presented theorem should be therefore valid for all known totally amenable systems. This in turn means that for a large class of coherent binary systems, the worst case effort of Algorithm 21.1 is really less than the effort given in (21.3) and, finally, Algorithm 21.1 is much faster than the use of standard methods as e.g. the inclusion-exclusion-method.

References

1. Agrawal, A. and Barlow, R. E. (1984). A survey of network reliability and domination theory, *Operations Research*, **32**, 478–491.

2. Barlow, R. E. and Iyer, S. (1985). Computational complexity of coherent systems and the reliability polynomial, *ORC-Report 85-6*, Operations Research Center, University of California, Berkeley.

3. Barlow, R. E. and Proschan, E. (1975). *Statistical Theory of Reliability and Life Testing*, New York: Holt Rinehart, and Wilson.

4. Chatterjee, P. (1974). Fault tree analysis: reliability theory and systems safety analysis, *ORC-Report 74-34*, University of California, Berkeley.

5. Garey, M. and Johnson, D. (1979). *Computers and Intractability*, New York: Freeman.

6. Huseby, A. B. (1989). Domination theory and the Crapo beta-invariant, *Networks*, **19**, 135–149.

7. Rosenthal, A. (1975). A computer scientist looks at reliability computations, In *Reliability and Fault Tree Analysis* (Eds., R. E. Barlow, J. B. Fussell and N. P. Singpurwalla), pp. 133–152, Philadelphia: SIAM.

8. Papastavridis, S. and Koutras, M. V. (1993). Consecutive k-out-of-n systems and their generalizations, In *New Trends in System Reliability Evaluation* (Ed., K. B. Misra), pp. 228–248, New York: Elsevier.

9. Satyanarayana, A. and Chang, M. K. (1983). Network reliability and the factoring theorem, *Networks*, **13**, 107–120.

10. White, N. (1986). *Theory of Matroids*, Cambridge, MA: Cambridge University Press.

22

The Computer-Assisted Analysis of the Semi-Markovian Stochastic Petri Nets and an Application

Anatoli Paul Ulmeanu and Dumitru Cezar Ionescu

Politehnica University of Bucharest, Bucharest, Romania

Abstract: A semi-Markovian Stochastic Petri Net is a powerful tool modelling a wide class of applications. Correspondingly, increasing its flexibility, it is more difficult to solve the mathematical model associated. Performance algorithms are proposed in order to develop an automated-computer state space generation and to perform the stochastic analysis of the large systems.

Keywords and phrases: Semi-Markov process, semi-Markovian stochastic Petri Net, extended stochastic Petri Net

22.1 Introduction

Petri Nets (PN) provide more and more to be a performing modeling tool applicable to many systems. Its powerfulness is given by both graphical and mathematical resources. As graphical tool, the PN operate with the common graph symbols: the nodes (two kind: places and transitions) and the directed arcs [Murata (1989)]. In addition, tokens are used to simulate the concurrent and dynamic activities of systems. A main goal was to developing the basic PN to model the deterministic and non-deterministic behaviour of systems, particularly the asynchronous and synchronous concurrent systems. One of the major steps in this direction was taking into account the timing considerations of these systems. Thus, an important PN class was defined and opened for study: the *timed* Petri net. Noe and Nutt (1973) associated a fixed delay to each transition between its time enabling and its time firing. Merlin and Farber (1976) introduced the timed PN for the analysis of protocol communications. A crucial moment in the timed PN history was marked by the definition of

Petri Nets, so called Stochastic Petri Nets (SPN). They were independently proposed by Natkin (1980) and Molloy (1982) by associating with each transition an exponentially distributed firing time. Molloy (1982) has shown that SPN are equivalent to continuous time Markov chains. Therefore, with the introduction of SPN, a bridge is established between two important classes of systems models: graph models (i.e., Petri net model) and the probabilistic models used to reliability and performance analysis. Further, on Generalized Stochastic Petri Nets (GSPN) proposed by Marsan, Balbo and Conte (1984) allow the immediate transitions (zero firing time), an exponentially distributed firing time being associated with timed transitions (as in SPN). The major limitation of the SPN models is the requirement of exponentially distributed firing times. Extended Stochastic Petri Nets (ESPN) introduced by Dugan (1984) allow the transition firing times to belong to any arbitrary probability distribution. In addition, the immediate transitions are allowed in ESPN. In this case, a reduction of the reachability graph must be done in the beginning of the analysis, by absorbing all the markings that enable an immediate transition (so called *vanishing markings*). Thus, in the reduced reachability marking graph there are only markings that enable timed transitions (so called, *tangible markings*). Clearly, the ESPN reflect more generally and more accurately the behaviour of the systems, but they are also more difficult to evaluate. Therefore, Dugan et al. (1984) underline the general conditions under which the reachability graph associated with an ESPN in the continuous time case can be solved as a semi-Markov process, so called semi-Markovian Stochastic Petri Nets (SMSPN).

The paper focuses on the assessment algorithms of the SMSPN that would permitted the computer-assisted analysis of large semi-Markov systems, especially in the reliability and maintainability fields.

This paper is organized as follows. Section 22.2 provides background results and important probabilistic definitions that described the PN stochastic behaviour. In Section 22.3, we develop a efficient computer-assisted analysis of the SMSPN. In Section 22.4, an application in reliability and maintainability field is presented. Finally, in Section 22.5, remarks and future developments are presented.

It is important to note that any reachability marking graph considered in this analysis is finite, already reduced and is obtained from a lived and bounded ordinary PN.

All the PN extensions (with the counter arc, counter-alternate arc, inhibitor arc) including the high-level PN (predicate and colored) may be transformed in an ordinary PN and, therefore, it will be not distinctly studied.

22.2 Background Material in the Stochastic Behavior of Petri Nets

A number of references [Dugan (1984), Florin (1985), Molloy (1982)] are available to provide background on PN stochastic analysis. The material presented in this section is a summary of the important definitions and theoretical results.

By definition [Baccelli (1992), David (1989), Murata (1989)], a Petri Net (PN) is a 4-tuple:

$$PN \equiv (P, T, A, M_0) \tag{22.1}$$

where $P = \{p_1, p_2, .., p_m\}$ is a finite set of places

$$\begin{aligned} T &= \{t_1, t_2, .., t_n\} \text{ is a finite set of transitions} \\ A &\subseteq (P \times T) \cup (T \times P) \text{ is a set of arcs} \\ M_0 &: P \to N \text{ is the initial marking } (N \text{ is the set of natural numbers}) \\ M_0(p_i) &= \mu_i \\ P \cap T &= \emptyset,\ P \neq \emptyset \text{ and } T \neq \emptyset. \end{aligned}$$

A marking M_k is said to be reachable from a marking M_0 if there is a sequence of firings that transforms M_0 in M_k. A firing or occurrence sequence is denoted by $\sigma = (t_{i_1}, t_{i_2}, .., t_{i_r})$ where t_{i_k} is the k-th firing transition (time ordering). We denote by $M_0 \stackrel{\sigma}{\rightarrow} M_k$. The set of *all* possible tangible markings reachable from M_0 is denoted by $R(M_0)$ and it is identified with the set of the PN reduced reachability marking graph nodes. The set of all possible firing timed transitions from a marking M is denoted by $F(M)$. Finally, the set of *all* the tangible markings reachable from any $M \in R(M_0)$ after a firing timed transition $t_k \in F(M)$ is denoted by $\mathcal{RF}(M)$.

We also use the following symbols for pre-set and post-set

$$\begin{aligned} t^- &= \{p \mid (p,t) \in A\} \text{ the set of input places of transition } t \in T \\ t^+ &= \{p \mid (t,p) \in A\} \text{ the set of output places of transition } t \in T. \end{aligned}$$

A Extended Stochastic Petri Net (ESPN) is a 5-tuple [Dugan (1984), Molloy (1982)]:

$$ESPN \equiv (P, T, A, M_{0,}r\,(t))$$

in the continuous time case, and

$$ESPN \equiv (P, T, A, M_{0,}\rho\,(k))$$

in the discrete time case, where

$r(t) = (r_1(t), r_2(t), .., r_n(t))$ is the firing hazard rate vector function associated with the set T of transitions, for any time $t \in R_+$ (the set of non negative real numbers) when the transition is enabling,

$\rho(p) = (\rho_1(p), \rho_2(n), .., \rho_n(p))$ is the firing step probability vector function associated with the same set T of transitions, for any $p \in N$ when the transition is enabling.

The both firing hazard rate and firing step probability vectors will be defined in respect of the firing time random vector X:

$$X = (X_1, X_2, .., X_n).$$

We consider a probability space in which our ESPN is held and we denote by $\mathcal{P}$ the probability measure.

For the continuous time case, we define the firing hazard rate associated with any transition $t_k \in T$, any time when it is enabling, by a non-negative and continuous function $x \to r_k(x)$, where $x \geq 0$ is the time elapsed after its enabling

$$r_k(x) = \lim_{\Delta x \to 0_+} \frac{\mathcal{P}\{x < X_k \leq x + \Delta x \mid X_k > x\}}{\Delta x}. \tag{22.2}$$

The quantity $r_k(x) \cdot \Delta x + o(\Delta x)$ represents the probability of firing the transition t_k in the interval $(x, x + \Delta x]$, given that the transition t_k has already been in its enabling marking for x units of time.

It is important to recall that in the ESPN and SPN, as well as any kind of PN, only one transition firing at time is allowed [David (1989)]. This property is full agreement with the main assumption of the semi-Markov and Markov systems: the probability of more than one system transition in $(x, x + \Delta x]$ is $o(\Delta x)$.

For the discrete time, we define the firing step probability associated with any transition $t_k \in T$, for any $n \in N$ when it is enabling

$$\rho_k(n) = \mathcal{P}\{X_k = n + 1 \mid X_k > n\}. \tag{22.3}$$

In the denumerable set $E = R(M_0)$ let us consider the r.v. $0 = S_0 < S_1 < ... < S_n < ...$ representing the ESPN firing times, and $J_0, J_1, ..., J_n, ..$ the r.v. which take values in E, referring the ESPN markings at the time of successive firings. If the ESPN is a SMSPN one, then $(J, S) = \{J_n, S_n; n \in N\}$ is a Markov renewal process (MRP). In this case, we denote by $Q(t)$ the kernel matrix $(Q_{ij}(t))_{i,j \in E}$.

$$Q_{ij}(t) = \mathcal{P}\{J_{n+1} = j, S_{n+1} - S_n \leq t \mid J_n = i\} \tag{22.4}$$

for all $n \in N, i \neq j$ and $Q_{ii}(t) = 0$.

Also, we denote by $G_{ij}(t)$ the conditional sojourn time distributions

$$G_{ij}(t) = \mathcal{P}\left\{S_{n+1} - S_n \leq t \mid J_n = i, J_{n+1} = j\right\}. \tag{22.5}$$

The first member of the equation (22.4) may be easily written as

$$Q_{ij}(t) = \int_0^t \prod_{k \neq i} (1 - G_{ik}(x)) dG_{ij}(x) \tag{22.6}$$

for any $i, j \in R(M_0)$, $i \neq j$, $k \in \mathcal{RF}(i)$, $i \xrightarrow{t_k} j$.

Remark. In the case when the firing hazard rate transition and the probability firing step transition are constant by respect of time, then the ESPN is a Markovian Petri Net, so called *Stochastic Petri Net* (SPN):

$$\begin{aligned} SPN &\equiv (P, T, A, M_{0,}r) \text{ continuous time} \\ SPN &\equiv (P, T, A, M_{0,}\rho) \text{ discrete time} \end{aligned}$$

In the first case, the sojourn time in any marking follows an exponential distribution , in the second case the sojourn time follows a geometrical distribution. Due to the memoryless property of the exponential distribution of sojourn time (firing delays), it has been shown that the reachability marking graph of a k-bounded SPN is equivalent to a finite Markov chain [Lin (1988)].

22.3 Computer-Assisted Analysis of the Semi-Markovian Petri Nets

In this section, we develop a performing computer-analysis of the semi-Markovian Stochastic Petri Nets, accordingly with the main algorithm presented in Figure 22.3.

The Petri Nets implementation on the computer-assisting analysis is generally based on pointer linked list, directed graph associated which any Petri nets (timed or not) actually consists of. Each node of the pointer list will have the following (minimum) associated information:

- the node symbol
- the number of tokens (only for the place type node)
- the type and parameters of the firing times distributions
- left-hand pointer and right-hand pointer which point to the next net node and to next son of net node, respectively.

Figure 22.2 is an illustration of the Petri net of Figure 22.1. It should be pointed out that the nodes sequence on the linked list (like t_1, p_1, p_2, p_3, p_4 as depicted in the Figure 22.2) is to be arbitrarily selected by the user.

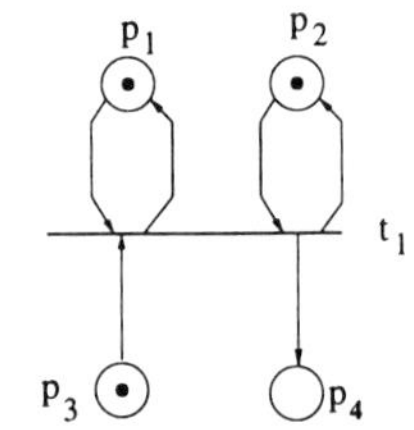

Figure 22.1: Example of a Petri net

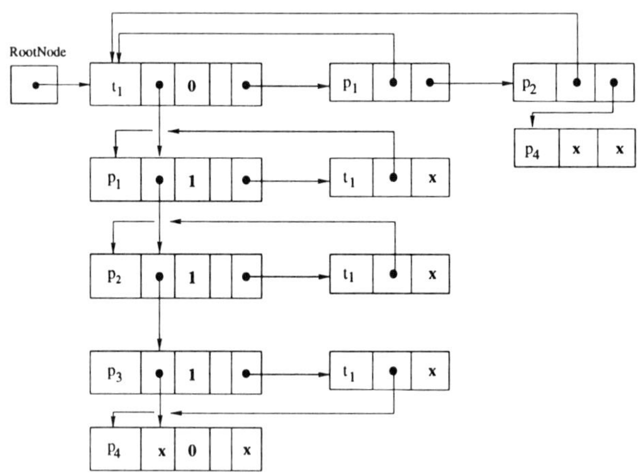

Figure 22.2: Pointer list modelling the Petri net of Figure 22.1

The automated generation of the markings of the Petri net may be achieved by means of the algorithm presented in Figure 22.4, implemented in the PAMS application [Ulmeanu (1995)], by calling the entitled procedure *Firing(m)* (the marking m is assigned to the initial marking M_0).

Figure 22.5 presents the effect of the $Firing(M_0)$ procedure calling which consists in the gradual building up of the Petri net reachability graph. The $Firing(m)$ procedure exit condition (the flag *Stop*) is double: no more enabled transitions to be fired and $NewMarking$ points to the initial marking M_0 (the initial marking was found out). It is worth mentioning that there are two important cases which lead to an infinite number of the recursive procedure callings, cases which should be carefully avoided:

- the total number of the Petri markings increases after each firing. Therefore new markings are indefinitely generated as the necessary exit condition from the recursive procedure has not been met

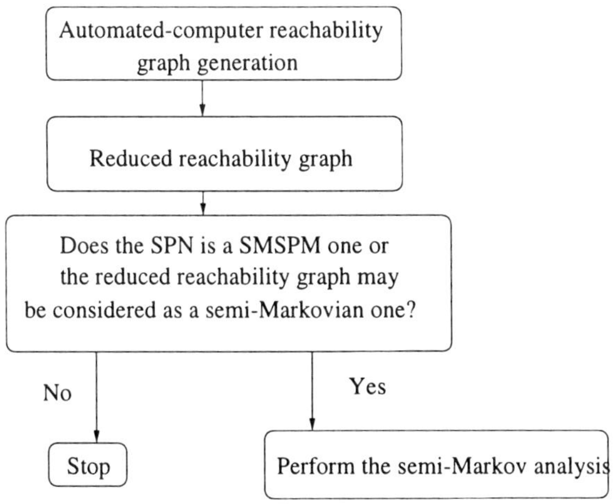

Figure 22.3: The main algorithm for the computer-assisted analysis of a SMSPN

```
Firing (m)        /* m is the current marking */

Firing_Enabled_Transitions (RootNode, m, n, t)        /*find F(m)*/

For each of the n Enabled transition i:=1 to n do

NewMarking:=Firing_Transition( t[i], m)

z:= Search_Marking_in_List ( t[i], NewMarking)

If z>0, then   /* marking does not on the list */
    Add_to_the_List (New_Marking)
    N:=N+1    /* N is the current number of resulting marking */
    New_Marking→ index :=N
    Make_the_Link (m, New_Marking)
    If Not_Stop then   Firing (New_Marking)
else
    Make_the_Link (m, New_Marking)
end   /* end of if */
end   /* end of for */
```

Figure 22.4: The "Firing" recursive procedure

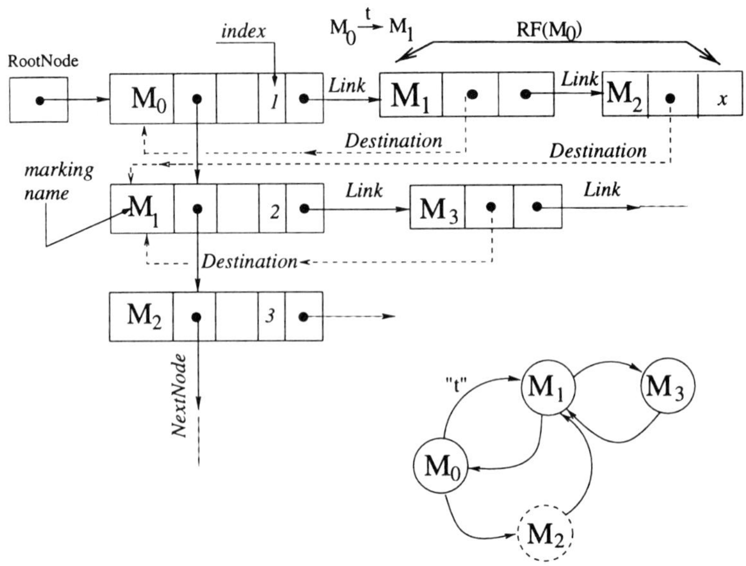

a) List of pointers modelling the reachability graph denoted by b)

b) Reachability graph (example)

Figure 22.5: Pointer list modelling the reachability graph of the Petri net

```
Kernel_Matrix_Function (t)

Q:=Allocate_Integrand_Space (N, N)
m :=Initial_Marking
While m differs from 0
Q [m -> index][m->index] :=0
m2 : =m->Link

for x:=0: step: t
prod :=diff G [ m->index][m2->destination->index](x)
pointer:=m2->destination->index

    While m2 differs from 0
    If m2->destination->index differs from pointer then
      prod *=1-G[m->index][m2->destination->index](x)
      End of if
      m2:=m2->Link
    End of While

End of For
Q[m->index][pointer]:=prod
m:=m->Next_Node
End of While
Return (Q)
```

Figure 22.6: Algorithm for generating the kernel matrix function

- the total number of Petri net tokens is preserved, but the Petri net gets locked because there are no more enabled transitions and the initial marking has not been found out.

In order to perform the stochastic analysis, the kernel matrix function is generating using the algorithm proposed in Figure 22.6.

22.4 Application

We shall consider five places and five transitions PN depicted in Figure 22.7, modelling a $2 \times 100\%$ system with a cold stand-by unit. When a failure affects the operating unit, the transition "t_1" is fired. Consequently, a token is removed from place "p_1" (labeled "Unit Up", a token is added to the place "p_3" (labeled "Units Failed") and a token is set to the place "p_5" (labeled "Recovering Demand"). Under these circumstances, if a token is placed in the place "p_2" (labeled "Units Stand-By"), the transition "t_2" is enabled and its firing sets the stand-by unit into operation state. On the other hand, the stand-by unit may fail when inactive, consequently a token may go from the place "p_2" to "p_4" (labeled "Undetected standy-unit unit failure"). If one assumes that no systematic preventive maintenance policy is enforced for detecting such an event, then it could be detected only by an eventual call to stand-by unit (e.g. a token in the place "p_5"). If this happens, the transition "t_4" is enabled, the tokens are removed from the places "p_4" and "p_5" and added in the places "p_3" and "p_5". Once the repair is completed, the unit becomes "Stand-by", and a token is added to "p_2", being removed from the place "p_3". Two repairing teams are available at any time.

The initial marking of the net contains one token in the place p_1 and one token in the place p_2. Using the PAMS (Performance Analysis Markov Systems) package tool [Ulmeanu (1995)], we generate automatically the nine states marking set $R(M_0)$ shown in Figure 22.7. The marking reachability graph is depicted in Figure 22.10 and we are assuming that is a semi-Markovian one, having its conditional sojourn distributions functions as presented in Figure 22.9.

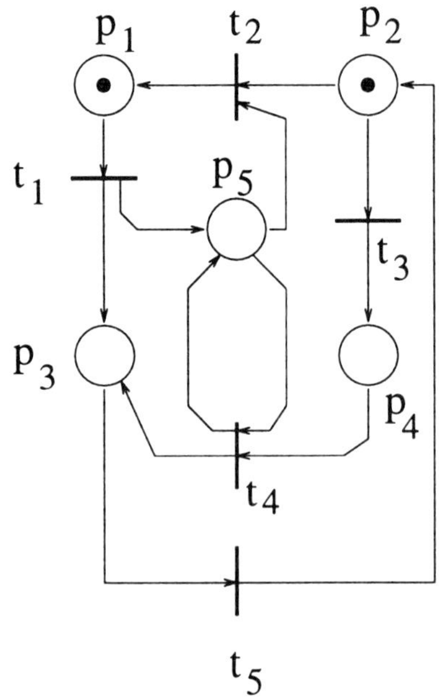

Figure 22.7: PN-1

State	p_1	p_2	p_3	p_4	p_5
M_0	1	1	0	0	0
M_1	1	0	1	0	0
M_2	1	0	0	1	0
M_3	0	1	1	0	1
M_4	0	0	2	0	1
M_5	0	0	1	1	1
M_6	0	1	0	1	1
M_7	0	0	0	2	1
M_8	0	2	0	0	1

Figure 22.8: State space of PN-1

Transition	*Conditional sojourn time distribution functions*
t_1	$Weib(\alpha = 1000\ h, \beta = 2.)$
t_2	$Exp(r_2 = 0.2\ h^{-1})$
t_3	$Exp(r_3 = 0.001\ h^{-1})$
t_4	$N(\mu = 10\ h, \sigma = 0.5\ h)$
t_5	$LOGLN(a = 6\ h; b = 1.16\ h)$

Figure 22.9: The conditional sojourn time distributions functions for the reachability graph depicted in Figure 22.10

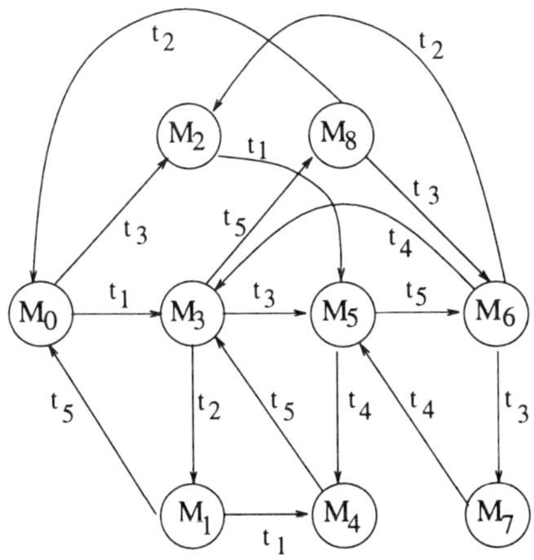

Figure 22.10: The marking reachability graph of PN-1

The states M_0, M_1 and M_2 are up states, the other M_3, M_4, M_5, M_6, M_7 and M_8 are down ones. In the Figure 22.9, $Exp(\lambda)$ means the exponential distribution of parameter λ, $Weib(\alpha, \beta)$ means the Weibull distribution of parameters α (the scale parameter) and β (the shape parameter), $N(\mu, \sigma)$ means the normal distribution of parameters μ (the expected value) and σ (the variance) and $LOGLN(a, b)$ means the log-normal distribution of parameters a and b, i.e. the probability density function is given by $f(t) = \frac{1}{b \cdot t \cdot \sqrt{2 \cdot \pi}} \cdot exp(-\frac{(ln(t)-a)^2}{2 \cdot b^2})$. Finally, according with the input parameters depicted in Figure 22.9, we can easy find out the asymptotic availability A_∞, the Mean Up Time MUT and the Mean Down Time MDT of the system, using the numerical method proposed in [Limnios (1993)]. Figure 22.11 shows the asymptotic availability of the system as a function of expected parameter μ. Figure 22.12 shows the asymptotic availability of the system as a function of exponential parameter λ_2, whereas the Figure 22.13 shows the Mean Up Time and the Mean Down Time of the system as function of exponential parameter λ_3.

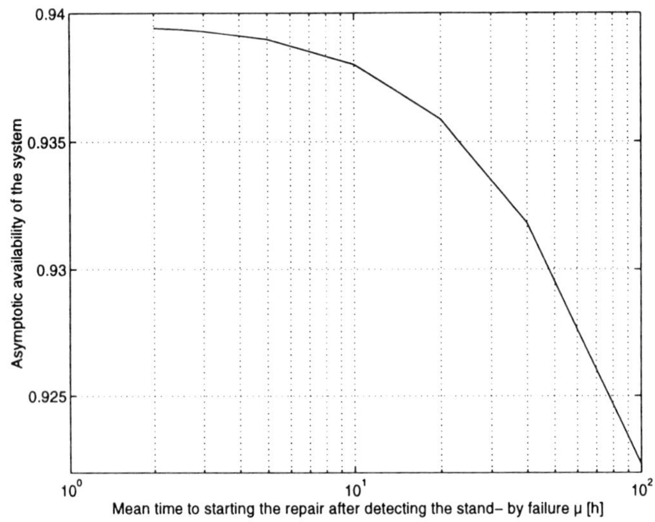

Figure 22.11: The asymptotic availability of the system versus the mean time μ to starting the repair after detecting the stand-by failure

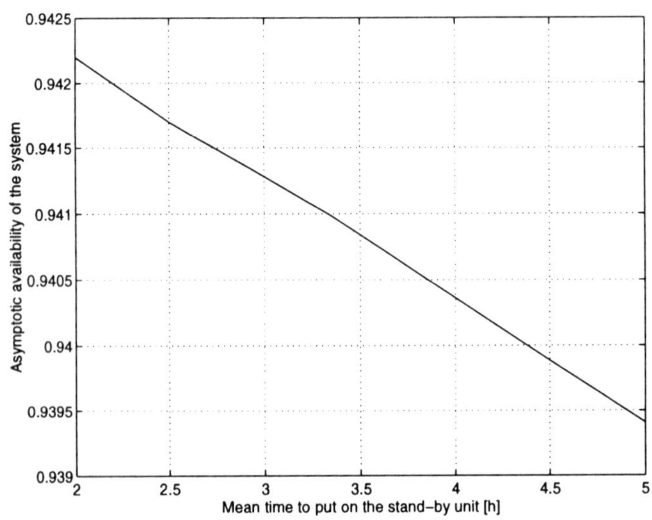

Figure 22.12: The asymptotic availability of the system versus the mean time $1/\lambda_2$ to put on the stand-by unit

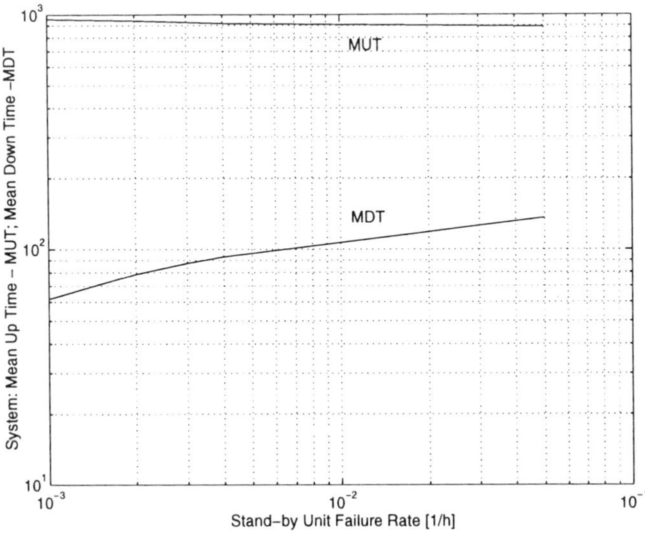

Figure 22.13: The Mean Up Time and the Mean Down Time of the system versus the mean time to failure $1/\lambda_3$ of the stand-by unit

22.5 Conclusions

The results that are given may be regarded as a very usefully bridge between a flexible tool modeling (Petri Nets) and a general mathematical tool (semi-Markov and Markov systems). The SMSPN models are often more compact and more easier to construct than using another models, such as a reliability diagram block or a fault tree, since the SMSPN graphical and mathematical resources match the real system behaviour, especially its competitive and concurrent activities. Moreover, the SMSPN models permit us the automatic generation of the marking reachability graph and of the semi-Markov or Markov state space. This is indispensable for moderate and complex systems. We point-out here only the semi-Markovian case, but the general improvement over that we are presented is well known [Baccelli (1992), Dugan (1984), Molloy (1982), Lin (1988)]. The rich body of references in the field of Petri nets [Murata (1989)] provide its growing interest in large domains, especially in the reliability and maintainability fields.

Acknowledgement. This work was initiated while the first author held a Ph.D. stage at Université de Technologie de Compiègne UTC. He would like to thank Professor Nikolaos Limnios for his hospitality and helpful support; and also express his gratitude to the Agence d'Environnement et de la Maîtrise de Energie ADEME for the grant awarded for this stage.

References

1. Baccelli, F., Cohen, G., Olsder, G. J. and Quadrat, J.-P. (1992). *Synchronization and linearity. An Algebra for Discrete Event Systems*, New York: John Wiley & Sons.

2. David, R. and Alla, H. (1989). *Du Grafcet aux réseaux de Petri*, Hermès edition.

3. Dugan, J. B., Trivedi, K., Geist, R. M. and Nicola, V. F. (1984). Extended Stochastic Petri Nets. *PERFORMANCE'84*, pp. 507–519. North-Holland: Elvesier Science Publishers B. V.

4. Florin, G. and Natkin, S. (1985). Les réseaux de Petri stochastiques, *Technique et Science Informatiques*, **0752-4072/85/01**, 143–160.

5. Limnios, N. (1993). A transient solution method for semi-Markov systems, *Statistics and Probability Letters*, **17**, 221–220.

6. Lin, C. and Marinescu, D. C. (1988). Stochastic high-level Petri nets and applications, *IEEE Transactions on Computers*, **7**.

7. Marsan, M. A., Balbo, G. and Conte, G. (1984). A class of generalized stochastic Petri nets for the performance evaluation of multiprocessor systems, *ACM Transactions on Computer Systems.*

8. Merlin, P. M. and Farber, D. J. (1976). Recoverability of communication protocols - implications of a theoretical study, *IEEE Transactions on Communications.*

9. Molloy, M. K. (1982). Performance analysis using stochastic Petri nets, *IEEE Transactions on Computers*, **31**, 913–917.

10. Murata, T. (1989). Petri nets: Proprieties, analysis and applications, *Proceedings of the IEEE*, **77**.

11. Natkin, S. (1980). *Réseaux de Petri Stochastiques, Thèse de Docteur Ingenieur.*

12. Noe, J. D. and Nutt, G. J. (1973). Macro e-nets representation of parallel systems, *IEEE Transactions on Computers.*

13. Papoulis, A. (1965). *Probability, Random Variables, and Stochastic Processes*, New York: McGraw-Hill Publishing Company.

14. Ulmeanu, A. P. (1995). *Manuel de l'utilisateur PAMS*, Université de Technologie Compiègne, France.

23

Incremental Approach for Building Stochastic Petri Nets for Dependability Modeling

Nicolae Fota, Mohamed Kaaniche and Karama Kanoun

SOFREAVIA, Issy les Moulineaux, France
LAAS-CNRS, Toulouse, France

Abstract: This paper presents a modeling approach based on Generalized Stochastic Petri Nets (GSPN) aimed at the dependability (reliability, availability and safety) evaluation of large hardware and software fault-tolerant systems. The model is built and validated in an incremental manner, according to a set of construction guidelines and rules, aimed to obtain models which are structurally valid, readable, compact and flexible.

Keywords and phrases: Dependability modeling, Generalized Stochastic Petri Nets, incremental approach

23.1 Introduction

Dependability evaluation of large fault-tolerant computing systems requires the modeling of the failure and repair behavior of hardware and software system components and the numerous interactions between them, resulting in complex models. Depending on the dependability measures to be evaluated, the modeling level of detail can furthermore increase this complexity. State-space models, in particular homogeneous Markov chains, are commonly used to model the dependability of fault-tolerant systems, because they are able to capture various functional and stochastic dependencies among components. To facilitate the specification and generation of large state-space models, higher level model types such as GSPNs (Generalized Stochastic Petri Nets with timed and immediate transitions) are generally used since they: a) allow a compact representation of the behavior of systems involving synchronization, concurrency and conflict phenomena, b) provide some structural verification of the model, and c) can be automatically converted to Markov models.

Several studies have been conducted to build models based on GSPNs or their offspring's, in a modular way, i.e., model composed of sub-models representing components of the system [see, e.g., Rojas (1996) and Sanders and Meyer (1991)]. To the best of our knowledge, some of the methods proposed are efficient when sub-models are loosely coupled (this is the case, for instance, when the models are built for performance evaluation), others become hard to implement when interactions are too complex. However, for the dependability modeling of fault-tolerant systems, multiple and complex interactions between system components have to be explicitly considered because of the dependencies induced by the component failures and repair [Kanoun and Borrel (1996)and Ziegler (1996)].

In this paper, we present a set of guidelines for the efficient and reliable construction of dependability models based on GSPN. To master complexity, a system model is progressively built and validated according to a *modular and incremental approach* taking into account the following criteria: 1) enhanced readability and compactness (via the use of a reduced number of places and transitions and well defined modeling constructs), 2) flexibility and reusability of parts of the model to ensure easy modification of the model when new assumptions are considered. Our guidelines aim to assist the user in the implementation of the system behavior and failure assumptions into the GSPN formalism, while mastering the model complexity and avoiding modeling errors that are inherent to this formalism (unbounded places, conflicts between immediate transitions, etc.). In the following, only a subset of these guidelines is outlined. A more detailed description of our approach is given in Fota (1997).

23.2 Presentation of the Incremental Approach

The model is built and validated in an *incremental* manner. At the initial step, the behavior of the system is described taking into account the failures of only one selected component, assuming that the others preserve their operational nominal state. The failures of the other components are then integrated progressively in the following steps of the modeling process. At each step: 1) a new component is added, and 2) the GSPN model is updated and validated (taking into account the impact of the additional assumptions on the behavior of the components already included in the model). At each integration step, the GSPN model is validated. The validation is carried out at the GSPN level (structural verifications) and also at the Markov level in order to check the different scenario represented by the model. When the Markov chain size increases, the exhaustive analysis of the Markov chain is impractical. In this case, sensitivity analyses are used to check the validity of the model assumptions.

With the incremental approach, only few additional assumptions are added at each step: the user has to model only the behavior of the new component and describe how the previous version of the model has to be modified to account for the interaction of this component with those already integrated. With respect to the order of integration, it is recommended to start the modeling with the components which behavior is highly dependent on the state of the other components. Thus assumptions corresponding to interactions between components are not modeled at once but progressively added at each step.

To ensure better control of the model evolution, we defined a set of rules and guidelines for modeling the components behavior and their interactions. These rules are not mandatory but facilitate significantly the construction of the model and particularly its validation. They allow to perform the modular construction of a structurally valid GSPN. Also, they promote reuse of some parts of the model as the components and the interactions are well identified.

23.3 Guidelines for Modular Construction of GSPN Models

The components' behavior is described by sub-models called *modules*, while interactions between components are modeled using *module coupling mechanisms*.

Modules. For each component, a GSPN *module* is built to describe the behavior of the component resulting from the occurrence of its internal events. Each *module* is built of internal places, characterizing the states of the component, and internal transitions, modeling the occurrence of failure and restoration events specific to that component. The *basic rule* for the module construction stipulates that its marking invariant should be equal to 1. Therefore, each module has a single token and each internal transition has a single internal input place. The latter cannot be connected to the internal transitions through inhibitor arcs or bi-directional arcs. To improve the compactness of the module, two additional rules (called *optimization rules*) are recommended: 1) avoidance of immediate internal transitions, and 2) module folding using memories.

We illustrate the latter rule with the example of Figure 23.1. When a module exhibits identical patterns, a folding can be done. In this case the states of the modeled component are given by the marking of one or several supplementary places called *memories* (which are not involved in the marking invariant of that module: place M_A in the example), combined to the marking of the rest of places internal to that module (places P1_A, P2_A and P3_A).

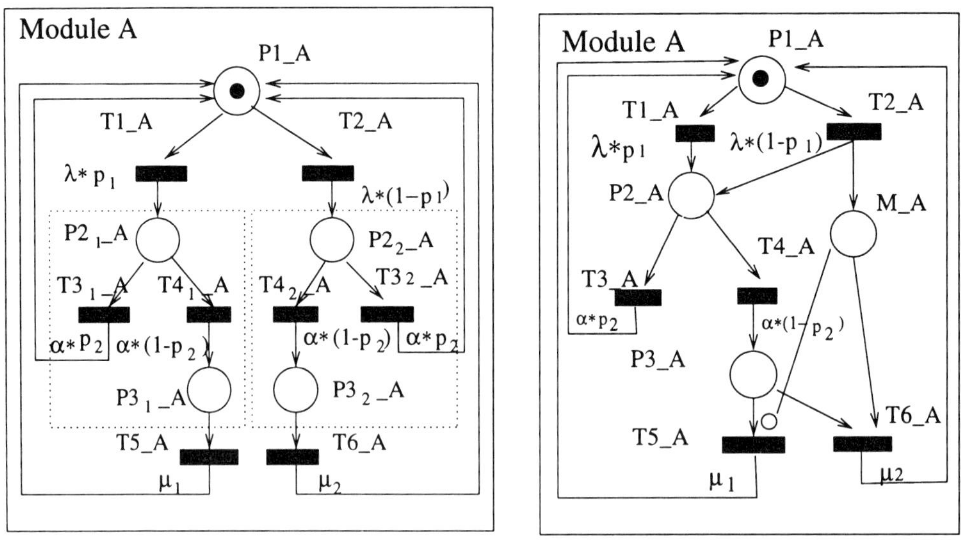

a. Initial GSPN b. Folded GSPN

Figure 23.1: Module folding using memories

Module coupling mechanisms. The different types of interactions among components, resulting from functional, structural or stochastic dependencies, are implemented through three basic module coupling mechanisms: marking tests, common transitions and interconnection blocks (see Figure 23.2). *Marking tests* are used when the occurrence of an event of a given component (internal transition T1_A of module A in Figure 23.2a) is conditioned upon the marking of other components modules. Only places of these modules that are not vanishing can be involved in the test (a place is said to be *vanishing* if it can contain tokens only in vanishing markings [1]). Inhibitor arcs and bi-directional enabling arcs are used to implement marking tests. Thus, when the internal transition fires, the marking of the places involved in the test remains unchanged.

Common transitions are shared by several modules (Figure 23.2b). They describe the occurrence of events which are common to different components, leading to the simultaneous marking evolution of the involved modules.

The *interconnection block* models the consequence of the occurrence of an event of a given component in terms of state changes of other components (Figure 23.2c). These consequences may depend on the state of components other than those modeled by the initializing and the target modules. A block connects one or several initializing modules to one or several target modules. It is built of one *input place* (IP) and a set of immediate *output transitions* (OT_k).

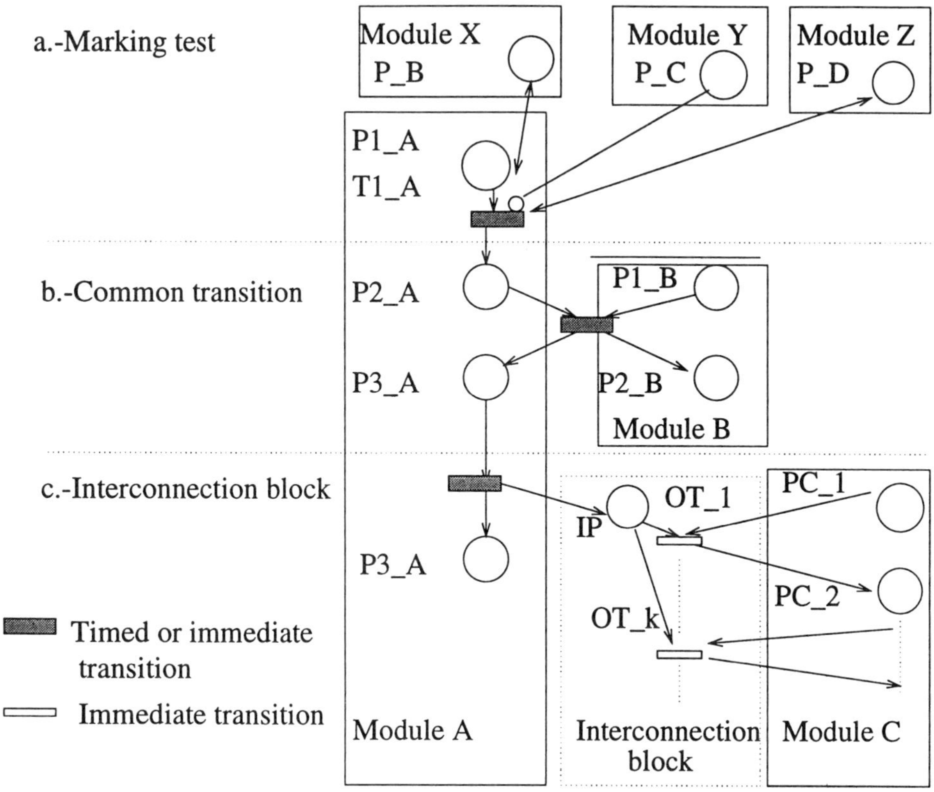

Figure 23.2: Module coupling mechanisms

Two types of blocks are defined, according to the type of consequences they model: *immediate blocks* and *deferred blocks*, and each one should satisfy specific construction criteria. Only immediate block construction will be presented here.

Immediate block construction rules. As the modeled consequences are immediate, the token which entered the block input place following the firing of the initializing module internal transition, should leave that place before a second event occurs. Thus, the input place should be a vanishing one.

In the example of Figure 23.3, B is an initializing module and A a (unique) target module. In order to ensure that the token placed in the block input place IP is immediately removed, all the possible states of module A should be tested and the evolution of this module, resulting from the occurrence of the initializing event, should be determined (we recall the one token per module rule). The set of the block output transitions should be built by assuring that IP is a vanishing place. When one of the output transitions fires, the token of module A might be transferred to another place of that module (the transfer may depend on the marking of other modules, as is the case for the subset OT1_j, j=1,2,...p, or not, as for the firing of OT2) or it might stay in the initial place. In the latter case, the token of the block input place is absorbed by means of a specific output transition, connected by a bi-directional arc to the corresponding place of the module (it is the case of OTk, aimed to empty IP when Pk_A is marked and the block is initialized).

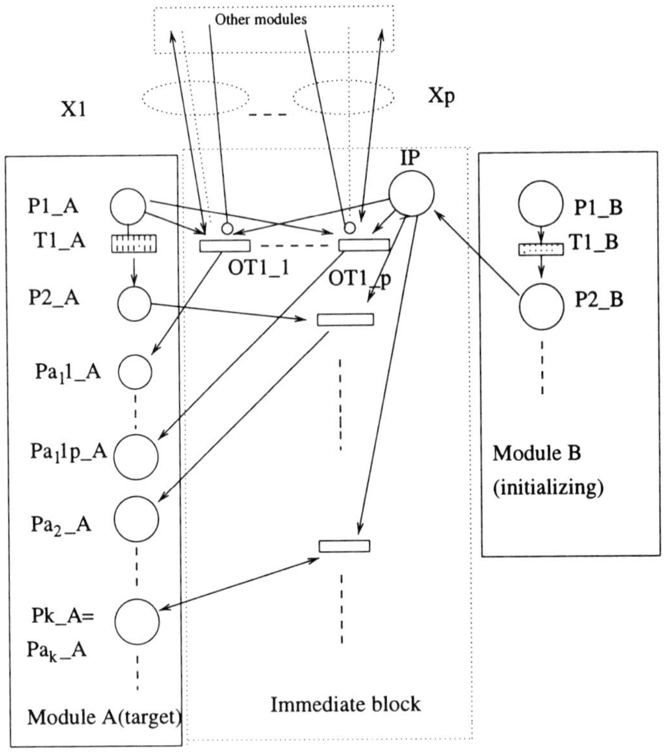

$n =$	number of places internal to module A
$m =$	number of module A not vanishing placed ($m \leq n$)
$X1, \ldots, Xp =$	marking tests on output transition OT1_1,...,OT_p, respectively
OT1_1,...,OT1_p=	output transitions provided for the cases where P1_A is marked
$a_1p\ (p \in (2, \ldots, n)) =$	index to places to which the token the module is transferred following the block activation in a context where X1,...,Xp respectively, become TRUE and P1_A is marked output transition provided for the case where Pj_A is marked
OTj (j = 2,...,m)=	output transmission provided for the case where P1_A is marked
Pa: A(i = 2,...,m)=	place in which the token of the module is transferred when Pj_A is marked (note that the token may remain in the departure place, e.g. Pk_A $\equiv Pa_k$_A associated with OTk)

Figure 23.3: Example of using an immediate block - case of one target module

Consequently, to build an immediate block it is necessary to ensure that for every possible marking of the model, there is one and only one output transition of the immediate block that fires when its input place is marked. As a result, all the possible markings of the modules involved in the module coupling described by that block should be identified before associating an output transition to each of these markings. To find the number of output transitions needed, one should compute the Cartesian product of the possible markings of the target module and of the other modules which are conditioning the evolution of the target module following the firing of the transition having initialized the block.

This construction rule is formalized in the following [see Fota (1997) for further details]. To each tangible marking k of the target modules and of other modules which may condition the consequences of the event occurrence, we associate a logical function B^k (called B-condition) defined as follows:

$$B^k = \bigcap_i b_i^k \text{ where } b_i^k = TRUE \text{ if } m(Pi) = 1 \text{ and } b_i^k = FALSE \text{ otherwise}$$

$Pi \in \{P^k\}$ = set of all places of target modules and of other modules which are not vanishing and are conditioning the consequences of an event occurrence; $m(Pi)$ is the marking of Pi.

To each possible B-condition B^k we associate an output transition OT_k. The input places of transition OT_k are defined by the places Pi involved in the B_condition which are not vanishing, in addition to the block input place IP.

The interconnection block composed of input place IP and output transitions OT_k is an immediate block if and only if the set of B-conditions associated to the output transitions OT_k is:

- complete: $\bigcup_k B^k = TRUE$ and
- exclusive: $\forall\ B^i, B^i \in \{B^k\}\ (i \neq j),\ B^i \wedge B^i = FALSE$

Block decomposition. To ensure re-usability and thus improve compactness and flexibility of the model, generic elementary blocks can be defined to describe specific interactions. These blocks can be connected according to a parallel-series topology to model complex interactions (Figure 23.4). Note that given the complexity of the B-conditions of certain immediate blocks, the block development is facilitated by the use of binary decision diagrams [see Fota (1997)].

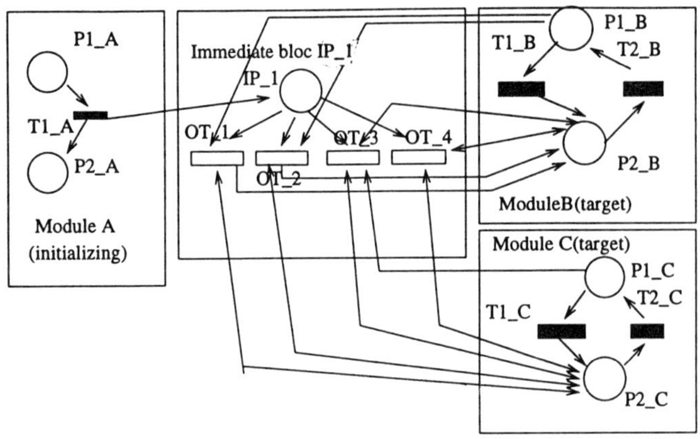

a:Interaction modeled by a single block

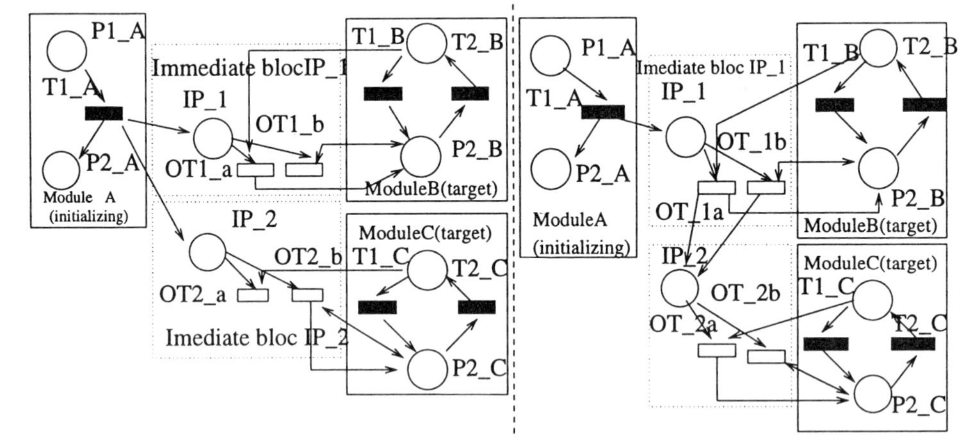

b. Parallel decomposition of the single block.

c. Series decomposition of the single block.

Figure 23.4: Block decomposition in parallel or series elementary blocks

The use of series-parallel sets of elementary immediate blocks may lead to the simultaneous enabling of several output transitions of different blocks. Priorities should be defined for the firing of these transitions to ovoid conflicts. It is the case if the firing of the output transitions of one block changes the marking of at least a not vanishing place involved in the B-conditions associated to output transitions of other block(s) of the same set. We established the following priority rule: a highest priority level should be assigned to the block whose B-conditions are changed by the firing of at least one of the output transitions of another block of the same set. If reciprocally, the B-conditions of the latter block change following the firing of one of the output transitions of the former, the corresponding interactions should be modeled by a single block.

To assist the user in the development of interconnection blocks, we defined several additional rules which are not presented here. For example, the modeling of components exhibiting identical behavior is possible by using several tokens in a module, providing that all B-conditions can still be expressed in a binary logic. Otherwise, an extension of the defined construction rules would be necessary, based on the relaxation of the one-token-per-module condition.

In the next section we illustrate the incremental approach and some of the construction rules on a simple architecture.

23.4 Example: Duplex System

The duplex system architecture (Figure 23.5) is derived from one of the subsystems of the French ATC computing system presented in Fota (1997). It is composed of two application software replicas (So the operational one and Ss the spare), running on two redundant computers (Co and Cs) and accessing two redundant disks (Do and Ds). So accesses Do for reading or writing data and periodically sends a copy of its current state to Ss through a local data link. Ss maintains the back-up data on Ds. Do and Ds can be accessed by either Co or Cs. Two configuration modes are possible: "local disk association" when Co is connected to Do, and "remote disk association" when Co is connected to Ds. The disks and the local data link are supposed to be failure-free since their failure rates are significantly small compared to those of the other components.

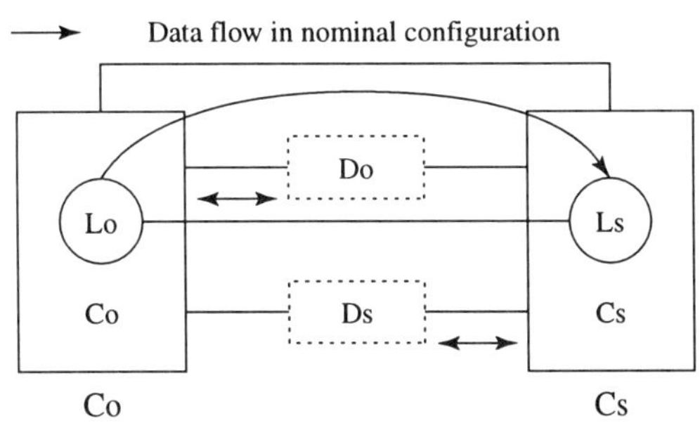

Figure 23.5: Duplex system architecture

The So replica exhibits two failure modes:

- "*So and Ss common mode failures, d_So_Ss*"; these failures are due to the corruption of part of the dynamic data processed by both replicas and are recovered by a manual procedure "*rec*" which involves So restart after elimination of the corrupted data;

- *"local mode, d_So_loc"*; these failures affect only So replica, the following recovery procedures are attempted consecutively: restart of So (*rst*), followed, in case of failure (*rst_n*: operational data corrupted), by a manual reconfiguration of Ss into the operational mode (*sw*: Co and Cs, Do and Ds are interchanged). If Cs is down but the back-up data is still available, So can be restarted on Co after having associated Ds to the computer ("*ras*"). Finally, if the back-up data is unavailable, in case of failure of "rst", procedure "rec" is performed.

With regards to the Ss replica, it is assumed that any failure (*d_Ss_loc* or *d_So_Ss*) leads to the loss of the dynamic data processed by this replica. Moreover, Ss can be restarted only if So is operational. A computer failure leads to the stop of the software replica hosted. When Co fails, Cs becomes the operational computer after the system reconfiguration. If both computers are down, data is lost.

The model is built in four steps, at each step a supplementary component being integrated. So is the first component to be modeled, due to the fact that its behavior is highly dependent on the state of the other components. For example, the failure of Co leads the So replica to be stopped whereas So failure does not affect Co. In the second step Ss is added, in the third one Cs is added and finally Co is integrated in the fourth step.

At the first step, the model describes the behavior of So assuming that Co, Cs and Ss are preserving their initial, nominal states (Figure 23.6). The evolution of So is modeled as a result of the occurrence of its own events only. Additionally, consequences that So might induce on the behavior of the other components are not taken into account at this step.

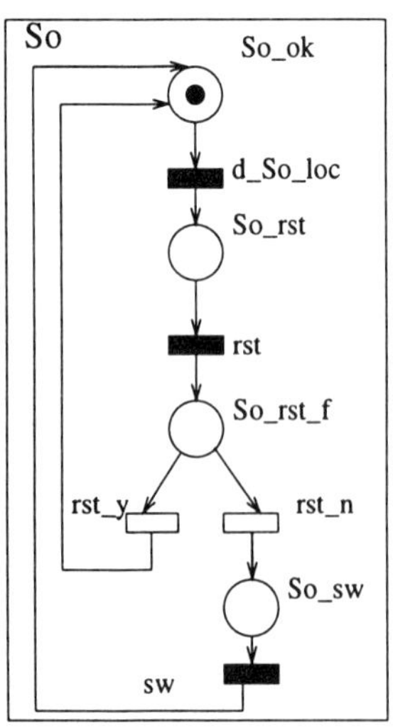

Place	Semantics
So_ok	So in "up"state
So_rst	So being restardes
So_rst_f	So at the end of restart
So_sw	So waiting for manual reconfiguation of Ss into operationalmode

Transation	Semantic
d_So	Local mode failure of So
rst	So restart
rst_y	Successful restart of So
rst_n	Failed restart of So
sw	Manual reconfiguration of Ss into operational mode

Figure 23.6: First step: So modeling

In Figure 23.6, the occurrence of the local mode failure (d_So_loc) leads to the restart of the replica (rst). If the restart is successful (rst_y and its associated probability), So returns to its initial state, otherwise (rst_n), the manual reconfiguration of Ss into the operational mode (sw) is performed and thus So enters a state where this procedure is progressing (replicas have changed their roles). In fact, this procedure involves to stop Ss, but this consequence is not modeled at this step (as Ss is assumed to preserve its nominal state, it will be considered available as soon as So and Ss change their roles). Common mode failure of So and Ss is taken into account only at the second step, when component Ss is integrated. By integrating component Ss at the second step, besides the So and Ss common mode failure, the Ss own local failure and restart are taken into account, as well as the procedure "rec", which is performed after unsuccessfully attempting to restart So while Ss is unavailable.

Figure 7 illustrates the modeling of step So-Ss, highlighting the model evolution from the first step to the current one. The shaded parts of the model illustrate how the GSPN model is updated. Module Ss is added to the initial model, while the already existing So module is enriched (places So_tst_sw, So_rec and transitions c_sw, c_rec, rec), to take into account the procedure "rec". Transitions c_sw and c_rec model the dependence of So evolution on the state of Ss after the failure of procedure "rst" (procedure "sw" is possible only if Ss is up). The interactions between So and Ss modules, resulting from the failure and repair assumptions, are implemented using marking testing (on transitions c_sw and c_rec), a common transition (d_So_Ss) and an interconnection block (StpSo_Ss, which models the stop of Ss due to the occurrence of "sw" in So).

At the next modeling step, component Cs is integrated (Figure 23.8). Besides adding module Cs and blocks StpCs_Ss (used to model Ss stop due to Cs failure) and Chg_data (managing the evolution of backup data state), already existing modules or blocks are enriched to take into account the impact of the failure and repair assumptions of Cs.

Module So gets additional places and transitions to model the procedure "ras", which is used to restart So when Cs is down but the back-up data is still available. Place Ss_hs_sp is added to module Ss to describe the Ss state where it is stopped while back-up data is still available on the associated disk. Finally, the block StpSo_Ss is enriched with an additional output transition (t2), to move the Ss token from place Ss_hs_sp to Ss_hs, following the firing of c_sw or c_ras (and given that Ss_hs_sp is marked). Note that the set of B-conditions Ss_ok and Ss_hs_sp associated to the output transitions of block StpSo_Ss is complete and exclusive (Ss_hs is never marked when the block is initialized).

Because of lack of space we don't present here the last step of the GSPN model construction. Note that in the case of this relative simple example chosen for the need of the presentation, the complexity of the final GSPN model (about 20 places and 40 transitions) is comparable to that of the Markov chain (15 states), but the interest of the GSPN modular and incremental modeling has been largely proven with complex applications Fota (1997) and Fota et al. (1997).

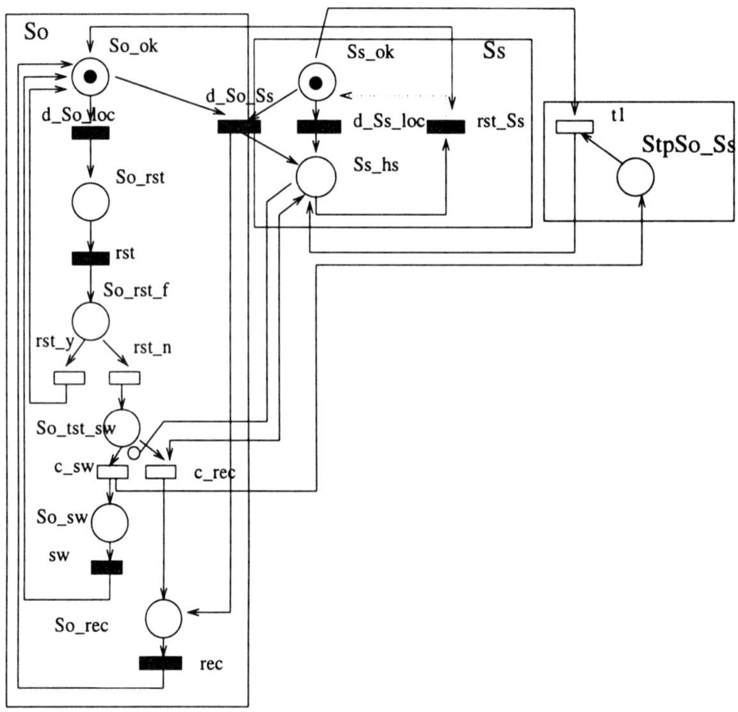

Additional places	Semantics
Ss_ok	Ss in “up” state
Ss_hs	Ss being restarted
StpSo_Ss	Input place of block modeling the stop of Ss due to So
So_tst_sw	So before the testing of state of Ss

Additional transition	Semantics
d_Ss_loc	Local model failure of Ss
rst_Ss	So restart
d_So_Ss	Common mode failure of So and Ss
c_sw	Choice of basic procedure
c_rec	Choice of rest procedure
rec	So restart after elimination of corrupted data
t1	Output transition of block StpSo_Ss

Figure 23.7: Second step: So-Ss modeling

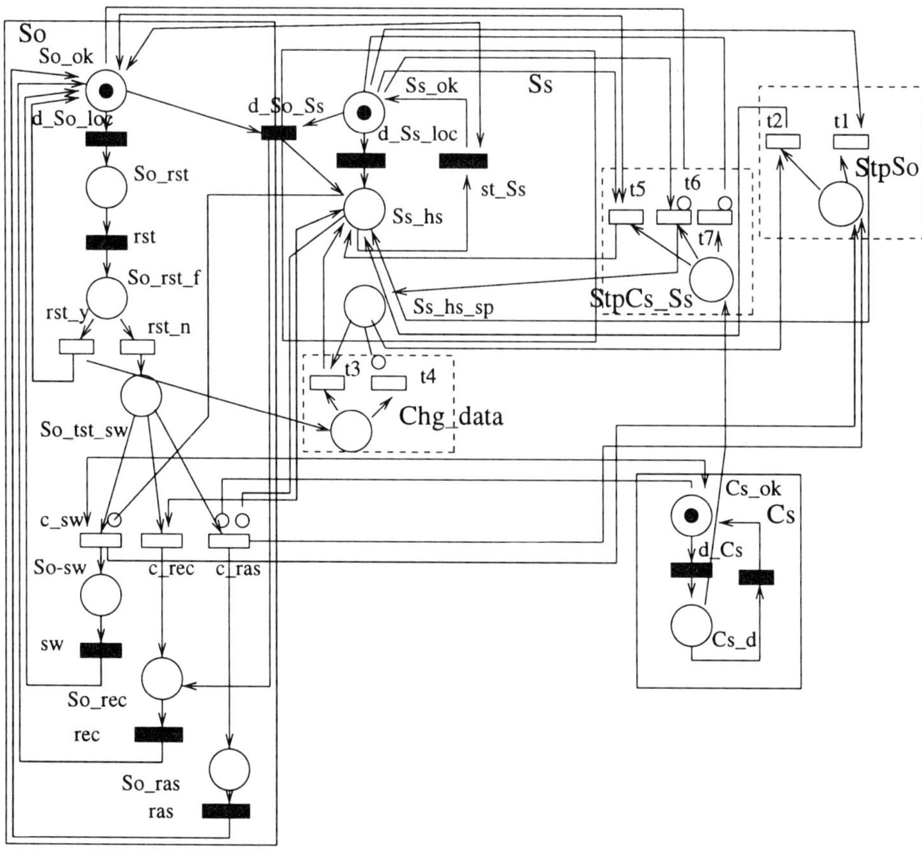

Additional places	Semantics
Ss_hs_sp	Ss "down", back-up data still available
So_ras	So being restarted on Co after remote disk association
Cs_ok	Cs in "up" state
Cs_d	Cs in "down" state
Chg_data	Input place of block modeling the state changing of back-up data
StpCs_Ss	Input place of block modeling the stop of Ss due to Cs

Additional transitions	Semantics
c_reas ras	Choice of reas procedure Restart of So on Co after remote disk association
d_Cs	Cs failure
r_Cs	Cs repair
t2_t7	Output transitions of blocks

Figure 23.8 Third step So-Ss-Cs modeling

23.5 Conclusions

In this paper, we proposed an incremental and modular approach to assist the users in the construction and validation of complex GSPN models. A GSPN model is structured as a set of modules coupled via three main mechanisms: marking tests, common transitions and interconnection blocks. The use of interconnection blocks enhances model readability and allows easy modification of the model when new assumptions need to be included. Moreover, we have defined a set of guidelines and rules for implementing the modules and their interconnections. These rules aim to obtain structurally valid models and reduce the number of modeling errors inherent to the GSPN formalism. Following this approach, we successfully modeled the dependability of the main subsystems of the French air traffic control computing system referred to as CAUTRA [Fota et al. (1997)]. The CAUTRA is implemented on fault-tolerant computers distributed over five control centers and one centralized operating center, connected through a WAN. The models constructed for the different subsystems are very complex. For example the GSPN of the system devoted to Radar and Flight Plan Data Processing has more than 100 places and 500 transitions and the corresponding reduced Markov chain is about 25 000 states Fota (1997) and Fota et al. (1997).

The proposed approach is suitable for the dependability modeling of computer systems with numerous hardware and software components, multiple interactions, and complex procedures for fault tolerance and restoration. Several extensions of this work are currently investigated: 1) the definition of rules for model folding to optimize the GSPN model for redundant components with similar behavior, 2) the definition of a higher level language for modules and interconnection mechanisms specification along with transformation rules for GSPN model generation.

Acknowledgements. This work was partially supported by CENA (Air Navigation Research Center), France, while Nicolae Fota was with LAAS - CNRS. The authors would like to thank Alain Peytavin at CENA for his helpful technical suggestions, support and confidence which made this work possible.

References

1. Marsan, A. M., Balbo, G., Conte, G., Donatelli, S., Franceschinis, G. (1995). *Modeling with Generalized Stochastic Petri Nets*, New York: John Wiley & Sons.

2. Fota, N. (1997). Incremental specification and construction of dependability models - Application to CAUTRA air traffic control computing system, *Ph.D. Thesis*, INP Toulouse (in French).

3. Fota, N. et al. (1997). Dependability evaluation of an air traffic control computing system, *Report LAAS 97-466* (in French).

4. Kanoun, K. and Borrel, M. (1996). Dependability of fault-tolerant systems - explicit modeling of the interactions between hardware and software components, *IEEE International Computer and Dependability Symposium IPDS '96*, pp. 252–261, Urbana-Champaign, Illinois.

5. Rojas, I. (1996). Compositional construction of SWN Models, *Report ECS-CSG-21-96*, Department of Computer Science, University of Edinburgh.

6. Sanders, W. H. and Meyer, J. (1991). Reduced base model construction methods for stochastic activity networks, *IEEE Journal on Selected Areas in Communications*, **9**, 25–36.

7. Ziegler, C. Dependability of automotive embedded architectures, *Ph.D. Thesis*, Report LAAS 96-289, INP Toulouse, France.

24

Lifetime of High Temperature Working Pipes

E. Alămoreanu, R. Iatan, R. Chiriţă and R. Ceauşu

Politehnica University of Bucharest, Bucharest, Romania

Abstract: This paper introduces a method for predicting lifetime of industrial pipe systems. Creep and fatigue effects of material are considered and a method of simulating reliability is developed. The estimated lifetime as well as remaining time to failure, expressed in operating hours or in number of loading cycles, are functions of several main parameters such as: creep strains, strength resource, internal damage state, material properties (hardness, ultimate strength, flow limit, density). For a system of high temperature working pipes, an algorithm of computing lifetime and reliability is applied.

Keywords and phrases: Lifetime, pipe, creep, fatigue, simulation, reliability

24.1 Introduction

Under static or quasi-static pressure load, a pipe is commonly supposed to exhibit a longitudinal failure, assuming that is made from a tenacious material, having a medium grade of internal damage. The spreading of fracture, the state and internal pressure at the moment the fracture occurs are related to metal wall temperature, critical fluid pressure being lower for higher temperatures. When analysing metal behaviour is tenacious or fragile, whether the applied loads and stresses are of elastic or plastic type, as well as the internal structure damage state and the rate of damage propagation. In case of long lasting thermal exposure, fracture propagation may shift from tenacious to fragile behaviour due to internal structure modifications.

24.2 Failure Risk

At temperatures above $450C^{\circ}$, forged or laminated steels exhibit important reductions of elasticity and flow limits. One can introduce the failure risk as

$$R_{\sigma} = \int_{V} f(\sigma)\, dv \tag{24.1}$$

where σ is the stress of an elementary volume dV. The integral is evaluated over the whole volume, each element exposed to thermal stress having its share to the failure risk. The stress distribution function may be introduced in the following form proposed by Weibull, that proved good agreement with experimental data:

$$f(\sigma) = \left[\frac{(\sigma - \sigma_0)}{\sigma_{\mu}}\right]^{m} \tag{24.2}$$

where σ_0 is a limit under of that the probability of fracture (P_R) is zero, σ_{μ} is an approximate value of the ultimate strength for a specimen of volume V_0; m is Weibull modulus - a parameter describing the homogeneity of material (a greater m results for a higher homogeneity). A common practice is to plot $\ln(\ln \frac{1}{1-P_R})$ over $\ln \sigma$. Such Weibull distribution proves if experimental data points plotted on these coordinates align themselves in a straight line with the slope m. For most of the situations one may assume $\sigma_0 = 0$, but for a higher precision σ_0 may be considered:

$$\log \sigma_0 = \log \left[\frac{v}{2(m+1)} \Big/ \ln \frac{1}{1-s}\right] \Big/ m \tag{24.3}$$

Parameters σ_0 and m are determined from tensile and bending tests on specimens of the same material, but of different volumes. The factor m yields from

$$m = \ln(V_2/V_1)/\ln(\sigma_{ri}/\sigma_{rt}) \tag{24.4}$$

where σ_{ri} is the bending ultimate strength; σ_{rt} is the tensile ultimate strength. For such computations, statistical analysis of experimental data is required. Admissible stress values drawn out directly from mechanical tests worked on fragile materials may be accurate only if the model reproduces exactly the real estate of loading, the volume subjected to stress and the stress volume distribution. Obviously, it is not possible to introduce on a specimen test all variables that may be taken into account for a real case. A statistical approach using an intrinsic admissible stress and a homogeneity factor may offer a better estimation.

24.3 Defining Reliability

Let the strength R_0 and the load S be represented by two random independent variables each following a normal distribution $R \in N_1(m_r, \sigma_r)$ and $S \in N_2(m_s, \sigma_s)$. R and S have the following densities of probability:

$$f_1(r) = \frac{1}{\sigma_r\sqrt{2\pi}} \exp\left(-\frac{(r-m_r)^2}{2\sigma_r^2}\right) \tag{24.5}$$

$$f_2(s) = \frac{1}{\sigma_s\sqrt{2\pi}} \exp\left(-\frac{(s-m_s)^2}{2\sigma_s^2}\right) \tag{24.6}$$

where m_r and m_s are the mean values of random variables r, s; σ_r, σ_s are standard deviations of r, s. The probability of successful operation, or reliability is:

$$R = P(R_0 > S) = P(Y > 0) = \frac{1}{\sigma_y\sqrt{2\pi}} \int_0^\infty \exp\left[-\frac{(y-m_y)^2}{2\cdot\sigma_y^2}\right] dy \tag{24.7}$$

where Y is another normal distributed variable $Y = R_0 - S, Y \in N(m_y, \sigma_y)$, which mean and standard deviation are: $m_y = m_r - m_s$ and $\sigma_y = \sqrt{\sigma_r^2 + \sigma_s^2}$. The reliability yields:

$$R = \frac{1}{2\pi} \int_{z_0}^\infty \exp\left(-\frac{z^2}{2}\right) dz \tag{24.8}$$

where $z = \frac{(y-m_y)}{\sigma_y}$ and for $y = 0$

$$z_0 = \frac{0 - m_y}{\sigma_y} = -\frac{m_r - m_s}{\sqrt{\sigma_r^2 + \sigma_s^2}} \tag{24.9}$$

24.4 Mathematical Model for Lifetime Estimations

When designing a pipe system, lifetime is refereed to as the period of time the system should work at normal operating parameters without any other intervention. After a long operating time, one could be interested in estimating the remaining time resource, for an external extension of initial lifetime estimations under specific risk assertions. The main mechanisms of a pipe being affected by high temperature operating conditions are related to creep and thermal fatigue. A procedure for estimating lifetime will therefore be close related to these phenomena. Determining strength and reliability parameters may involve the following limit states: fracture state (tenacious or fragile); plastic joint state; environmental induced internal damage under statistical loading; limit geometric displacement state. A first step in modelling the actual structure consists of defining geometrical and strength parameters. Computational methods such as finite element method are involved that have to be provided with data on geometrical configuration (such as the thickness of pipe walls) and with the following strength parameters of pipe's material: R_m ultimate strength; $R_{C0.2}$ technical flow limit, at normal temperature; $R_{r/t/T_e}$ technical long time strength; $R_{1/t/T}$ technical creep limit. At this stage, the operating time and parameters are to be determined by a correct evaluation of the actual design and of mechanical characteristics of materials. Maximum equivalent normal stress is computed according to the limit state of maximum tangent stress (Tresca) by taking into account the effects of internal pressure, of the component own weight, of squeezed bearings, and also considering a third party of thermal induced stresses. This approach allows to determine time to failure under creep effects and under fatigue efforts, respectively. For the points where intense efforts are expected, non-destructive testing procedures are performed at certain time intervals during system operation. The procedures involved may be: dynamic response analysis, magnetic powders analysis, penetrating radiation tests, strain measurements. The aim of such tests is tracking out data on cumulating strains that are useful in predicting the actual time resource. For estimating variable loading strength, one way introduce the non-linear cumulating damage criteria that adds effects of statistical and dynamical loads. The yielding operating time is

$$\sum_{i=1}^{n} \frac{N_i}{[N]} + \left(1.25 \frac{R}{R_{r/t/T}}\right)^m \leq D \tag{24.10}$$

where: N_i is the number of loading cycles at the same asymmetry factor; n is the number of such set cycles; $[N]$ is the limit number of cycles that is tracked by experiments for an asymmetry factor $\frac{\sigma_{min}}{\sigma_{max}} = \frac{1}{3}$; R is the stress determined from creep data; $R_{r/t/T}$ technical endurance strength for an extra t hours of operating at a T temperature; D - limit damage factor in respect of creep exposure and fatigue; m - a factor depending on material and type of loading.

The limit number of cycles could be expressed as:

$$[N] = \frac{1}{n_N} \left(\frac{0.25 E_T \ln \frac{100}{100 - Z_T}}{\sigma_a - \sigma_{-1}} \right)^{\frac{1}{\alpha}} \tag{24.11}$$

where n_N is the strength resource after 5 loading cycles %; E_T is the longitudinal modulus of elasticity at a given temperature T; Z_T is the constriction related to the technical endurance strength; σ_a is the limit magnitude of the variable stress cycle; σ_{-1} fatigue strength under symmetric loading cycle; α coefficient of material behaviour.

One should use a simple criterion for asserting availability that is:

$$\sum_{i=1}^{n} \frac{N_i}{[N]} \leq 1 \tag{24.12}$$

where $[N]$ yields from equation (24.11) for a value of σ_a calculated as

$$\sigma_\alpha = \frac{2.3\eta_g E_T}{4\sqrt{N}} \log \frac{100}{100 - Z_m} + 0.8R_{r/t/T} \tag{24.13}$$

where Z_m is the average value of tensile constriction related to long time strength; η_g is a coefficient of creep and relaxing influence; N the number of loading cycles up to fracture.

24.5 Simulating Reliability

Simulation, as a substitute for experimentation, introduces mathematical and logical models describing the behaviour of real system over a certain operating time. Geometrical characteristics (thickness of pipe's wall) and strength parameters of material, as described above, may be derived by computer "experiments" on a "physical model" satisfying the operating conditions of the real component. One may use Monte Carlo Simulation where the solution of the problem is represented by a stochastic variable or a stochastic process and based on generating a uniformly distributed random number, a selection of the stochastic variable or process is obtained allowing an estimation of the interesting parameter. A block diagram of the simulation model reveals the logical dependencies between input and output parameters. If dealing with random variables, the operating characteristics, that is hypothesis and equations connecting input variables of system states and output variables, take the form of probability density functions. As well one may encounter statistical parameters of operating characteristics among the input parameters. Such parameters have to be defined by statistical observations of the real system.

24.6 Algorithm of Simulation

The sequence of steps performed on a computer for simulating reliability of a pipe system is described. As input parameters for reliability prediction one can assume: values of creep strains; strength resource; internal damage occurrence;

long time plasticity; status of strength characteristics; material density. As for reliability on availability, one has to make a distinction among several situation of defining it: when designing a system (prediction of reliability), when experimenting on a system (technical reliability), during industrial operating (real state reliability). There also may be a nominal reliability that is indicated by technical standards. The pipe system could be represented as a graph in which the nodal points are elements of the system and arcs are connections between these elements. Connections may be of the serial type, parallel type or combinations of the two. A numerical approximation of the reliability function may be obtained by generating possible states of the system (that require knowledge of reliability functions of the elements) and evaluating for each state the system performance. The ratio of the number of successful events to the total number of events may represent an estimation for the reliability function of the system, the accuracy being related to the number of events generated. The simulation algorithm is presented in Figure 24.1.

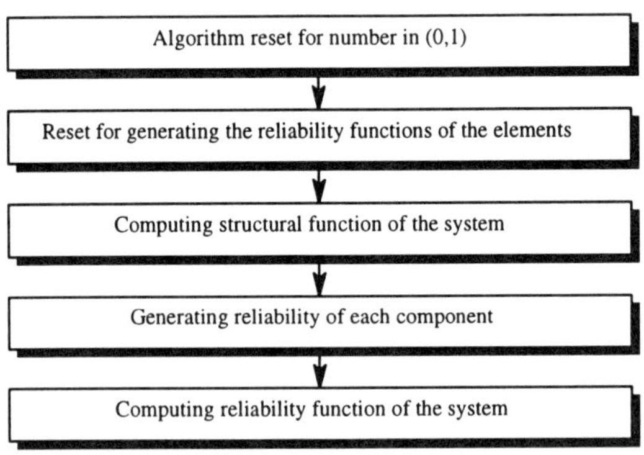

Figure 24.1: Simulation algorithm

For a example system including n components with serial or parallel links, a structural function may be introduced Figure 24.2.

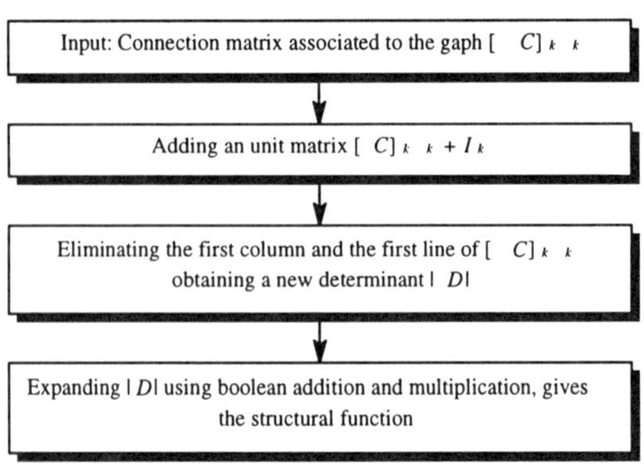

Figure 24.2: Structural function for a system with serial or parallel links

24.7 Simulating Reliability for Components

Based on data from endurance tests of each component concerning time and load dependence of strength parameters, one can deduce the probability density associated to parameters of components. An algorithm presented in Figure 24.3 may be used.

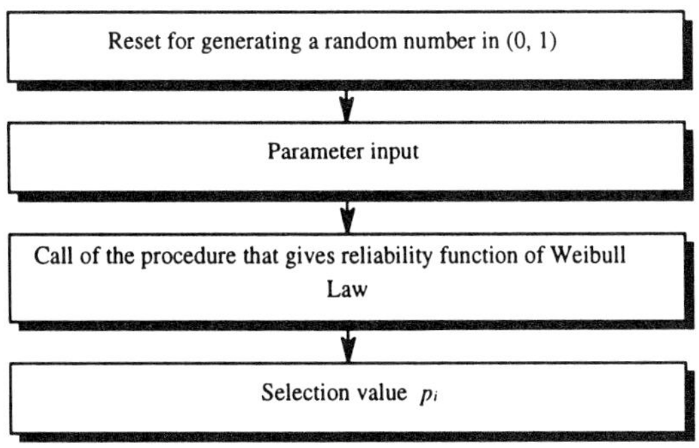

Figure 24.3: Algorithm based on endurance tests

24.8 Simulating System Reliability

Numerical computation of reliability function may be performed using a simulation algorithm based on Monte Carlo method. Each component may have one of two states: working or failure. An algorithm for system reliability function is presented in Figure 24.4.

If introducing a selected parameter into the computational pattern described, the yielding reliability allows to establish reliability for the system with more parameters successively selected. On the other hand it allows to establish a lifetime prediction as well as the remaining time up to failure. The chart in Figures 24.5 and 24.6 presents a general procedure used for an evaluation of lifetime, where: t_R is time up to failure of a pipe under certain stress and temperature; t is time of exposure; ϵ cumulated specific train; $F(\frac{t}{t_R})$ function of creep damage. Real damage is indicated by nondestructive techniques.

The technique described above requires a monitoring system integrated to a data acquisition system and measurement and control units. Software analysis on acquired data should provide information on system behaviour at specific operating stages that will allow decisions regarding system management and maintenance.

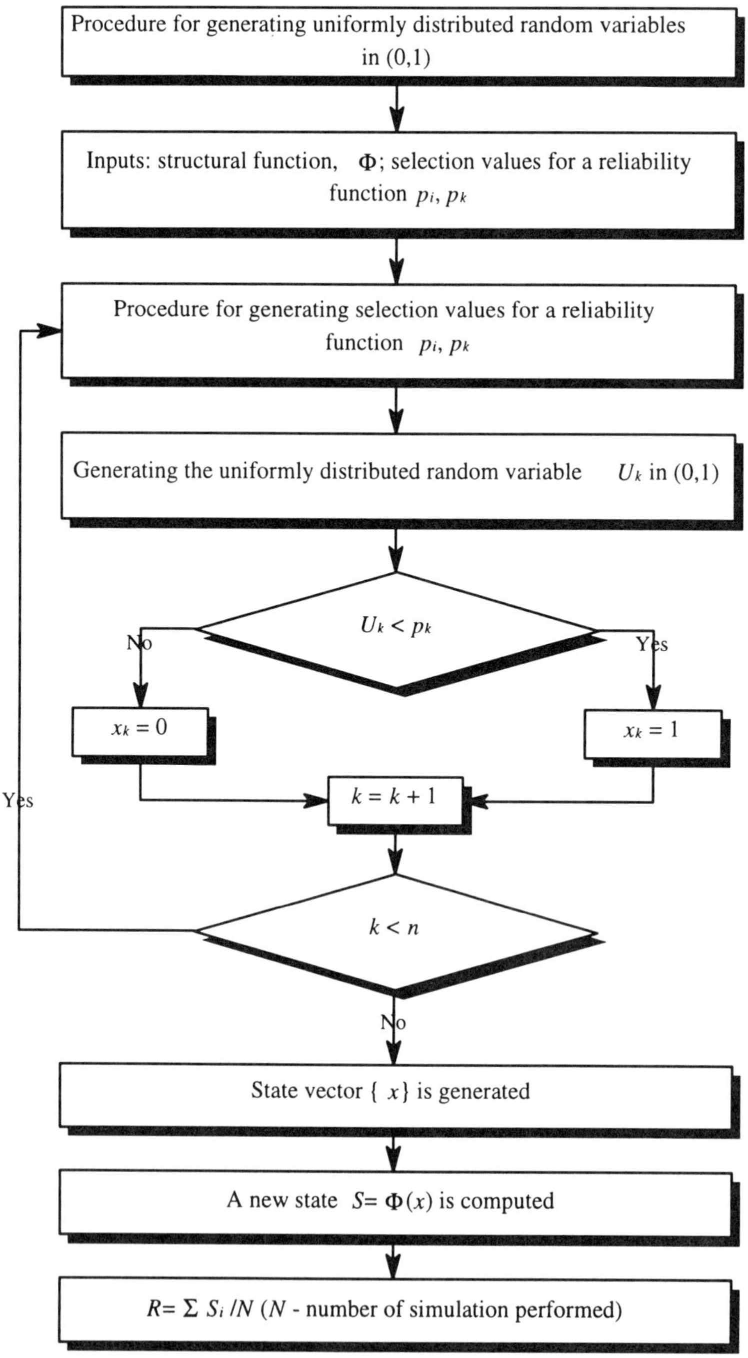

Figure 24.4: Algorithm for system reliability function

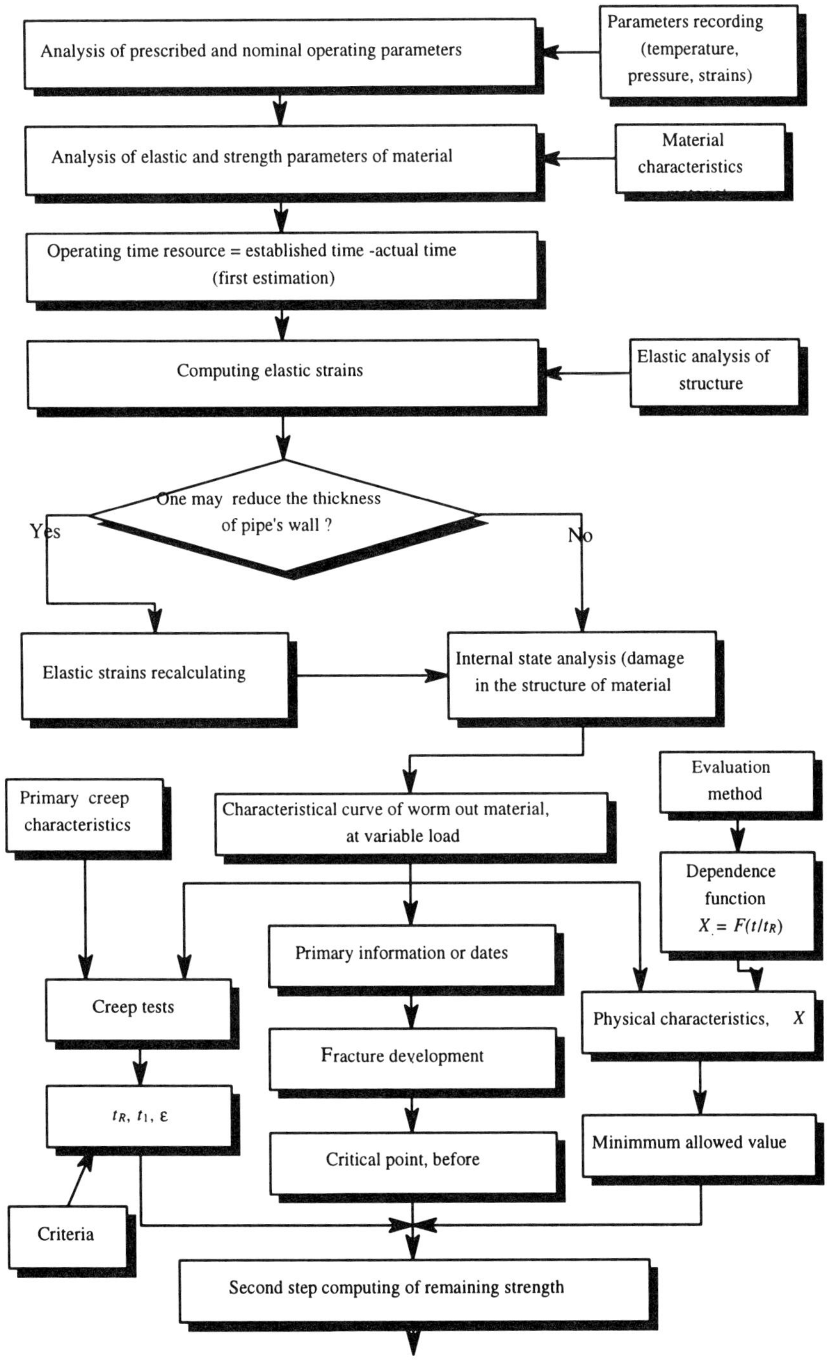

Figure 24.5: General procedure (continued in Figure 24.6)

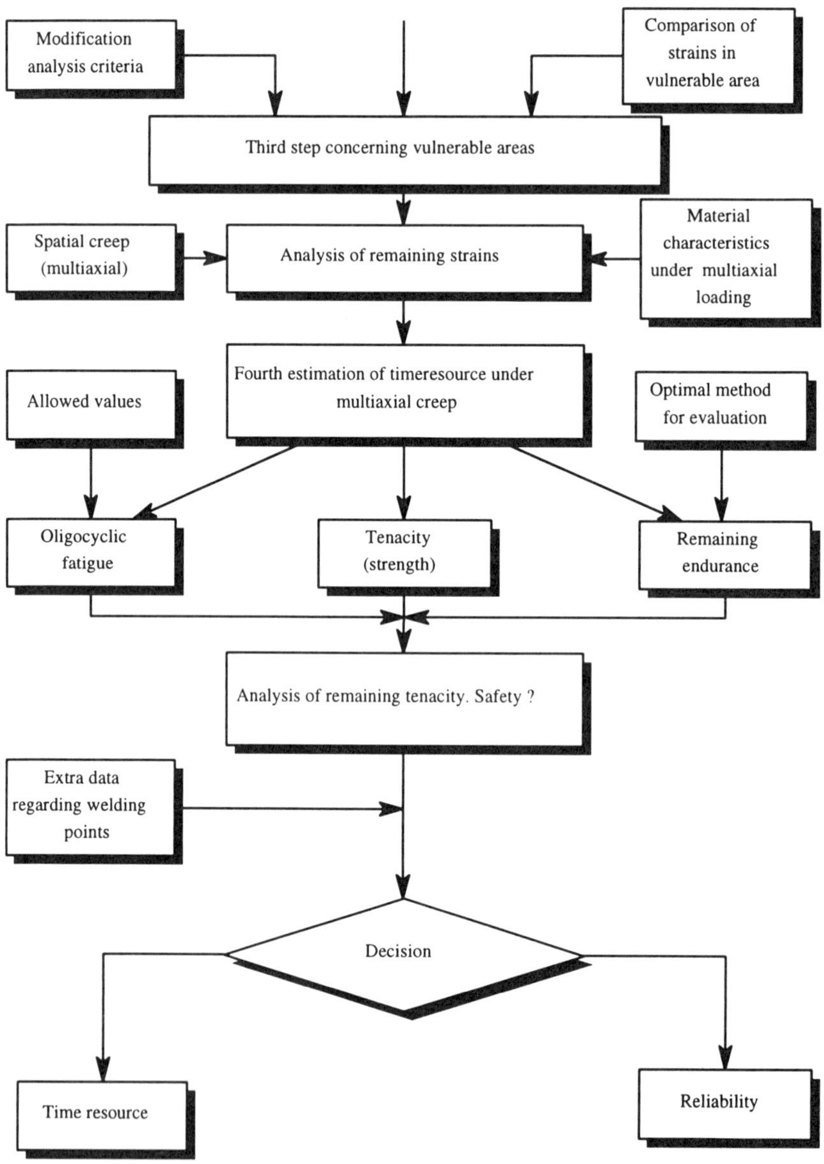

Figure 24.6: General procedure (continued from Figure 24.5)

References

1. Barbu, Gh. (1992). *Simulating Models in Reliability*, Bucureşti:Technical Publishing House.

2. Lee, D. R. and Bloom, J. M. (1993). Assessment of fully plastic J and C^* integral solution for application to elastic plastic fracture and creep crack growth, *J.P.V.T.*, **115**.

3. Gnedenko, B., Beliaev, Y. and Soloviev, A. (1972). *Methodes mathematiques en theorie de la fiabilite*, Moscow: Mir Publishing House.

4. Coffin, L. F. Jr., A Study of Cyclic - Thermal Stresses in a Ductile Metal, *Trans. ASME*, **76**, 931–950.

5. Fleşer, T. (1991). Possible evaluation of remaining lifetime in power and chemical plants, *O.I.D.I.C.M*, Bucureşti.

6. Stasenko, I. V. (1986). *Pipes computation to creep*, Bibl. Rascetika, Moscow.

Index